NUCLEIC ACID ANALYSIS

NUCLEIC ACID ANALYSIS
PRINCIPLES AND BIOAPPLICATIONS

Editor

CHARLES A. DANGLER
Division of Comparative Medicine
Massachusetts Institute of Technology
Cambridge, Massachusetts

WILEY-LISS

A JOHN WILEY & SONS, INC., PUBLICATION
New York • Chichester • Brisbane • Toronto • Singapore

Address All Inquiries to the Publisher
Wiley-Liss, Inc., 605 Third Avenue, New York, NY 10158-0012

Printed in the United States of America

Library of Congress Cataloging-in-Publication Data

Nucleic acid analysis : principles and bioapplications / edited by
   Charles A. Dangler.
     p.   cm.
   Includes index.
   ISBN 0-471-10358-6 (cloth : alk. paper)
   1. Nucleic acid probes.   2. Nucleic acid probes—Diagnostic use.
   I. Dangler, Charles A., 1957–   .
   QP624.5.D73N82   1996
   591.8′8—dc20   95-47202
                                       CIP

The text of this book is printed on acid-free paper.

10 9 8 7 6 5 4 3 2 1

CONTENTS

CONTRIBUTORS

Asim K. Bej, Department of Biology, University of Alabama at Birmingham, Birmingham, AL 35294-1170 **[1,231]**

Bruce Budowle, Forensic Science Research and Training Center, FBI Academy, Quantico, VA 22135 **[79]**

Jarasvech Chinsangaram, Department of Veterinary Pathology, Microbiology, and Immunology, School of Veterinary Medicine, University of California, Davis, CA 95616 **[131]**

Catherine Theisen Comey, Forensic Science Research and Training Center, FBI Academy, Quantico, VA 22135 **[79]**

Charles A. Dangler, Department of Veterinary Science, The Pennsylvania State University, University Park, PA 16802; present address: Division of Comparative Medicine, Massachusetts Institute of Technology, Cambridge, MA 02139 **[31]**

Chitrita Debroy, Supelco, Inc., Supelco Park, Bellefonte, PA 16823 **[31]**

Peter Feng, Center for Food Safety and Applied Nutrition, FDA, Washington, DC 20204 **[203]**

Walter E. Hill, Seafood Products Research Center, FDA, Bothell, WA 98041-3012 **[203]**

Keith A. Lampel, Center for Food Safety and Applied Nutrition, FDA, Washington, DC 20204 **[203]**

James R. Lupski, Department of Molecular Human Genetics and Department of Pediatrics, Baylor College of Medicine, Houston, TX 77030 **[157]**

Meena H. Mahbubani, Miles College, Birmingham, AL 35208 **[231]**

Edward R. B. McCabe, Department of Pediatrics, University of California, Los Angeles, CA 90024-1752 **[67]**

The numbers in brackets are the opening page numbers of the contributors' articles.

Linda L. McCabe, Department of Pediatrics, University of California, Los Angeles, CA 90024-1752 **[67]**

Bennie I. Osburn, Department of Veterinary Pathology, Microbiology, and Immunology, School of Veterinary Medicine, University of California, Davis, CA 95616 **[131]**

Calvin H. Vary, Maine Medical Center Research Institute, South Portland, ME 04106 **[47]**

James Versalovic, Department of Molecular Human Genetics, Baylor College of Medicine, Houston, TX 77030 **[157]**

William C. Wilson, Arthropod-Borne Animal Diseases Research Laboratory, USDA, Agricultural Research Service, Laramie, WY 82071-3965 **[105]**

PREFACE

This text is intended for two types of readers. The first group is the relatively uninitiated, those persons not familiar with molecular biology, but who find themselves drawn into the world of nucleic acid technology because of the more applied scientific and commercial aspects evident on the horizon. Within this first group, scientists not already familiar with nucleic acid interactions and the development of nucleic acid assays will hopefully find this text valuable in helping them to become familiar with these topics. The text describes several approaches for generating nucleic acid probes and offers some advice on selecting labeling systems and assay formats. The second audience consists of persons already familiar with basic nucleic acid hybridization assays, as well as those interested members of the first group. This group will benefit especially from the sections on applications. The chapters on applications address developments in veterinary and human medicine, as well as the often neglected markets of agriculture and environmental sciences, to shed light on the current and future direction of commercial and applied research interests.

This book was originally conceived as a primer on applications of nucleic acid technology, with an emphasis on the hybridization of nucleic acids, because of the central importance of this molecular interaction in defining the specificity and sensitivity of nucleic acid techniques in general. Obviously, other techniques besides hybridization protocols must be considered when developing applications, including those for commercial use. Various approaches to address sample handling and processing, user interface, and signal detection may be coupled to the basic hybridization reaction to design or to customize an efficient and potentially profitable application. The fashioning of a nucleic acid-based application is defined by utility and driven by ingenuity. Recognizing the broad spectrum of possible strategies, and the availability of several excellent techniques-oriented books (such as *Current Protocols in Molecular Biology*, or the abridged volume, *Short Protocols in Molecular Biology*, edited by F. M. Ausubel et al. and published by John Wiley & Sons), this text does not attempt to present detailed protocols for specific applications, but instead serves as a menu of options with examples describing representative applications. It is hoped that by illustrating different fields that might benefit from the development of assays, and by clarifying some of the basic technology, this text will help stimulate the development of novel and commercially successful applications of nucleic acid hybridization techniques.

The contributions, cooperation, and patience of many authors were required in the preparation of this volume, and I am indebted to all of them. Dr. Bennie I. Osburn and Peter Prescott were integral in the initial conception and framing of this book. I am deeply grateful for their advice and assistance. Susan King and Colette Bean, both of John Wiley & Sons, shepherded this project to completion. I thank them for their endurance, guidance, and support.

CHARLES A. DANGLER

CHAPTER 1

NUCLEIC ACID HYBRIDIZATIONS: PRINCIPLES AND STRATEGIES

ASIM K. BEJ
Department of Biology,
University of Alabama, at Birmingham,
Birmingham, AL 35294-1170

1.1. INTRODUCTION

The study of the myriad arrangements of the four nucleotides that constitute the genomes of each and every organism in this diversified living world has had a profound influence in virtually all areas of biology. The genetic make-up of each living organism contains a vast library of information. The analysis of these immense genetic messages requires the isolation and characterization of particular DNA sequences of interest. Nucleic acid hybridization is one of the primary approaches by which a great deal of information about the genetic structures and functions in all living organisms is being unfolded. The process that underlines all approaches involving nucleic acid hybridization, first described by Marmur and Doty [1], is the formation of the double helix from two strands of DNA that are complementary to each other. Successful nucleic acid hybridization is based on two fundamental criteria: (1) the two DNA strands involved in the hybridization process must have some degree of complementarity and (2) following hybridization, the extent of complementarity will determine the stability of the duplex DNA. Based on these two underlying principles of nucleic acid hybridization, a variety of avenues have been established involving this unprecedented approach to unveil the complexity of the underlying genetic principles of the living world and to address numerous enigmatic biological problems. This chapter will portray the fundamental principles of the

Nucleic Acid Analysis: Principles and Bioapplications, pages 1–29
© 1996 Wiley-Liss, Inc.

nucleic acid hybridization method and highlight its role in some of the portentous areas of the biological sciences.

1.2. WHAT IS NUCLEIC ACID HYBRIDIZATION?

Nucleic acid hybridization is a process in which inconsonant nucleic acid strands with specific organization of nucleotide bases exhibiting complementary pairing with each other under specific given reaction conditions, thus forming a stable duplex molecule. This phenomenon is possible because of the biochemical property of base-pairing, which allows fragments of known sequences to find complementary matching sequences in an unknown DNA sample. The known sequence is called the probe, which is allowed to bind and hybridize with a complementary target sequence under specific conditions and provide a signal to confirm the presence of a specific sequence of interest in a sample. For hybridization, both the probe and the target DNA have to be in single-stranded form. Since the DNA molecule *in vivo* or *in vitro* exists in double-stranded form, the strands must be "denatured." By reversing the denaturation process, the probe is allowed to bind to its complementary target strand by a process called "reassociation" or "annealing." During duplex formation between the probe and the target strands, by the annealing process, a hybrid double-stranded molecule is formed, and this reaction is called "hybridization." Conceptually, all nucleic acid hybridization methods appear to be the same. However, actual performance and interpretation of each hybridization reaction involving binding of a unique probe with its complementary target DNA requires evaluation and optimization of a set of parameters to resolve the reaction conditions.

Since both DNA and RNA share the same biochemical principles of base-pairing, the hybrid formation during the nucleic acid hybridization process can occur between DNA–DNA, DNA–RNA, or RNA–RNA. Also, a number of approaches such as dot blot, colony or plaque lifts, Southern blot (DNA–DNA), Northern blot (RNA–DNA), *in situ,* and solution hybridization (all based on the basic principles of nucleic acid hybridization) are used.

1.3. SAMPLE PREPARATION FOR NUCLEIC ACID HYBRIDIZATION

The preparation of the target nucleic acids for the probe to bind is particularly important for *in vitro* hybrizations. The target DNA is intimately associated with complex proteins, lipids, carbohydrates, and other nucleic acids within the cell. Therefore, the release of the nucleic acids from the cells, with successive purification from complex cellular matrices, followed by denaturation, is the first and primary step for a successful hybridization reaction. In general, the cells are lysed by treating with detergent, alkali, or simply boiling [2,3]. In some hybridization approaches, such as dot blot, Southern blot, Northern blot, or solution hybridization, purification of the nucleic acids following lysis is recommended. However, in many hybridization methods such as colony or plaque lifts, the cells are lysed, nucleic acids are denatured, and hybridization is performed without further purification. The *in situ* hybridization approach allows the probe to bind with previously denatured target nucleic acids without disrupting the tissues, cells, or chromosome [4].

1.4. VARIOUS APPROACHES AND KINETICS OF NUCLEIC ACID HYBRIDIZATIONS

There are three approaches to nucleic acid hybridization: hybridization in liquid phase (solution hybridization), hybridization on solid supports, and *in situ* hybridization. They are commonly used depending on the type of information that is expected to be generated from the hybridization result. These three hybridization approaches are described below.

1.4.1. Solution Hybridization

In the liquid hybridization, both the target nucleic acid and the probe interact in an optimized hybridization solution. The hybridization in the liquid phase has the maximum kinetics. Therefore, most of the time, the kinetics of a hybridization reaction is determined in liquid phase. Also, analysis of genome organizations and genetic complexity are generally done in liquid phase hybridizations. In liquid hybridization, the nucleic acids are purified from the target cells or tissues and transferred to a hybridization solution in which the single-stranded probe is added for hybridization. The nucleic acids are then denatured either by heating the sample or by treatment with alkali, and the annealing of the probes to their complementary target nucleic acids is performed. Hybridization may be followed by monitoring continuous decrease in the optical density, in a spectrophotometer at 260 nm wavelength, as the annealing of the probe to its complementary target nucleic acid occurs. The rate of the reaction is a measure of the concentration of the complementary sequences.

In liquid phase and possibly other hybridization processes, the initial event in the annealing of the two complementary strands is a nucleation reaction followed by a rapid process of "zippering" of the two single-stranded nucleic acids forming a duplex molecule. A mathematical model based on the assumption that the two strands are unbroken and have 100% complementarity was derived by Wetmur and Davidson [5]. However, in practice, unless the complexity of the genome is simple and the size is small such as in viruses, the complex nucleic acid molecules from eukaryotes or other higher organisms are sheared to a desired size range before being subjected to solution hybridization. As a result, during hybrid formation, numerous single-stranded tails are available for binding with other molecules, forming complex concatamers. The remaining single-stranded forms are one of the key features for quantitative determination of the hybrid molecules. Conventionally, following hybridization the sample is subjected to S1 nuclease digestion, which destroys the single-stranded molecules. The sample is then passed through a hydroxyapatite chromatography column in which the double-stranded duplex molecules are selectively bound and the single-stranded molecules are eluted by using a buffer with specific ionic strength. Later, the double-stranded molecules are eluted using another buffer with a different ionic strength, and quantitation is determined by spectrophotometric, flurometric, colorimetric, or radioactive methods. An alternative approach following hybridization and S1 nuclease treatment is selective precipitation of the duplex molecules using trichloroacetic acid (TCA). Hydroxyapatite binding measures the fraction of the DNA that is linked to structures containing duplexes and follows the equation [6]

$$H = (1 + kC_0t)^{-1}$$

where H is fraction of DNA bound to the hydroxyapatite, k is observed rate constant ($M \cdot sec^{-1}$), C_0 is original concentration of the nucleotides ($mol \cdot L^{-1}$), and t is time (in seconds). In 0.14 M sodium phosphate buffer with a pH of 6.8, the double-stranded DNA will bind to hydroxyapatite, but the single-stranded DNA will not.

During the S1 nuclease digestion the single- and double-stranded nucleic acids will be distinguished, and the kinetics of the nucleic acid hybridization will follow the equation

$$S = (1 + kC_0t)^{-0.44}$$

where S is fraction of nucleotides that remain unpaired, and k, C_0, and t are the same as above.

If the DNA is sheared to a desired size range, the concentration of the fragments that contain a particular sequence is inversely proportional to the genome size. As a result, the rate of the reannealing reaction between these strands is inversely proportional to the genome size. This is more pronounced in eukaryotic organisms because of the presence of a large number of repetitive sequences in the genome [5,7]. The relationship between the complexity of the genome and the hybridization reaction rate in solution is fundamental to both DNA–DNA and DNA–RNA hybridizations. In the RNA–DNA hybridization, the base sequence complexity of a species of RNA can be defined as the base sequence complexity of the DNA from which the RNA is transcribed. Therefore, fundamentally the DNA–DNA hybridization can provide a comparative picture of the genome complexity of a group of related or unrelated organisms when performed in solution. On the other hand, a quantitative evaluation of the expression of a specific gene requires identification of the specific message by hybridization followed by measurement of the total amount of the duplex molecules.

In solution hybridization, the complexity of the genome can be measured quantitatively by calculating the C_0t (for DNA) or R_0t (for RNA) value. C_0t (moles of nucleotide per liter) is the product of DNA concentration (R_0t for the RNA concentration) and time (in seconds) of incubation in a reannealing reaction. The complexity is presented as a measure of $C_0t_{1/2}$, which is a value when one-half (50%) of the total DNA in the reaction has reannealed. The C_0t value can be determined from the equation

$$C/C_0 = 1/1 + kC_0 t_{1/2}$$

where C is concentration of nucleotides at time t (mole per liter), C_0 is original concentration of nucleotides at time 0 (mole per liter), k is reassociation rate constant, and t is time (in seconds). Therefore, $C_0 t_{1/2}$, i.e., when the reannealing reaction is half completed at $t_{1/2}$, can be defined mathematically as

$$C/C_0 = 1/2 = 1/1 + kC_0 t_{1/2}$$

so that

$$C_0 t_{1/2} = 1/k.$$

Since the $C_0 t_{1/2}$ is the product of the concentration and the time required to proceed halfway, a greater $C_0 t_{1/2}$ implies a slower reaction, which indicates the degree of complexity of the genome. Recently, high performance liquid chromatography (HPLC) has been used to separate and quantitate double-stranded hybrids from the single-stranded molecules following solution hybridization rapidly and accurately.

The complexity of any DNA can be determined by comparing its $C_0 t_{1/2}$ with that of a standard DNA of known complexity. Usually *Escherichia coli* DNA is used as a standard. Its complexity is taken to be identical with the length of a genome that implies that every sequence in the *E. coli* genome of 4.2×10^6 is unique. Therefore, the complexity of any DNA can be determined by the equation

$C_0 t_{1/2}$ (DNA of any genome)/$C_0 t_{1/2}$ (*E. coli* DNA) =
Complexity of any genome/4.2×10^6 bp

Comprehensive descriptions of methods for $C_0 t$ [8,9] and $R_0 t$ [10] are given elsewhere.

The primary advantage of the solution hybridization approach is that it provides kinetic simplicity with maximum rate of hybridization. One of the disadvantages of this approach is that the nucleic acids should be free of proteins, lipids, carbohydrates, and other contaminants that may interfere with the annealing process. Also, because of its faster reassociation rate, higher self-annealing may occur. To prevent this problem, excess amount of probe may be used in a solution hybridization reaction. This approach may not be efficient if a large number of samples have to be analyzed.

1.4.2. Hybridizations on Solid Support

In this approach, the single-stranded DNA or RNA is immobilized on a solid support such as nitrocellulose, nylon, or polystyrene and is available for hybridization with the probe in a liquid phase. After it was first described by Gillespie and Speigelman [11], the hybridization method on solid support became a popular analytical hybridization tool in molecular biology despite the fact that hybridization in the two phases (solid and liquid, sometimes referred to as "mixed phase") can be slow. Since the target nucleic acids are bound on the solid support, one of the advantages of this approach is that self-annealing is prevented. The probe molecules are labeled radioactively or nonradioactively so that following the hybridization process the hybrids can be detected either by autoradiography or by colorimetry. Also, a quantitative assay of the hybrids can be determined by using a scintillation counter or a densitometer.

Types of Solid Supports. A variety of solid supports for DNA–DNA or RNA–DNA hybridizations are available. Among some of the membrane solid supports, nitrocellulose, nylon, or chemically activated papers are common. The nature and type of membrane solid supports are selected depending on the purpose of the hybridization.

Nitrocellulose membrane binds with DNA and RNA with high efficiencies (80 $\mu g/cm^2$). However, nucleic acid fragments below 500 nucleotides in length are bound poorly. Also, the nitrocellulose membrane is relatively fragile and becomes brittle. Therefore, it is difficult to handle, especially when used multiple times.

The introduction of nylon membrane for solid support hybridizations has overcome this difficulty. The texture of the nylon membranes is more flexible, and they do not disintegrate with repeated usage. Moreover, nucleic acids of sizes lower than 500 nucleotides can be immobilized efficiently on the nylon membranes. The sensitivity of hybridization detection on various commercially available nylon membranes is claimed to be much higher than on the nitrocellulose membranes. For Southern and Northern hybridizations, both nylon and nitrocellulose membranes provide excellent hybridizations. For colony lifts, a nylon membrane with a pore size of 1.2 μm is recommended. For DNA dot blots, a membrane with a 0.45 μm pore size is recommended for large (>500 nucleotide) molecules, and a 0.22 μm pore size membrane is preferred for small (<500 nucleotide) molecules. For RNA dot blots, membranes with pore sizes of 0.1–0.22 μm provide excellent hybridization.

Recently, the use of a polystyrene surface (e.g., microtiter plates) has been used for hybridizations. Although this approach has not become popular to date, some of the advantages are (1) multiple samples can be handled more easily on a microtiter plate than on a nylon or nitrocellulose membrane, (2) automation is possible in this approach, (3) less amounts of reagents are used in a microtiter plate hybridization, and (4) the hybridization time can be shortened by several hours compared with membrane hybridization. The efficiency of hybridization in a microtiter plate has not been shown to be higher than in the membrane hybridization.

Immobilization of Nucleic Acids on Solid Supports. As a first step the nucleic acids are denatured before immobilizing onto solid supports. The conventional methods of immobilization do not allow the nucleic acids to bind to the nitrocellulose or nylon membranes covalently. The noncovalent immobilization of the nucleic acids may pose some problems such as at high temperature the nucleic acids may leach out from the membrane if hybridization is carried on for a long period of time. In addition, if the probe binds to the immobilized target nucleic acid by its entire length, the hybrid molecule may detach from the membrane and stay in the liquid phase [12]. This phenomenon may lead to no or unexpectedly poor hybridization signals with time. Therefore, it is necessary to bind the immobilized nucleic acids onto the membrane covalently. Most of the commercially available membranes are designed to bind to the nucleic acids covalently either by UV irradiation or by chemical treatment [13]. The chemically activated membranes bind with the nucleic acids covalently. Therefore, these membranes can be used for immobilization of both large and small nucleic acids. However, the binding capacity of the chemically activated membranes is much lower ($1–2$ $\mu g/cm^2$) than that of nitrocellulose membranes and many of the commercially available nylon membranes, and the procedures for binding are more complicated. The chemically activated membranes are commonly used for hybrid selection for the enrichment of specific RNA sequences [10].

Various Types of Solid Support Hybridizations. The solid support hybridization approach has broad applications, including (1) dot blot or slot blot hybridization, (2) bacterial colony or plaque hybridization, (3) Southern hybridization, (4) Northern hybridization, (5) *in situ* hybridization, and (6) hybrid selection by sandwich hybridization.

Dot Blot or Slot Blot Hybridization. The dot or slot blot hybridization approach is relatively rapid with the advantage that a large number of samples can be analyzed simultaneously against one probe under the same hybridization conditions. The method involves fixation of the single-stranded target nucleic acid molecules onto a solid support, which is a nitrocellulose or a nylon membrane containing free primary amine groups available for covalent binding with the nucleic acid molecules. The nucleic acid samples are immobilized with specific geometric arrays manually or by using commercially available dot or slot blot apparatuses that are capable of analyzing 96 or more nucleic acid samples simultaneously on a single solid support. The name of this hybridization approach is derived from the geometric arrays of the samples on the solid support, which can be in the form of solid small circles (dot blot) or solid moderately elongated wells (slot blot). The rest of the solid support that does not contain any target nucleic acid to be tested is inactivated by treatment with a blocking reagent followed by hybridization with a specific nucleic acid probe under specific hybridization conditions. Following hybridization, the solid support is processed by washing several times with several solutions to remove unbound probes and/or binding of the probe with nontarget nucleic acids.

Although purified nucleic acids were initially used for dot or slot blot hybridization assays, a number of studies showed that unpurified nucleic acid from the environmental or clinical samples can be used successfully for hybridization simply by lysing the cells in the samples with a strong base or a chaotropic agent such as sodium iodide, denaturing the nucleic acid with alkali such as sodium hydroxide, and denaturing the cellular proteins by protein-degrading enzymes such as proteinase K or treatment with an anionic detergent such as sodium dodecyl sulfate. The success of this modified sample preparation approach has made the dot or slot blot hybridization method popular. The disadvantage of using the unpurified nucleic acid sample is the binding of the probe to nonspecific target nucleic acids, thus providing false-positive hybridization signal and/or high background signal.

The hybridization signal generated by the dot or slot blot approach can be expounded qualitatively as well as quantitatively. The intensity of the hybridization signal is generally compared with a positive control in which a known amount of nucleic acid is immobilized to determine the quantity of nucleic acids present in the test samples. A negative control sample is necessary in this approach, which is used to determine a weak signal represents the result of the hybridization with a very small quantity of the target nucleic acid or the interaction with a large quantity of nucleic acids from nonspecific organisms. To eliminate any background hybridization signal, which is considered as "noise," appropriate controls should be considered in each experiment. If the hybridization is performed with the radiolabeled probe, the dots or the slots can be cut out and a quantitative evaluation of the hybridization can be determined by using a liquid scintillation counter.

A modified approach of the dot or slot blot nucleic acid hybridization has been described by immobilizing various probes onto solid support, each probe being specific for detecting a specific target nucleic acid molecule [14]. The solid support is exposed for hybridization to an unknown sample that is a mixture of various target or nontarget nucleic acids. Since each probe is specific for hybridization with a specific target nucleic acid, any positive hybridization signal to a specific spot determines the presence of the target nucleic acid in a complex mixed sample. This approach is called the reverse blot or immobilized capture probe hybridization [14,15], and has the advantage of simultaneous screening of environmental and clinical samples containing a diverse group of microbial species within a relatively short period of time (Fig. 1.1).

Colony or Plaque Hybridization. This hybridization approach is the simplest application of all hybridizations and useful when a large number of microbial colonies or bacteriophages are screened for a cloned gene or a DNA segment from a gene library or for screening environmental samples to determine the presence of a specific microorganism. In this procedure, the bacterial colonies or the plaques are transferred onto a nylon or a nitrocellulose solid support by the replica plating method. The colonies or the plaques are lysed on the solid support, and the nucleic acids are released on the membrane. The membrane is subjected to hybridization with the labeled probe as elaborated in the previous section.

Direct colony hybridization on primary cultivation is desirable to (1) avoid a cultivation bias encountered by selective media that may underestimate total abundance of a given genotype, (2) ensure that a given genotype is represented in the population, (3) provide optimum permissive growth conditions for stressed organisms that may be nonculturable on a selective media, and (4) reduce the analysis time for cultivation, presumptive quantification, and confirmation of a genotype or a phenotype.

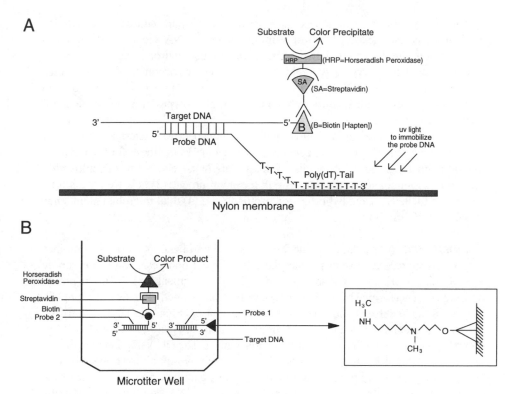

Fig. 1.1. Schematic representation of immobilized capture probe hybridization on a solid support. **A**: The solid support is a nylon or a nitrocellulose membrane on which the oligonucleotide probe is immobilized. Prior to immobilization, the probe DNA is treated with deoxyribonucleotidyl terminal transferase (TdT) to generate a poly(dT)-tail that is cross-linked on the membrane by using uv light. The target DNA is denatured and hybridized to the probe DNA. The biotin label at the 5'-end of the target DNA is then conjugated with HRP-SA, and color development is performed by exposing the HRP enzyme to its substrate. **B**: The solid support is a coated polystyrene surface on which the single-stranded probe DNA molecule is immobilized on a Covalink NH® microtiter well covalently by using a phosphoramidate bond (*Nunc*, Roskilde, Denmark). The amplified target DNA is denatured and hybridized with the first probe DNA, which is immobilized on the surface. Following hybridization, a second probe that is labeled with a biotin (hapten) or a radionucleotide is hybridized to the target DNA at a different segment. The color development and detection is performed as described above. The hapten-labeled hybridized second probe is then used for enzymatic reaction to colorimetrically detect the identity of the target DNA. The use of two-probe DNA for hybridization detection of the target DNA is called "sandwich hybridization," in which the target DNA is sandwiched between the two-probe DNA.

Colony hybridization on a secondary cultivation of pure cultures is usually to confirm a specific genotype or test a unique gene or DNA sequence. Colony hybridization is popular and widely used in characterizing environmental samples and screening colonies carrying recombinant DNAs. For environmental applications, this hybridization approach has been used for detection, enumeration, and isolation of microorganisms with specific genotype and/or phenotypes and for the development of gene probes.

The positive hybridization signal with the nonspecific bacterial colonies or plaques is the primary disadvantage of this procedure. In addition, for the environmental applications some of the disadvantages are (1) selectivity of the cultivation medium, (2) abundance of the organism in the environment, (3) how well the colony will adapt to the lysis and hybridization protocol, and (4) total microbial abundance at the time of sampling.

Southern Blot Hybridization. The Southern blot hybridization approach combines with conventional Agarose or polyacrylamide gel electrophoresis and is commonly used for characterization of a gene or a specific DNA segment (Fig. 1.2). The procedure was first performed by E.M. Southern in 1976 and named after the inventor [16]. The entire approach is conducted in two major phases. First, the DNA is purified from the target organisms and treated with restriction enzyme(s), and the DNA fragments are separated by the gel electrophoresis method. The DNA fragments in the gel are then denatured by treatment with alkali followed by immobilization onto solid support, i.e., nylon or a nitrocellulose membrane by capillary transfer, or one of the commercially available apparatuses that transfers the DNA onto the membrane by electroblotting or vacuum blotting. Second, the membrane is exposed to an appropriate labeled probe for hybridization under hybridization conditions. Generally, a known DNA complementary to the probe DNA is used as a positive control, and another DNA that is unrelated to the probe is used as a negative control. If the probe is radiolabeled, an autoradiogram is performed on a x-ray film. Positive hybridization signal(s) represent the specific restriction-digested DNA fragment(s) complementary to the probe. The Southern blot hybridization method is useful in determining the presence of a gene or a specific DNA sequence in a pool of DNA and its location, orientation, and molecular weight. Also, changes in the molecular weight of a gene due to mutation such as deletion, and identification of a single base mutation of a gene fragment, which may be an indication of a disease state, can be determined. One of the primary disadvantages of this method is that it may sometimes take days to perform unless a relatively expensive, commercially available, partially automated device is used. Most of the time the Southern hybridization is used for qualitative analyses of a DNA or a gene of interest. In the case of a single gene copy, the amount of target DNA must be very high for positive hybridization signal. Moreover, a limited amount of DNA can be bound on the membrane. Therefore, analysis of the signal-to-"noise" ratio can sometimes be difficult because of the low hybridization signal to the target DNA. However, the intensity of the hybridization with different restriction enzyme–treated DNA fragments can be useful to determine a quantitative analysis such as the number of gene copies.

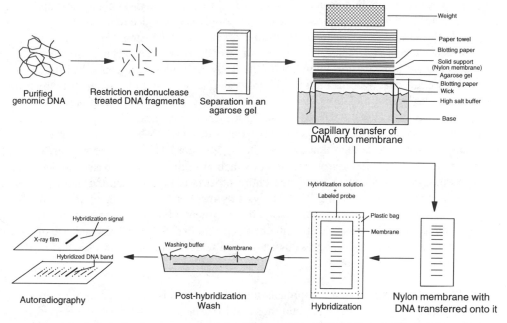

Purified
genomic DNA

Restriction endonuclease
treated DNA fragments

Separation in an
agarose gel

Weight
Paper towel
Blotting paper
Solid support
(Nylon membrane)
Agarose gel
Blotting paper
Wick
High salt buffer
Base

Capillary transfer of
DNA onto membrane

Hybridization solution
+
Labeled probe

Plastic bag
Membrane

Hybridization

Nylon membrane with
DNA transferred onto it

Hybridization signal

X-ray film

Hybridized DNA band

Autoradiography

Washing buffer   Membrane

Post-hybridization
Wash

Fig. 1.2. Schematic representation of the Southern blot DNA-DNA hybridization showing various essential steps of the process. Purified genomic DNA is treated with restriction endonuclease and separated in an agarose gel. The double-stranded DNA is then denatured by treatment with alkali (NaOH) and salt solution (NaCl). If the DNA fragments are too large ($\geq$3.0 kbp), they are treated with HCl to depurinate. The DNA on the gel is then transferred onto the nylon membrane by capillary action using a high-salt buffer. The nylon membrane is then subjected to treatment with blocking reagents and hybridized with a single-stranded DNA probe tagged with a radionucleotide for autoradiography or with a hapten molecule (biotin) for colorimetric detection. Following hybridization, the membrane is washed in a series of washing buffer to remove excess probes. An X-ray film is placed on the membrane and exposed for autoradiography.

Northern Blot Hybridization. In Northern blot hybridization, the target nucleic acid is the RNA, and detection of a specific RNA or mRNA is performed by using a DNA or a RNA probe. The fundamental principle of the Northern blot approach is the same as the Southern blot hybridization except that some procedural modifications have been introduced because of some of the different properties of the RNA molecules. The Northern blot hybridization is used primarily to investigate the gene expression through identification of a specific mRNA by hybridization with a specific probe. Since most mRNA is shorter in size than the genomic DNA, enzymatic digestion is not required prior to electrophoresis. Moreover, the size of the mRNA can be determined simply by comparing molecular weight with the appropriate RNA size standards in the gel and the hybridized segment of the total RNA. The RNA molecules may exist in fold-back conditions with one or more hairpin structures. Therefore, the RNA molecules in the sample are denatured simply by heating. Also, a denaturing agent such as formaldehyde or glyoxal is added to the gel to keep the RNA molecules in denatured condition. Since the double-stranded and single-stranded nucleic acid molecules migrate differently from each other in the electrophoretic gel, this ensures migration of the RNA molecules with their actual molecular weight. The hybridization procedures using a DNA or an RNA probe is the same as the Southern blotting method, and the result is visualized on an autoradiograph or colorimetrically directly on the hybridization membrane.

One of the disadvantages of this approach is that the RNA molecules are extremely susceptible to digestion by ubiquitously present RNAses. Therefore, great care must always be taken, from the RNA extraction from the cells or tissues to performing hybridization steps. Also, the entire process is time consuming and may take days to obtain any result. RNA analysis by the Northern blotting approach is popularly used to determine qualitatively the expression of a specific gene. In addition, the Northern blot hybridization method is useful to determine the actual size of the message, any problem of cross-hybridization with other mRNA species, and to the size of an unknown message.

In situ Hybridization. In this approach, the hybridization is performed in the morphologically intact cell, tissue, or chromosome [4]. The *in situ* hybridization is used to demonstrate the presence of a specific gene in a specific location of a tissue, cell, or chromosome (see Figure 11.11 in Chapter 11). This approach represents the best combination of nucleic acid target available for hybridization in a morphologically intact tissue, cell, or chromosome without damaging its morphological characteristics. To make the probe available for hybridization with the chromosomal DNA, the intact cells or the tissues are treated with various enzymes. The penetration of the probe molecules in the test tissue or in the cell is one of the most important aspects of this approach and is generally performed by using a tissue permeabilizer such as various detergents, protease, DNAse, or RNAse in appropriate situations. An *in situ* hybridization result is analyzed in a morphological context, and the sensitivity of detection of the target nucleic acid sequence by a labeled probe is greatly influenced by the number of copies of the target sequence and their distribution. Following hybridization with an appropriate probe, the morphology of the tissue or the cell is subjected to counterstain. Although a quantitative approach to the hybrid molecules has been reported, sometimes it can be complicated and confusing. Therefore, *in situ* hybridization is not commonly used for qualitative hybridization assays.

In *in situ* hybridization methods, each sample must be processed individually, which makes this approach tedious and relatively time consuming. This difficulty is the major drawback of the approach.

Sandwich Hybridization. Sandwich hybridization was developed to overcome the problem of nonspecific hybridization signals generated by the other conventional hybridization methods. Another advantage of this method is that the sample processing is minimum and thus the total time is enormously reduced. A portion of the semipurified test DNA is hybridized with a target probe that was previously immobilized onto a solid support. Following hybridization, the rest of the "free" test DNA is subjected to hybridization with a probe DNA that is labeled either with radionucleotide or by colorimetric reagent. After appropriate washing of the nonspecifically bound probe, the detection is performed through autoradiography or by colorimetric result. This hybridization approach will only generate positive signal if there is a "sandwich" between the immobilized target probe and the labeled probe (see Fig. 1.1B). This approach has eliminated most of the background signals associated with many of the hybridizations.

1.4.3. Quantitative Aspects of Solid Support Hybridizations

The random collision between the two complementary nucleic acid sequences is the fundamental principle in hybridization reactions. Two major phenomena are involved in the hybridization process: (1) the time course of the reaction in solution is determined by the concentration of the reacting species and by the second-order rate constant, k, and (2) the melting temperature (T_m) of the duplex nucleic acids is the key factor for the hybrid stability during and following hybridization processes [10]. The kinetics and factors affecting quantitative aspects of the solid support hybridization have been less extensively studied than solution hybridization because of the complexity of the reaction. Hybridization of a single-stranded nucleic acid probe to the membrane-bound complementary DNA or RNA involves two reactions that compete with each other: In one reaction, the single-stranded probe molecules in the liquid phase tend to reassociate to form partially duplex "hyperpolymers," and in the other reaction the probe molecules interact with the complementary nucleic acids bound on the membrane. As the hybridization reaction proceeds, the number of single-stranded molecules disappears due to the formation of hybrids, which can be expressed by the following equation:

$$-d[C_s]/dt = k_1[C_f][C_s] + k_2[C_s]^2$$

where C_s is concentration of single-stranded nucleic acids in liquid phase, C_f is concentration of single-stranded nucleic acids in the solid phase (i.e., on the membrane), k_1 is rate constant for the hybridization reaction on the filter, k_2 is rate constant for the reassociation constant in the liquid, $k_1[C_f][C_s]$ represents the hybridization on solid support, and $k_2[C_s]^2$ represents the reassociation in the liquid phase.

Although a variety of factors affect the rate constants during hybridization, two factors contribute to keep the rate constants equal, i.e., $k_1 = k_2$. These factors are (1) the rates of the zippering phenomenon following nucleation, at the filter, and in the solution are the same, and (2) the molecular weight of the single-stranded nucleic acid in solution (generally the probe) is less than the molecular weight of the nucleic acids bound on the solid support (i.e., on the membrane) [17–19]. In addition, successful filter hybridization depends on two factors: The probe in the solution must diffuse effectively to the solid support, and the probes must interact and hybridize with the nucleic acids on the solid support in the given hybridization conditions. Therefore, at low values of [C_f], the initial rate of hybrid formation is proportional to the $[C_f]$, but the rate does not increase linearly at higher values [10,20–22]. The reassociation is a rate-limiting process when the value of $[C_f]$ is low, whereas a high value of $[C_f]$ increases the hybridization rate rapidly. As a result, the solution surrounding the membrane consumes the probe molecules due to the diffusion of the probe to the membrane for hybrid formation. Therefore, the diffusion of the probe is an important factor for solid support hybridization processes and can be expressed with by a term "J." The term J is a function of the diffusion coefficient and the concentration gradient of the probe in the liquid phase. When $k_1[C_f][C_s]>J>0$, the rate of disappearance of single-stranded nucleic acids in a hybridization reaction can be expressed by the equation

$$-d[C_s]/dt = k_2[C_s]^2$$

In addition, the reassociation of the probe in a hybridization reaction is significant when J is $k_2[C_8]^2$ [10]. From the above discussions, it is clear that if the diffusion of the probe is faster, then the overall hybridization reaction can be accelerated. The factors that help to increase the diffusion of the probe in the liquid phase are (1) smaller probe (15–30 nucleotide long oligonucleotide probes are useful), (2) high incubation temperature (>50°C), (3) gentle shaking of the reaction bag for efficient collision between the probe and the target nucleic acid on the membrane, and (4) minimum amount of hybridization solution in the bag. Therefore, it is important to provide optimum conditions considering the above-mentioned factors for the hybridization reaction, where the probe in the liquid phase diffuses to the solid phase and interacts with the membrane-bound target nucleic acid molecules to form hybrids rather than reassociate with themselves to form "hyperpolymers" or "concatemers." It has been demonstrated by Flavell et al. [21] that approximately 20%–30% of the probe molecules in the liquid phase of a hybridization reaction will interact with themselves and reassociate and will be unavailable for hybridization with the membrane-bound nucleic acids. Also, it has been shown that 10% of a denatured double-stranded probe added in a hybridization reaction are homologous rather than complementary sequences [21].

1.5. FACTORS AFFECTING THE KINETICS OF NUCLEIC ACID HYBRIDIZATIONS

A number of important physical and chemical factors affect each hybridization reaction, which involves a specific probe and target nucleic acid. These factors function during different events such as rate of reassociation and hybrid stability of the nucleic acid hybridizations.

1.5.1. Factor Affecting the Rate of Reassociation

Temperature. The maximum rate of reassociation of a duplex DNA is generally 25°C below the melting temperature (T_m), which is considered to be the optimum temperature for reassociation. Salt concentration in the hybridization reaction plays an important role in the reassociation of a duplex DNA at an optimum temperature [5,23,24]. For example, in the presence of Mg^{2+} ions, the rate at which phage lambda DNA reassociates is reduced by two orders of magnitude when the reassociation is carried out at a temperature 55°C below the T_m value [6].

Salt Concentration. The rate of hybridization increases with higher salt (NaCl) concentrations up to 1.2 M NaCl, which is approximately seven times higher than the standard salt concentration of 0.18 M [24]. Divalent cations have a more pronounced effect than the monovalent cations (Na^+) in the reassociation kinetics of a hybridization reaction. To eliminate the interference of the divalent cations that may be present in many of the reagents, a reasonable amount of a chelating agent such as EDTA is generally added to the hybridization cocktail [6].

Base Mismatches. Given specific hybridization conditions, the reassociation between two single-stranded nucleic acids may occur even when there are mismatches. It has been estimated that if the reassociation is carried out at a temperature that is optimal for the mismatched sequences, that is, 25°C below the T_m value, then the rate of reassociation is reduced at least twofold [6,23].

Fragment Lengths. In solid phase membrane hybridization, the effect of length of probe versus the membrane-bound target is not fully understood. However, the concentration and the secondary structures may play important roles in membrane hybridization. In solution hybridization the rate of zippering is much faster than the nucleation. Once the probe finds its complementary sequence, the first step is the nucleation. Immediately after the nucleation, zippering of the two complementary strands occurs. When the complementary strands are of the same size, the rate of reassociation rises as the square root of the length [25]. This can be expressed by the equation

$$k(L_1) = k(L_2) \times (L_1/L_2)^{1/2}$$

where k is rate constant ($M^{-1} \cdot sec^{-1}$) and L_1, L_2 are average lengths of DNA fragments.

When the size of the probe molecules is smaller than the target molecules, the rate of reassociation increases in proportion to the target fragment length. In contrast, when the target molecules are small and the probe molecules are large a significant reduction in rate of reassociation results.

For the RNA–DNA hybrids, the rate of reassociation is the same as the DNA–DNA hybridization under standard conditions (~0.18 M NaCl). However, if the salt concentration is raised, the rate of reassociation does not rise as fast as the reassociation rate of DNA–DNA hybrids under similar conditions. This is possibly because the presence of higher concentrations of salt stabilizes the secondary structures of RNA molecules caused by random self-interactions.

Complexity of the Nucleic Acids. The repetitive sequences in any genome play an important role in the kinetics of nucleic acid reassociation. The reassociation rate of the genome from prokaryotic organisms, where repetitive sequences are less numerous than in eukaryotic organisms, is inversely proportional to the complexity. The reassociation kinetics of any nucleic acid is largely dependent on the complexity of the genome. This follows the equation

$$C_0 t_{1/2} = 1/k$$

where k is rate constant and $C_0 t_{1/2}$ is 50% reassociation of the total DNA at time t (in seconds). Also, the rate of reassociation is inversely proportional to the length of the reassociating nucleic acids. Therefore, one can express this as

Complexity = Reassociation $\propto$ 1/Complexity (or the genome size)

Base Composition. The presence of total %GC affects the rate of nucleic acid hybridization. Since the GC base pairs have greater thermal stability than the AT pairs, the rate increases as the %GC content in the nucleic acid increases [5,10].

Formamide. Use of formamide in a hybridization reaction between 30% and 50% does not have any profound effect on the rate of reassociation [10]. However, a concentration of 20% formamide reduces the rate only up to one-third in solid phase membrane hybridizations [26]. On the other hand, use of formamide to a concentration of 80% in DNA–DNA hybridization reduces reassociation rate constant by a factor of 3 and for RNA–DNA hybridization by a factor of 12 in solution and possibly in solid phase membrane hybridizations [27].

Dextran Sulfate. Use of an inert polymer such as dextran sulfate to a concentration of 10% in a hybridization reaction increases the rate of reassociation approximately 10-fold [19]. The inert polymer helps to concentrate the DNA in the liquid phase of both solution and solid phase membrane hybridization reactions rather than be dispersed in the volume with less interaction with each other. It has been reported that in the case of single-stranded oligonucleotide probes the use of dextran sulfate increased the reassociation rate by up to three- to fourfold, whereas in the case of double-stranded probes the rate increased by up to 100-fold. As a result, in the double-stranded probes the yield of hybrids increased. This increase in the hybrid molecules was mostly caused by the formation of hyperpolymers and concatemers among the probes, creating extensive networks of reassociated probes with several single-stranded areas open for hybrid formation with the target nucleic acids. Hybridization of these reassociated probes with the target nucleic acids increases the hybridization signals. This leads to an overestimation of the extent of hybridization. For purposes of qualitative determination of a hybridization reaction, the use of dextran sulfate may help when a few target nucleic acid molecules are present in the sample by increasing the hybridization signal per hybridized molecule. However, in the case of a quantitative hybridization assay, this phenomenon should be avoided; therefore the use of dextran sulfate may not be highly recommended. Also, for the same reasons, in many hybridizations dextran sulfate may produce high background signals.

Ionic Strength. The reassociation rate increases as the ionic strength in a hybridization reaction is increased. When the concentration of the sodium in a hybridization reaction is increased twofold above <0.1 M, the rate increases 5–10-fold [5]. However, with a further increase in the concentration of the sodium, the rate increases slowly up to about 1.5 M [24].

Base Mismatches. Many hybridization reactions involve interaction between the probe and the target molecules that are not 100% complementary. Hybridization with mismatched base pairs can lower the rate of hybridization and the melting temperature (T_m) of the hybrids [28]. If a hybridization reaction is carried out at a temperature that is about 25°C below the T_m, the rate of reassociation decreases by a factor of 2 for every 10% mismatch nucleotides between the probe and the target nucleic acids [23,28].

pH. Most hybridization reactions are performed at a pH between 6.8 and 7.4. However, if hybridization is carried out at 0.4 M sodium with a pH between 5 and 9, the rate of hybridization is not affected [5].

Viscosity. The rate of reassociation decreases as the viscosity of the hybridization solution increases. Therefore, when a hybridization solution is used with viscous reagent(s) such as a high concentration (7%) of SDS, a relatively high (60°–65°C) hybridization temperature is recommended.

1.5.2. Factors Affecting the Hybrid Stability

The hydrogen bonds between the nitrogenous bases are the key elements in keeping two single strands together. The thermal stability of the hydrogen bonds in a duplex DNA primarily depends on the melting temperature (T_m) of that DNA. The following factors can affect the stability of a duplex DNA in a hybridization reaction.

T_m *of the DNA–DNA Hybrids.* In a hybridization reaction where the probe and the target DNA manifest 100% complementarity, the T_m value can vary depending on (1) base composition, (2) ionic strength, and (3) denaturing agent [23,29,30]. This can be expressed in an equation as described by Howley et al. [26]:

$$T_m = 81.5 + 16.6(\log M) + 0.41(\% \text{ G} + \text{C}) - 0.72 (\% \text{ formamide})$$

where M is molarity of the monovalent cation (usually sodium) and (% G + C) is percentage of the total guanine and cytosine bases on the DNA.

If an aqueous solution of 1 M NaCl is used in the hybridization mixture, the above equation can be simplified to

$$T_m = 81.5 + 0.41(\%G + C)$$

It has been reported by Howley et al. [26] that the T_m of a duplex DNA in solution hybridization is higher than when used in solid phase membrane hybridization.

Also, in this equation the monovalent cation amount is 0.01–0.4 M NaCl [29], the %G + C is 30%–75% (50% is ideal) [1], and with the use of every 1% formamide the T_m decreases 0.75°C for poly(dA:dT) and 0.5°C for poly(dG:dC) [27].

T_m of the RNA–DNA Hybrids. In the case of RNA–DNA hybrids, it has been reported that the increase in the concentration of formamide and the decrease in the T_m are not linear [10]. Depending on nucleotide sequences (i.e., the %G + C) in an RNA–DNA hybrid, the presence of 80% formamide provides more stability than in the DNA–DNA duplex [27,31]. The advantage of using 80% formamide in RNA–DNA hybrids is that at this concentration formamide represses the DNA–DNA hybrid formation; therefore, the RNA–DNA hybrids are selected [27,32].

Formamide. Since the use of formamide at a concentration between 30% and 50% decreases the T_m of the nucleic acid hybrids in a hybridization reaction, the stability of the probe can be increased by lowering the hybridization temperature [10]. As a result, there will be better retention of the noncovalently bound hybrids on the membrane and less possibility of damaging the nitrocellulose membrane.

Ionic Strength. The stability of a duplex molecule depends on the salt concentrations in the hybridization solution. In general, the higher the salt concentration, the more stable is the hybrid. In the case of mismatched hybrids, a higher salt concentration is required for stability of the duplex molecules than in 100% matched hybrids.

Mismatched Hybrids. In a hybridization reaction, it is important to know the extent of complementarity between the probe and the target nucleic acids. During the hybridization process it is desirable to provide conditions that will stabilize the duplex molecules of interest and discard the nonspecific hybrids. The parameters that affect the rate of reassociation described previously may vary depending on the degree of mismatch between the two nucleic acids. It has been reported that although the exact figure depends on the total G + C content, in general, with every 1% of mismatched bases in a DNA duplex the T_m is reduced by 1°C [23]. Also, the stability of the hybrids will depend on the distribution of the mismatched bases. For example, if 20% of the mismatched bases are concentrated in one region, and the rest of the nucleic acid has perfect matching, the T_m will be relatively higher and the stability of the hybrid can be well maintained. On the other hand, if there is a mismatched base at every fifth position along the hybrid, the duplex will be extremely unstable [10].

1.6. NUCLEIC ACID PROBES

1.6.1. Types of Probes

In nucleic acid hybridization, both single- and double-stranded DNA or RNA are used as probes for the detection of the target DNA. Generally, the probe molecule is labeled either with a radiochemical or with a hapten molecule to provide a signal upon successful base-pairing to the target nucleic acid molecule. Recently, the use of synthetic oligonucleotides became popular in nucleic acid hybridizations.

Double-Stranded DNA Probe. The double-stranded DNA probes have a wide range of applications, especially in gene cloning, screening of a gene library, or colony blot hybridization. Generally, the double-stranded probe is a cloned DNA fragment of a gene or a segment of a DNA with low complexity. Before adding to the hybridization reaction, the double stranded probes are required to be denatured either by boiling followed by quick plunging of the reaction tube on ice or by alkali treatment. In the hybridization reaction, the complementary single-stranded probe sequences undergo two competing reactions; the reassociation of the probes in the solution and hybridization with the target nucleic acids. Therefore, use of single-stranded DNA (especially synthetic oligonucleotide probes) instead of double-stranded DNA is preferable.

Single-Stranded DNA Probe. One of the advantages of using a single-stranded probe is that there is no competing reaction in the solution with the probe itself. Therefore, the interaction is only between the probe and the target nucleic acid molecules, and possibly the reassociation rate is more efficient. Single-stranded synthetic oligonucleotide probes, 16–30 oligonucleotides long, are used for nucleic acid hybridizations and are synthesized in a commercially available DNA synthesizer. Other single-stranded probes, such as genomic DNA or a cloned DNA fragment, can be generated by strand separation of double-stranded DNA from a bacteriophage recombinant, such as a M13 phage recombinant.

RNA Probe. The widespread presence of ribonucleases makes it extremely difficult to use RNA as a probe in nucleic acid hybridizations. The use of a specific mRNA or synthesized cDNA as a nucleic acid probe may not be very useful because they contain a complex mixture of sequences due to the relatively short half-life of most mRNAs and synthesis of incomplete cDNAs. As a result, the entire pool of probe may contain a very small quantity of the actual sequence of interest. The use of a low quantity of the RNA sequences that represent the probe of interest will result in an inefficient hybridization rate. Therefore, it may be almost impossible to get a signal from the hybridization reaction. Also, using RNA polymerase specific for a promoter can be used to generate single-stranded labeled RNA probes from either one of the two DNA strands (Fig. 1.3).

However, it has recently become possible to generate a large quantity of the desired RNA from a cloned gene fragment in a number of specifically designed re-

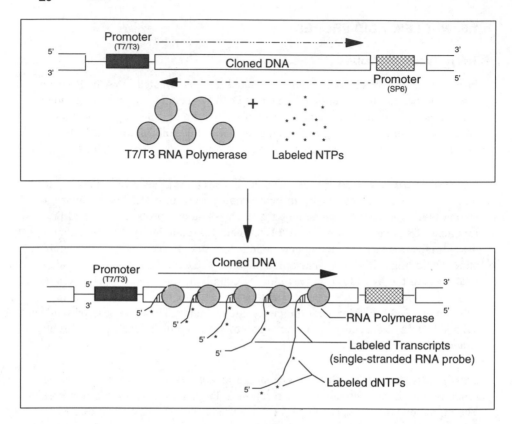

Fig. 1.3. Schematic representation of generation of single-stranded labeled RNA probes. First, the DNA segment from which the probe is to be generated is cloned into a plasmid vector under the control of a T7, T3 or a SP6 promoter. The cloned DNA is subjected to RNA synthesis by adding T7, T3 or SP6 RNA polymerase, labeled NTPs, and a salt buffer. Transcription of the cloned DNA by the appropriate polymerase enzyme generates labeled single-stranded RNA probe, which can directly be used for hybridization. Generally, the DNA segment is cloned flanked by two heterologous promoters. As a result, the labeled RNA probe can be generated either one of the two DNA strands as necessary.

combinant plasmids. These RNAs are single stranded and possess low complexity. As a result, there is little or no competition with itself in the liquid phase of a hybridization reaction, and this increases the reassociation rate with the target nucleic acids.

Polymerase Chain Reaction (PCR)–Generated Probes. Recently, the use of an *in vitro* DNA amplification method called PCR [15] was shown to generate double- or single-stranded DNA probes within a short period of time. To apply this approach, one must know the sequences at the two ends of the nucleic acid of interest. Two oligonucleotide primers are designed and PCR amplification of the DNA sequence of interest is performed in the presence of a very small quantity of the target DNA [15,33,34]. Using unequimolar quantities of two oligonucleotide primers, an asymmetric PCR approach can be used to generate a single-stranded DNA probe of the two strands [8,14,33,35–37] (Fig. 1.4). Also, the labeling of the probe molecules with either radiochemicals or one of the hapten molecules (e.g., biotin) can be achieved during the PCR amplification process [38]. Generation of single- or double-stranded probes by the PCR method is easy and rapid. However, since the fidelity of the thermostable DNA polymerases differs, the choice of polymerase should be considered carefully so as to achieve amplified probe DNA strands with maximum complementarity to the template DNA.

1.6.2. Labeling the Probes

Approaches of Labeling. Labeling of the probe with an indicator molecule is essential to determine successful hybrids following the hybridization reaction. It is necessary to label the probe with such a substance that will not interfere with the hybridization reaction. The sensitivity of the hybridization signal depends on the method of label incorporation, the type of label, and the detection assay. Most of the labeling reactions are performed by various nucleic acid modifying enzymes. Some

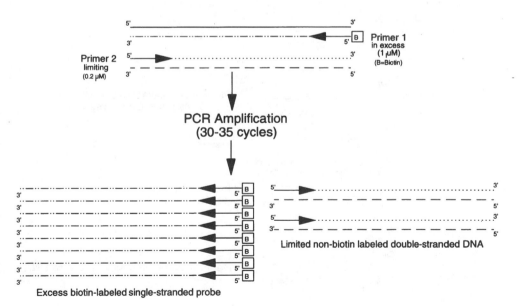

Fig. 1.4. Schematic diagram of generation of probe molecules by the PCR DNA amplification method. The PCR amplification is performed in the presence of a radionucleotide or a hapten molecule that is incorporated during DNA synthesis. Single-stranded DNA probes can be directly generated by the asymmetric PCR method in which one of the two primers is added in less quantity.

of the enzymes commonly used for incorporation of labels in the probe are as follows.

Bacterial Polynucleotide Kinase (PNK). Bacterial PNK incorporates a single labeled nucleotide at the γ-position of the 5'-phosphate of each of the probe molecules [3].

Terminal Deoxyribonucleotidyl Transferase (TdT). TdT incorporates multiple labeled nucleotides as a tail at the 3'-hydroxyl ends of each of the probe molecules [3]. The multiple incorporation of the labeled nucleotides increases the probe signal [13].

Nick Translation. This approach utilizes two DNA modifying enzymes. First, the DNAse I creates a nick at random on the double-stranded DNA probes. Then the DNA polymerase I (DNA pol I) enzyme, with its exonuclease activity, removes nucleotides from 5' ends of the nick. The polymerase activity of the DNA pol I adds the labeled nucleotides to the 3' end of the nick [3] (Fig. 1.5).

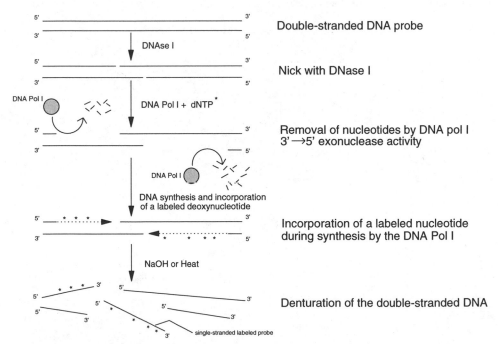

Fig. 1.5. Schematic representation of the "nick-translation" method of labeling DNA probe. First, random nicks on the double-stranded DNA are generated by DNAse I enzyme. Then removal and resynthesis of the DNA strand are performed by DNA Polymerase I enzyme. During DNA synthesis, labeling of the probe is done by incorporating radionucleotides or hapten molecules. Following incorporation of the label, the double-stranded DNA is denatured by treatment with NaOH or boiling water, followed by rapid cooling on ice.

Random Primer Extension Approach. This approach utilizes approximately 6-nucleotides-long synthetic oligonucleotides that bind to the probe molecules randomly and synthesizes the complementary strands by extending the primers with DNA pol I. During primer extension, the labeled nucleotides are incorporated [3]. RNA probes can be generated directly from the DNA template by incorporating labeled ribonucleotides during synthesis [3] (Fig. 1.6).

Direct Labeling. This approach does not utilize any of the nucleic acid modifying enzymes. Photochemically activated labels (e.g., photobiotin) can be mixed with

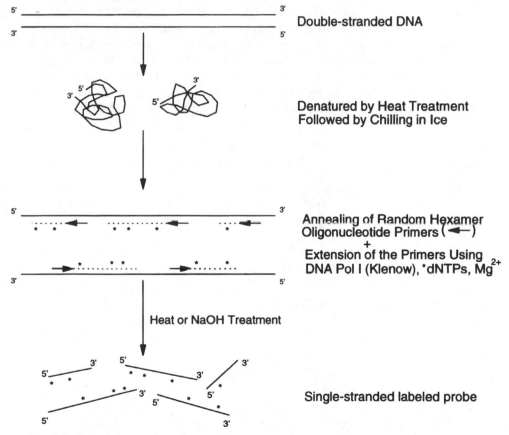

Fig. 1.6. Schematic representations of the "random-priming" procedure by which labeled probes are generated. First the double-stranded DNA molecules are boiled with small (generally six nucleotides long) primers and quickly cooled by placement into ice. The small primers base pair randomly with the target DNA, and the Klenow fragment of the DNA polymerase I synthesizes the new strands on the template during when the radionucleotides or the hapten molecules are incorporated. Following synthesis, the double-stranded DNA is denatured by treatment with NaOH or in boiling water, followed by rapid cooling by placement on ice.

DNA or RNA, and exposure to visible light forms chemical bonds with the probe molecules. A variety of chemical linkers have been developed to bind the reporter molecules such as biotin, or fluorescent or chemiluminescent molecules, to the probe DNA for rapid and sensitive detection of the hybridization assays.

Types of Labels. Two broad categories of labels are commonly used in hybridization assays. The conventional and most sensitive probes are labeled with one of the many radioisotopes such as phosphorus 32 and 33 (^{32}P, ^{33}P), iodine 125 (^{125}I), tritium (^{3}H), sulfur 35 (^{35}S), or carbon 14 (^{14}C). The detection of the labels following hybridization is commonly performed either by autoradiography or by a scintillation counter. The sensitivity of radiolabeled probe hybridization is <1 pg to 1–10 fg of target DNA. The other labels are nonradioactive in nature and are nontoxic and safe. These labels are nucleotides that are chemically bonded with one of the hapten molecules such as biotin or fluorescent or chemiluminescent substances. Upon successful hybridization, generally the biotin-labeled probes are analyzed by color development with peroxidase or alkaline phosphatase enzymes. The fluorescent and chemiluminescent labeled probes are analyzed by autoradiography. The sensitivity of the nonradioactively labeled probe hybridization is between 1 and 5 pg of the target nucleic acids [14].

1.6.3. Determination of T_m Values of the Probes

One of the critical factors in the hybridization reaction is the determination of the T_m of the nucleic acid probes. The choice of a specific hybridization temperature largely depends on the T_m value of the probe to be used in a specific hybridization reaction.

1. For the DNA–DNA hybrids, the T_m is calculated by the equation [28]

$$T_m = 81.5 + 16.6 \times \log [Na^+] - 0.65 \times (\% \text{ formamide}) + 41 \times (G + C)$$

2. For the RNA–DNA hybrids, the T_m is calculated by the equation [39,40]

$$T_m = 79.8 + 18.5 \times \log [Na^+] - 0.35 \times (\% \text{ formamide}) + 58.4 \times (G + C) + 11.8 \times (G + C)$$

3. For synthetic oligonucleotide probes between 16 and 30 nucleotides long, the T_m is calculated by the following equation in the presence of 1 M Na^+:

$$T_m = T_H - 5°C = 2 \times (\text{No. of A} + T) + 4 \times (\text{No. of G} + C) - 5°C$$

The melting temperature is dependent on the length of the probes. The melting profile is broad when probes with variable lengths are used [10]. This is most apparent at short average lengths of hybridized probe in accordance with empirical relationship [24]:

$$T_n - T_m = 650/L$$

where T_n is melting temperature of the long DNA molecules, T_m is melting temperature of the short hybrids, and L is length of the probe in nucleotides. Generally, a stepwise melting of the hybrids using constant temperature with a variable salt concentration and vice versa is performed to determine a precise T_m of a specific probe [10].

1.7. DETERMINATION OF OPTIMAL HYBRIDIZATION CONDITIONS

The optimal hybridization conditions depend on the purpose of the experiment and on the extent of complementarity of the probe with the target nucleic acid. If the probe and the target nucleic acid are expected to have a high degree of complementarity, which will allow only well-matched hybrids to form, the hybridization can be performed in "high stringent" conditions. On the other hand, if the probe and the target nucleic acid have a high degree of mismatch, the hybridization reaction is performed in "low stringent" conditions. In general, the hybridization reaction is performed in three steps: prehybridization, hybridization, and posthybridization wash.

1. **Prehybridization:** Following transfer of the nucleic acid on the solid support, it is subjected to prehybridization with appropriate blocking reagents to precoat the surfaces to prevent the labeled probe from binding nonspecifically to the solid support. Improper blocking may lead to high background signals.
2. **Hybridization:** In the hybridization reaction it is essential that the probes are denatured completely. Appropriate ionic strength, temperature, and time of hybridization are essential for successful hybridization. Hybridization is generally carried out either in aqueous solution or in the presence of formamide.
3. **Posthybridization wash:** The posthybridization wash is performed to remove unhybridized probe from the solid support and to dissociate the probe DNA that is loosely bound to the target nucleic acids. The salt concentration, washing temperature, and the time of treatment are important factors to achieve a stringent or nonstringent hybridization result in a specific hybridization assay.

Before each hybridization reaction it is important to optimize the buffer composition, temperature, and time of incubation. If the hybridization is performed for well-matched hybrids, high temperatures such as 65°–68°C in aqueous solution and 42°C in the presence of 50% formamide, in combination with the washing conditions of low salt concentration ($0.1\times$ SSC) and high temperature (5°–25°C), are recommended [28,41,42]. If the hybridization is to be performed with a poorly matched probe, lower formamide concentration (25%) at 35°–42°C, with washing solution containing high salt concentration (6 × SSC at 37°–55°C) is recommended [28,31,41,42].

In general, for closely related hybrids, either high temperature and low salt concentration or low temperature and high salt concentration are used during the hybridization and washing steps. The time of incubation is found to have profound effects on discriminating the closely related and distantly related sequences in a hybridization reaction. For double-stranded DNA probes, the increase in hybrid formation during incubation longer than three times $C_0t_{1/2}$ is negligible. To determine the time in hours needed to achieve $C_0t_{1/2}$ for any other probe, the following equation is used:

$$n = 1/A \times B/5 \times C/10 \times 2$$

where A is the weight of probe added (in μg), B is the complexity of the probe (which is proportional to the length of the probe in kb), and C is the volume of the reaction (in ml).

1.8. DISCUSSION

The use of nucleic acid hybridization alone or in combination with other biochemical methods has provided a key to solve many problems in almost all areas of the biological sciences. The broad applications of the nucleic acid hybridization techniques include cloning and screening of cloned genes, identification of specific microbial pathogens in clinical, food, and environmental samples, diagnosis of human genetic diseases, and in forensics. It is noteworthy that although introduction of the PCR method in biological sciences has brought about a revolution, the hybridization method is routinely used as a diagnostic tool following PCR DNA amplification to determine the identity of the amplified DNA. Although most of the underlying principles of the nucleic acid hybridization methods are well understood, with the emergence of new technologies, improvement of this approach is necessary to solve many of the unanswered questions in biological and biomedical sciences.

REFERENCES

1. Marmur, J., Doty, P.: Determination of the base composition of deoxyribonucleic acid from its thermal denaturation temperature. J. Mol. Biol. 5:109–118 (1962).

2. Ausubel, F.M., Brent, R., Kingston, R.E., Moore, D.D., Smith, J.A., Sideman, J.G., Struhl, K.: Current Protocols in Molecular Biology. New York: John Wiley & Sons, Inc., 1987.

3. Sambrook, J., Fritsch, E.F., Maniatis, T.: Molecular Cloning: A Laboratory Manual, 2nd ed. Cold Spring Harbor, NY: Cold Spring Harbor Laboratory Press, 1989.

 4. Pardue, M.L.: *In situ* hybridization. In Hames, B.D., Higgins, S.J. (eds.); Nucleic Acid Hybridization: A Practical Approach. Washington, D.C.: IRL Press, 1985.

 5. Wetmer, J.G., Davidson, N.: Kinetics of renaturation of DNA. J. Mol. Biol. 31:349–370 (1968).

 6. Britten, R.J., Davidson, E.H.: Hybridization strategy. In Hames, B.D., Higgins, S.J. (eds.); Nucleic Acid Hybridization: A Practical Approach. Washington, D.C.: IRL Press, 1985.

 7. Britten, R.J., Kohne, D.E.: Repeated sequences in DNA. Science 161:529–540 (1968).

 8. Bej, A.K., Perlin, M.H., Atlas, R.M.: Effect of introducing engineered microorganisms on soil microbial community diversity. FEMS Microbiol. Ecol. 86:169–176 (1991).

 9. Torsvik, V.L., Goksoyr, J., Daae, F.L.: High diversity of DNA in soil bacteria. Appl. Environ. Microbiol. 56:782–787 (1990).

10. Young, B.D., Anderson, M.L.M.: Quantitative analysis of solution hybridization. In Hames, B.D., Higgins, S.J. (eds.); Nucleic Acid Hybridization: A Practical Approach. Washington, D.C.: IRL Press, 1985.

11. Gillespie, D., Speigelman, S.: A quantitative assay for DNA–RNA hybrids with DNA immobilized on a membrane. J. Mol. Biol. 12:829–842 (1965).

12. Hass, M., Vogt, M., Delbecco, R.: Loss of siminial virus 40 DNA–RNA hybrids from nitrocellulose membrane; implications for the study of virus-host DNA interactions. Proc. Natl. Acad. Sci. U.S.A. 69:2160–2164 (1972).

13. Church, G.M., Gilbert, W.: Genomic sequencing. Proc. Natl. Acad. Sci. U.S.A. 81:1991–1995 (1984).

14. Bej, A.K., Mahbubani, M.H., Miller, R., DiCesare, J., Haff, L., Atlas, R.M.: Multiplex PCR amplification and immobilized capture probe for detection of bacterial pathogens and indicators in water. Mol. Cell. Probe 4:353–365 (1990).

15. Saiki, R.K., Gelfand, D.H., Stoffel, S., Scharf, S.J., Higuchi, R., Horn, G.T., Mullis, K.B., Erlich, H.A.: Primer-directed enzymatic amplification of DNA with a thermostable DNA polymerase. Science 239:487–494 (1988).

16. Southern, E.M.: Detection of specific sequences among DNA fragments separated by gel electrophoresis. J. Mol. Biol. 98:503–517 (1976).

17. Lee, C.H., Wetmer, J.G.: Independence of length and temperature effects on the rate of helix formation between complementary ribopolymers. Biopolymers 11:549–561 (1972).

18. Wetmur, J.G.: Excluded volume effects on the rates of renaturation of DNA. Biopolymers 10:601–613 (1971).

19. Wetmur, J.G.: Acceleration of DNA renaturation rates. Biopolymers 14:2517–2524 (1975).

20. Brinsteil, M.L., Sells, B.H., Purdom, I.F.: Kinetic complexity of RNA molecules. J. Mol. Biol. 63:21–39 (1972).

21. Flavell, R.A., Birfelder, E.J., Sanders, J.P., Borat, P.: DNA–DNA hybrid on nitrocellulose filters: General considerations and non-ideal kinetics. Eur. J. Biochem. 47:535–543 (1974).

22. McCarthy, B.J., McConaughy, B.L.: Related base sequences in the DNA of simple and complex organisms. I. DNA/DNA duplex formation and the incidence of partially related base sequences in DNA. Biochem. Genet. 2:37–42 (1968).

23. Bonner, T.I., Brenner, D.J., Neufield, B.R., Britten, R.J.: Reduction in the rate of DNA reassociation by sequence divergence. J. Mol. Biol. 81:123–135 (1973).

24. Britten, R.J., Graham, D.E., Neufeld, B.R.: Analysis of repeating DNA sequences by reassociation. Methods Enzymol. 27E:363–406 (1974).

25. Britten, R.J., Davidson, E.H.: Studies on nucleic acid reassociation kinetics: Empirical equations describing DNA reassociation. Proc. Natl. Acad. Sci. U.S.A. 73:415–419 (1976).

26. Howley, P.M., Isreal, M.F., Law, M.F., Martin, M.A.: A rapid method for detecting and mapping homology between heterologus DNAs. J. Biol. Chem. 254:4876–4883 (1979).

27. Casey, J., Davidson, N.: Rates of formation and thermal stabilities of RNA:DNA and DNA:DNA duplexes as high concentrations of formamide. Nucleic Acids Res. 4:1539–1552 (1977).

28. Beltz, G.A., Jacobs, K.A., Eickbush, T.H., Cherbas, P.T., Kafatos, F.C.: Isolation of multigene families and determination of homologies by filter hybridization methods. Methods Enzymol. 100:266–284 (1983).

29. Schildkraut, C., Lifson, S.: Dependence of the melting temperature of DNA on salt concentrations. Biopolymers 3:195–208 (1965).

30. McConaughy, B.L., Laird, C.D., McCarthy, B.J.: Nucleic acid reassociation in formamide. Biochemistry 8:3289–3295 (1969).

31. Kafatos, F.C., Jones, C.W., Efstratiadis, A.: Determination of nucleic acid sequence homologies and relative concentration by a dot blot hybridization procedure. Nucleic Acids Res. 7:1541–1552 (1979).

32. Schmeckpepper, B.J., Smith, K.D.: Use of formamide in nucleic acid reassociation. Biochemistry 11:1319–1326 (1972).

33. Bej, A.K., Mahbubani, M.H., Atlas, R.M.: Amplification of nucleic acids by polymerase chain reaction (PCR) and other methods. Crit. Rev. Biochem. Mol. Biol. 26:301–334 (1991).

34. Mullis, K.B.: The unusual origin of the polymerase chain reaction. Sci. Am. 262:56–65 (1990).

35. Bej, A.K., Mahbubani, M.H.: Applications of the polymerase chain reaction in environmental microbiology. PCR Methods Appl. 1:151–159 (1992).

36. Chamberlin, J.S., Gibbs, R.A., Ranier, J.E., Nguyen, P.N., Caskey, C.T.: Deletion screening of Duchenne muscular dystrophy locus via multiplex DNA amplification. Nucleic Acids Res. 16:11141–11156 (1988).

37. McCabe, P.C.: Production of single-stranded DNA by asymmetric PCR. In Innis, M.A., Gelfand, D.H., Sninsky, J.J., White, T.J. (eds.); PCR Protocols: A Guide to Methods and Applications. San Diego: Academic Press. 1989.

38. Levenson, C., Chang, C.: Nonisotopically labeled probes and primers. In Innis, M.A., Gelfand, D.H., Sninsky, J.J., White, T.J. (eds.); PCR Protocols: A Guide to Methods and Applications. San Diego: Academic Press. 1989.

39. Bodkin, D.K., Knudson, D.L.: Sequence relatedness of palyam virus genes to cognates of the palyam serogroup viruses by RNA–RNA blot hybridization. Virology 143:55–62 (1985).

40. Suggs, S.V., Hirose, T., Miyake, T., Kawashime, E.H., Johnson, M.J., Itakura, K.: In Brown, D.D. (ed.); ICN–UCLA Symposia of Developmental Biology. San Diego: Academic Press. 1986.

41. Sim, G.K., Kafatos, F.C., Jones, C.W., Koehler, M.D., Efstratuadis, A., Maniatis, T.: Use of the cDNA library for studies on evolution and developmental expression of the chorion multigene families. Cell 18:1303–1316 (1979).

42. Meinkoth, J., Wahl, G.: Hybridization of nucleic acid immobilized on solid support. Anal. Biochem. 138:267–284 (1984).

CHAPTER 2

GENERATION OF NUCLEIC ACID PROBE MOLECULES

CHITRITA DEBROY and CHARLES A. DANGLER
Supelco, Inc., Supelco Park, Bellefonte, PA 16823 (C.D.);
Department of Veterinary Science, The Pennsylvania State University,
University Park, PA 16802 (C.A.D.)

2.1. INTRODUCTION

Nucleic acid probe molecules are an essential feature of almost all systems designed to detect specific nucleic acid sequences. The base sequence of a probe molecule defines the specificity of the corresponding detection reaction. The incorporation of functional chemical moieties into the probe molecule permits a variety of signaling systems to be coupled to the hybridization reaction, permitting the user to monitor and quantify the hybridization reaction. The final chemical structure of the probe molecule may affect the efficiency of the hybridization reaction by influencing the thermal and chemical stability of the probe–target sequence duplexes.

2.1.1. Selection of Appropriate Probes

To develop any nucleic acid probe–based detection system, the primary step is to generate a nucleic acid probe that has the appropriate specificity (i.e., complementary base sequence) for the desired target sequence. Clearly, the strongest hybrid duplex forms when the probe and target sequences are completely complementary. The occurrence of base mismatches in the probe–target hybrid undermines the thermal stability of the duplex. However, in some instances the use of nucleic acid probe sequences that are not perfectly complementary to the target sequence is tolerated or even desirable. This happens particularly when there is only partial conser-

Nucleic Acid Analysis: Principles and Bioapplications, pages 31–45
© 1996 Wiley-Liss, Inc.

vation of the target sequence within a family of common genes and the use of a single, common hybridization probe is preferred. Also, probe sequences with some base pair mismatch can be used when the target sequence is unknown, but is suspected to share homology with another defined target sequence and, therefore, also suspected to be partially complementary to the probe sequence that detects the known target sequence. In both of these cases, the lack of perfect base-pairing between probe and target nucleic acid strands is compensated by reducing the stringency of the hybridization conditions (e.g., reducing hybridization temperature, increasing the salt concentration, reducing formamide concentration) to maintain the stability of the hybrid duplex.

Specificity of the hybridization reaction may also be influenced by the length of the probe molecule. Genetic probes generated by cloning are typically much longer than probes generated by oligonucleotide synthesis. Accordingly, the sequences of the longer probe molecules are more complex, and complementary sequences are less likely to occur randomly. However, oligonucleotide probes typically demonstrate much higher specificity of detection for single base mismatches than the longer probe molecules. This property is a result of the lower melting temperature (T_m) of oligonucleotide probe–target sequence hybrid duplexes in comparison to longer nucleic acid probe molecules with the same (%G + C) content. Consequently, the oligonucleotide probe–target hybrid pair is more easily destabilized by single base mismatches.

Longer nucleic acid probe molecules offer some advantages to oligonucleotide probes. To exploit the longer chain structure of nucleic acid molecules, probe molecules may be designed to hybridize specifically only in defined regions of the molecule, while other regions of the molecule are used to couple functional units (e.g., polyA tracts, restriction site linkers) to the probe molecule. This has been used with great versatility in the polymerase chain reaction (PCR) to direct primer-dependent synthesis of modified target gene sequences. In this case, specific hybridization is only maintained by the 3' end of the probe (i.e., primer) molecules to direct the 5'–3' DNA polymerase activity of the reaction. Longer nucleic acid chains are also capable of accommodating the insertion or incorporation of multiple molecular labels, thereby increasing the density of label per probe molecule, the intensity of signal generated per hybrid pair, and the sensitivity of probe-based detection. In contrast, oligonucleotide probes, due to shorter length, are restricted in label content or label accessibility by steric factors.

In addition to the specific sequence or length of the probe molecule, the chemical characteristics of the probe molecule must be considered. As suggested above, incorporation of modified nucleotide analogues into the probe molecules can be used to couple radioactive and nonradioactive signal detection systems to the probe, which in turn can be used to monitor the progress of the hybridization reaction. Incorporation of biotinylated thymidine analogues has been a common method of capitalizing on nonradioactive, enzyme-based detection systems in nucleic acid detection systems. Chemical moieties to be used as detection labels can also be conjugated to the probe molecule after the nucleotide chain has been synthesized (e.g., photobiotinylation, sulfonation). While chemically modified or labeled bases may be useful as detection markers, it should be noted that modified bases can alter the hybridization reaction between probe and target sequences. Typically, interference in hybrid duplex stability occurs with modified probes due to effects on hydrogen bonding, commonly associated with the use of base analogues, and steric factors, resulting from bulky side chains (e.g., linker arms).

The selection of RNA versus DNA probe molecules deserves special mention. The higher thermal stability of hybrid duplexes containing RNA strands (i.e., RNA–RNA or RNA–DNA duplexes) can be either an asset or a liability, depending on whether the duplexes represent the desired probe–target interaction or nonspecific background. In either case, the use of RNA probes does require higher hybridization stringency conditions to avoid mismatched hybrids. The greater lability of RNA molecules, relative to DNA, also necessitates an extra measure of caution when storing and handling to avoid degradation of the probe.

The use of RNA probes does have several benefits that deserve mention. The typical method of generating RNA probes is through *in vitro* transcription from DNA templates, isolated by cloning or PCR amplification. The transcription reaction can often yield up to 30 μg of RNA transcripts from 1 μg of DNA template. The increased amount of probe can be used to enhance the efficiency of hybridization reactions or to perform more reactions. Background signals generated by non-hybridized probe may also be eliminated by ribonuclease (RNAse) treatment to remove unbound RNA probe during posthybridization processing, thereby increasing the specific signal-to-background ratio and facilitating interpretation of results. A related strategy has been employed in RNase cleavage protection assays.

2.1.2. Resources for Nucleic Acid Probe Sequences

The base sequence of the nucleic acid probe is a primary concern for the construction of a nucleic acid detection system. Obviously, one method of obtaining a specific gene sequence of interest is to sequence the gene in one's own laboratory. Fortunately, the current scientific environment is rich with published sequences from a wide spectrum of origins that are readily accessible for analysis. Computerized databanks (e.g., Genbank, EMBL) are a valuable source of nucleic acid sequences and their corresponding references. These resources are available to investigators with Internet access through Gopher or Worldwide Web servers. Hopefully, the controversial attempts to patent and moderate the use of specific gene sequences will not impair the future usefulness of these databases. The information available on the computerized systems has the added advantage of being readily accessible to the countless programs that allow users to examine and manipulate the raw sequence data. Sequence data can be found in journal articles and occasionally patent descriptions; however, in contrast to computerized databases, sequence data recovered from conventionally published text must be manually entered, often with considerable eyestrain, into a computer-accessible format to take advantage of available software.

2.2. ISOLATION OF NUCLEIC ACID SEQUENCES

Nucleic acid probes may be prepared directly from high molecular weight genomic DNA extracted directly from animal, plant, and bacterial cells by procedures that are essentially similar. This approach is relatively unsophisticated by present standards; however, it may have some utility for crude screening experiments to detect homology between nucleic acid samples, such as might be employed for genomic homology studies. Extracted genomic DNA often must be subjected to some selective refinement, such as restriction endonuclease digestion, to limit the specificity of the probes. However, individual gene sequences for use as probes are poor candidates for preparation by genomic extraction because they represent such a small percentage of the total genomic size. More refined and consistent nucleic acid probe molecules for individual genes are obtained in larger quantities by amplification, cloning, or synthesis.

For a clean DNA preparation from cells and tissues [1–4] conventional procedures are used that commonly include organic extraction of the cell lysate with detergents, proteases, and phenol, which degrade or render the proteins insoluble [5,6]. The nucleic acids remain soluble in an aqueous phase and are ultimately precipitated by ethanol in the presence of high salt concentration. Clean nucleic acid samples, considerably free of cellular contaminants, are required when they are used as substrates for specific enzymes such as reverse transcriptase for cDNA synthesis from extracted RNA or for reactions using various DNA polymerases. These enzymatic processes may serve as the basis for incorporating radioactive or nonradioactive nucleotide markers into the probe sequences to facilitate detection of hybridization results.

2.3. CLONED NUCLEIC ACID SEQUENCES

For preparation of DNA or RNA probes for hybridization, it is frequently desirable to clone the nucleic acid sequences of interest, in the form of DNA, into vectors, which facilitate the copious replication of the desired genetic sequence *in vitro*. This approach also limits the presence of contaminant sequences from the genome of origin, which increase background hybridization resulting in difficulty in interpreting the results. The process of cloning requires fractionating the original genome to isolate a desired specific genetic sequence, which after the cloning procedure will be propagated in the absence of its original cohort genetic sequences. The desired genetic sequence to be cloned is typically isolated by restriction endonuclease digestion from the genomic DNA, reverse transcription from isolated mRNA species to form copy DNA (cDNA), or PCR amplifications of either genomic DNA or cDNA. The latter two approaches in general are more amenable for generating probes for specific gene targets.

The cloned product may be generated efficiently and represents a consistent and stable reagent for use as a probe. Cloned genetic sequences are a clear improvement over isolating probe molecules directly from original genomic sources when needed.

However, this approach has been partially supplanted by PCR amplification technology, which permits the user to generate specific probe copies as needed from a reference source of template DNA or RNA. Clone-derived probes remain especially useful for producing specific probes for genes with uncharacterized sequences (hence, specific PCR primers cannot be defined) or for preparing probes for gene sequences that exist as rare copies or are derived from rare sources (therefore, they make poor candidates for reliable amplification on demand). A further benefit of cloned genetic probes is the ability to prepare large batches (e.g., milligram quantities) of probe sequences using relatively inexpensive means, far below the expense associated with preparing like amounts of DNA through PCR amplification alone.

If the probe sequences are derived from prokaryotic organisms with relatively small genomes, such as viruses or bacteria, purification is relatively simple since the desired sequence represents a significant proportion of the total nucleic acid. The sequence of interest can be inserted into the vector DNA to form a recombinant construct, and the resulting recombinant DNA can be propagated by transfection typically into a fast-growing bacterial system [7,8]. To develop probes from eukaryotic organisms where the genome size is relatively large, the task of isolating a specific probe sequence is of a correspondingly larger magnitude. The solution to this is to generate an intermediate library of cloned gene sequences in which each clone is represented by a distinct recombinant bacterial colony. The library of cloned sequences is then screened to determine which recombinant colony contains the desired cloned gene. The recombinant clone can be identified by the presence of either its DNA or the DNA-associated gene product. Specific mRNA sequences may also be cloned for use as probes; however, cDNA is first synthesized by reverse transcription before cloning. Details of the cloning procedures are available elsewhere [1,4,9].

2.3.1. DNA Probes From Cloned Sequences

For preparing DNA probes from cloned DNA, the cloned recombinant DNA is first isolated by a procedure similar to that described by Birnboim and Doly [10]. The DNA is often digested with appropriate restriction enzymes to cleave the probe DNA from the vector. The DNA probe is separated from the vector DNA by gel electrophoresis [1,4] and recovered from the gel [1]. It is subsequently labeled with nonradioactive or radioactive markers by the procedures described below.

Although standard cloning procedures yield double-stranded DNA, single-stranded DNA may also be used as probes. For generating single-stranded DNA probes, which consist of only one of the two complementary strands of the probe sequence, a double-stranded DNA fragment is inserted into the polycloning site of the double-stranded replication form of M13 bacteriophage or of a vector that contains homologous functional components [11]. Single-stranded DNA for use as probe is generated by replication and release of single-stranded recombinant bacteriophage by infected bacterial cultures. Single-stranded, labeled DNA probes complementary to the sequences encoded by the single-stranded bacteriophage M13 vector are synthesized *in vitro* using synthetic oligonucleotide primers complementary to M13 sequences. The primer is then extended by a DNA polymerase that incorporates the labeled nucleotides into the complementary strand, which is synthesized. The labeled single-stranded DNA probes can be separated in polyacrylamide gels under denaturing conditions as described by Sambrook et al. [1].

2.3.2. RNA Probes From Cloned Sequences

Plasmid vectors have been developed containing polycloning sites downstream from powerful RNA transcription promoters derived from bacteriophages SP6 [12], T7, and T3 [13–15]. This advancement in vector development has made it possible to synthesize labeled single-stranded RNA probes of high specific activity (i.e., the quantity of label per unit amount of probe). The DNA sequence of interest is inserted into the polycloning site of the vector carrying the RNA transcr ption promoter sites. To generate the RNA probe, the plasmid is linearized and incubated *in vitro* with an appropriate DNA-dependent RNA polymerase and nucleotide triphosphates (NTP). The initiation of RNA starts at the bacteriophage promoter, resulting in transcripts complementary to one of the two strands of the template. RNA probes generated by *in vitro* transcription have many advantages over single-stranded DNA probes. They can be labeled to high specific activity directly during the transcription reaction. Because of the higher stability of RNA-containing hybrid pairs, RNA probes can hybridize to their target sequences far more efficiently than homologous DNA probes and yield stronger signals in hybridization reactions than do DNA probes of equal specific activity [16]. Transcription reactions also yield higher numbers of RNA copies per molecule of DNA template than DNA labeling reactions. Consequently, more probe is synthesized (e.g., microgram quantities per reaction) to support more hybridization reactions or higher concentrations of probe per reaction.

2.3.3. Vector Selection

There are many commercially available vectors for generating DNA and RNA probes. Vectors may be selected, in part, based on the size of the fragment to be inserted, the desired cloning sites, the desired orientation of the DNA to be cloned, and the kind of nucleic acid probe to be generated (e.g., single-stranded DNA or RNA). Other useful selection criteria include the presence of selectable markers, which facilitate identification of recombinant clones, and the host system that will be used to propagate the recombinant vector [17].

2.4. OLIGONUCLEOTIDE SYNTHESIS

Synthetic oligonucleotides are also used as probes. These are synthesized either individually or in pools by automated instruments. These instruments can synthesize nanomole amounts of single-stranded oligomers, which can be as long as 70–100 nucleotides. Fortunately, because oligonucleotide synthesizers can be quite expensive, oligonucleotides can be synthesized by various companies both inexpensively and with a rapid turnaround time on orders.

The synthetic oligonucleotide can be either DNA or RNA. The oligomers are assembled one base at a time on solid supports contained in reaction columns. The solid support may have one of the four nucleotide bases attached when purchased. Reagents are delivered to the column to effect addition of successive bases. Five different chemical reactions are performed for each base that is coupled to the growing chain. Intervening wash steps remove excess reagents and reaction byproducts. The polymerization of bases is controlled in part by special derivatives of the nucleotide bases that are protected at key functional groups to prevent side reactions and to provide activation sites. Labeled molecules may be incorporated during the oligomerization process (e.g., biotinylated nucleotides) or subsequent to synthesis (e.g., conjugation with alkaline phosphatase, ^{32}P 5′-end labeling with T4 polynucleotide kinase [1,18]).

A consideration in the design and use of oligonucleotide probes is the necessity of having prior knowledge of the desired target gene sequence in order to specify the exact base sequence of the probe. This differs from the design and use of cloned probes, which may be used without prior knowledge of probe sequence. Because of the short length of the probe, oligonucleotide probes can be susceptible to several confounding molecular interactions that may impair their usefulness in a hybridization reaction. Computer-assisted sequence analyses may be useful to determine whether candidate oligonucleotide probes are free of intramolecular interactions (i.e., self-homology) that might impair hybridization to target molecules and to minimize the likelihood of interaction with nontarget sequences due to cross-homology. If homologies to the nontarget sequence appear greater than 70%, the probe sequence should not be used.

There are several advantages of using synthetic oligonucleotide probes. Because of their typically short length (e.g., 18–30 bases) compared with cloned probes, oligonucleotide probes have low molecular weights and a lower sequence complexity. Large quantities of oligonucleotide probes can be obtained from a single synthesis, making these probes very cost effective. Because of the shorter length and high concentration of probe molecules per hybridization reaction, shorter hybridization times are required to cover certain target sites than with the cloned probes.

A disadvantage of oligonucleotide probes are the weaker signals produced by labeled hybrid pairs than by cloned probes. Although specific activity per unit mass of oligonucleotide probe can be high, label concentration per molecule may be low because of the short length of the molecules. Consequently, if a target sequence hybridizes equimolar quantities of oligonucleotide or cloned probes, the latter will juxtapose a larger amount of label to the site of hybridization and offer greater sensitivity.

2.5. LABELING OF PROBE MOLECULES

2.5.1. Polymerase-Driven Labeling

The following approaches share the common feature of generating labeled probe molecules during polymerase-driven extension of a nucleic acid molecule. Label is incorporated into probe molecules by virtue of labeled nucleotide triphosphates serving as base substrates for the polymerase. Although only one labeled nucleotide base (i.e., ACGT) is typically used, it is possible to use multiple labeled nucleotide bases to enhance the specific activity of the probe. The ratio of radioactive and non-radioactive species of each base may be used to regulate the total yield and specific activity of the probe to be synthesized.

Nick Translation. This is one of the oldest methods for producing labeled double-stranded DNA of high specific activity. The procedure involves limited digestion of the double-stranded DNA template with DNase I to create single-stranded nicks in the DNA. *Escherichia coli* DNA polymerase I acts within the nicked sites to add a nucleotide to the free 3' end using the complementary strand as a template and concurrently remove the free 5' nucleotide. This series of enzymatic activities advances the polymerase in the 5' → 3' direction and incorporates labeled nucleotide into the newly synthesized chain [19,20]. The labeled probe molecules can be separated from the unincorporated nucleotides by procedures described later in the chapter. Standard kit reactions usually employ 1 μg amounts of DNA template. Because the nick translation process essentially results in replacement of original strands with labeled strands, no overall increase in probe mass is achieved. The nick translation process also results in random nicking of the DNA strands that remain unrepaired. When the probe molecules are denatured for use in hybridization reactions, they represent a heterogeneously sized population that appears as a smear if electrophoresed rather than as a single band. The size of these fragments may be important for some hybridization systems, such as *in situ* hybridization. A drawback of this labeling approach is the requirement for incubation at subambient temperature (i.e., 15°–16°C) to avoid the formation of hairpin turns.

Labeling DNA by Random Priming. An alternative to nick translation and developed by Feinberg and Vogelstein [21,22], this method offers a number of advantages. Oligonucleotides (6–12 nucleotides in length) serve as primers for initiation of DNA synthesis on single-stranded templates by DNA polymerases [23]. The oligonucleotides used are heterogeneous in sequence and will form hybrids in many positions along the template DNA molecule. In the standard procedure, a double-stranded DNA template is denatured and hybridized to the random primers. DNA polymerase extends the random primers and synthesizes single-stranded DNA from both template strands [1]. However, because the primer molecules are supplied separate from the template molecules, any nucleic acid molecule may serve as a candidate template for random-primer labeling, including single-stranded DNA and RNA. If RNA is used as the template, RNA-dependent DNA polymerase or reverse transcriptase is used. Probes can be labeled to very high specific activity; however, the reaction is usually conducted with lower amounts of template nucleic acid than is nick translation (e.g., 25 ng vs. 1 μg, respectively). Increases in template DNA content may result in concomitant reduction of specific activity. Because the site of initial strand synthesis is determined by random primer hybridization to the template strands, the probe strands are heterogeneous in size.

Polymerase Chain Reaction. PCR is used to amplify a segment of DNA that is delineated between two regions of known sequence. The approach is discussed in greater depth elsewhere in this volume. Rather than using random primers, as described above, primer pairs of known sequence are selected that flank the desired gene sequence to be amplified. The primers are oriented at the 5' end of complementary strands such that a polymerase-driven extension of these primers results in the synthesis of the intervening genetic sequence [24,25].

The template DNA is first denatured by heating in the presence of a large molar excess of each of the two synthetic oligonucleotide primers and the four dNTPs. The oligonucleotide primers anneal to their target sequence as the temperature is lowered, after which the primers are extended with DNA polymerase. The cycle of denaturation, annealing, and DNA synthesis is then repeated many times. The products of one round of amplification serve as templates for the next; as a result, each cycle theoretically doubles the amount of DNA product, resulting in an exponential increase in the yield of DNA. Although the original protocols for PCR [26–28] used the Klenow fragment of *E. coli* DNA polymerase I, it is now replaced commonly by the more thermostable enzyme *Taq* polymerase [29], which is purified from thermophilic bacterium *Thermus aquaticus* [30]. This enzyme can survive extended incubation at 95°C, is not inactivated by the heat denaturation, and is very efficient during amplification steps.

PCR amplification has been used extensively in the diagnosis of genetic disorders [31,32], for detection of nucleic acid sequences of pathogenic organisms in clinical samples [33,34], for genetic identification of forensic samples including DNA extracted from individual hairs [35,36] or a single sperm [37], and for analysis of mutations in activated oncogenes [38,39]. PCR is used to carry out a variety of tasks, including generation of specific sequences of cloned double-stranded DNA for use as probes. Although this process may be used to isolate gene sequences for the purposes of cloning or diagnosis, our interest here is the use of this approach to produce labeled probe from reference templates. The templates are usually readily available genomic or cloned DNA, which can be prepared to serve as template for the amplification reaction. By controlling the amount of labeled nucleotide and the timing of its addition to the amplification cycles, one can modulate the amount and specific activity of the labeled amplified product synthesized. Because the site of initial strand synthesis is dictated by the known primer sequences, the probe strands are homogeneous in size, unlike probes labeled by random priming or nick translation. The amplified DNA produced after the reaction may have high specific activity and may be used as a probe directly after purifying the product [24].

RNA Transcription. As noted earlier, labeled RNA probes can be made from DNA sequences cloned into one of the commercially available transcription vectors. These plasmid vectors contain polycloning sites downstream from powerful RNA transcription promoter sites for SP6, T3, or T7 RNA polymerases [13–15]. The addition of RNA polymerase in an *in vitro* transcription system in the presence of labeled nucleotide triphosphate precursors allows the synthesis of multiple RNA copies of the DNA insert from either direction to be used as sense or antisense probes [40,41]. RNA molecules generated by *in vitro* transcription have many advantages, as mentioned earlier. For example, they can be synthesized to very high specific activity, and they hybridize to the targets far more efficiently than homologous DNA probes. The RNA yield of transcription reactions usually exceeds the template DNA mass by at least 10-fold. One major disadvantage is the lability of RNA probes; they are highly susceptible to alkaline hydrolysis and degradation by ubiquitous ribonucleases. Caution must be practiced especially to minimize contamination by the latter.

2.5.2. Labeling of Pre-Existing Nucleic Acid Molecules

In some instances it may be useful to label pre-existing molecules, although the value of this approach may be limited for producing hybridization probes because labeled probes generated in this fashion tend to have lower specific activities. Arguably, nick translation could be considered here; however, this process has already been considered above because it involves the degradation of the original strand, with concurrent resynthesis of a replacement strand. In this section, methods that permit the attachment of labeled nucleotides to the ends of preformed nucleic acid strands are discussed. Unlike the methods that incorporate labeled nucleotides during synthesis of the nucleic acid, in which specific activity per molecule is proportional to the length of the molecule, in end-labeling techniques the specific activity per molecule remains the same despite variation in length. By its nature, end-labeling does not create an even distribution of label over the length of the molecule, and under some conditions end-labeled probes may be uniquely susceptible to inactivation by the presence of exonucleases.

Labeling the 5' End of DNA. This procedure is a method of choice for generating labeled DNA fragments to be used as gene probes, especially when oligonucleotides are used as the probe. The transfer of an isotopically labeled phosphate to the 5' terminus of a DNA strand is catalyzed by bacteriophage T4 polynucleotide kinase. Attachment of a labeled phosphate group to the 5' end of the probe strand has no impact on the base sequence of the probe. This approach has special utility because it allows the labeled molecule to be used as a primer and extended from its 3' end by various polymerase-driven reactions.

Labeling the 3' End of DNA. The 3' end of the DNA fragments that have been digested with restriction enzymes that produce protruding 5' termini can be labeled using the Klenow fragment of *E. coli* DNA polymerase I. In this reaction, in the process of generating a blunt end the recessed 3' end is filled using labeled dNTP. However, the selection of labeled nucleotide base is determined by the sequence of the protruding complementary 5' overhang, which may be predicted from the restriction endonuclease used to cleave the DNA.

When the 3' end of the DNA fragment is either blunt or protruding, bacteriophage T4 DNA polymerase may be used [1,4]. This enzyme carries a much more powerful 3' to 5' exonuclease activity than *E. coli* DNA polymerase I and is therefore the enzyme of choice for end-labeling DNA molecules with protruding 3' tails. By virtue of replacement synthesis the enzyme incorporates labeled nucleotide bases. The 3'–5' exonuclease activity of the enzyme first digests double-stranded DNA to produce molecules with recessed 3' termini. On addition of labeled nucleotides, the partially digested DNA serves as primer templates that are regenerated by the T4 DNA polymerase into blunt-end double-stranded DNA. The 3'–5' exonuclease activity degrades the single-stranded DNA much faster than double-stranded DNA so that after a molecule has been digested to its midpoint it will dissociate into half length single strands that will be rapidly degraded. It is therefore, important to stop the exonuclease reaction before the enzyme reaches the center of the molecule. The rate of exonucleolytic digestion is affected by the sequence of the DNA, and, for, statistical reasons, different molecules in the population will be digested to different extents. Consequently, the replacement synthesis method yields a population of molecules that are fully labeled at their ends but that contain progressively decreasing quantities of label in the center.

Terminal deoxynucleotidyl transferase (TDT) catalyzes the addition of nucleotides to the 3' end of nucleic acid strands. The process is not dependent on a complementary template, unlike the DNA polymerase-driven reactions. Consequently, an absolute lengthening of the strands from the 3' ends may occur. The composition of the elongated ends is determined by the free nucleotide bases in the reaction. The user may employ any labeled deoxynucleotide base; however, the use of dTTP, with the subsequent synthesis of poly-dT tails, is discouraged because high background resulting from nonspecific hybridization to polyA tracts on mRNA molecules may be encountered.

T4 RNA ligase may be used to label the 3' end of RNA molecules. However, this approach is far from standard in generating RNA probes for use in hybridization reactions. As noted earlier, RNA molecules are labeled to high specific activity typically during *in vitro* transcription.

2.5.3. Purification and Quantification of Labeled Probes

To eliminate high backgrounds, it is often necessary to remove unincorporated labeled nucleotides from the reaction mixture prior to using the newly labeled probe. Fortunately, some methods for probe synthesis, such as random-primer labeling and RNA transcription, have high incorporation rates, obviating the need for purification. When necessary, this typically is achieved by selective precipitation of polynucleotides, filtration, or chromatographic means.

Ethanol Precipitation. This method of nucleic acid purification is the traditional method for purifying probe molecules from labeling reactions, as well as for concentrating nucleic acid samples. The various conditions suggested for salt concentration, incubation temperature and duration, and centrifugation specifications by different laboratories are a strong argument for superstitious behavior among scientists. With this in mind, the example given below should be recognized as a general approach favored by the authors. The approach works best for DNA and oligonucleotides that are more than 18 nucleotides long. DNA after labeling can be heat treated to inactivate the enzymes that are present in the reaction mixture or extracted with organic solvents such as phenol:chloroform:isoamyl alcohol (50:49:1). The nucleic acid in the aqueous phase can be precipitated by ice-cold ethanol in the presence of 5 M ammonium acetate (pH 5.3) at –20°C [1]. For dilute solutions of DNA (e.g., less than 50 μg/ml), carrier molecules can be added prior to the addition of ethanol. The carrier molecules are unlabeled and may be nonspecific sheared DNA, RNA, or glycogen, which do not interfere with the hybridization reaction. The nucleic acids can also be precipitated from aqueous solution with the cationic detergent cetylpyridinium bromide [1]. The detergent is then removed from the nucleic acid by precipitating with ethanol. The nucleic acid is finally redissolved in the buffer of choice. The method is extremely rapid and efficient and works well with DNA labeled by nick translation, RNA labeled by *in vitro* transcription, and for oligonucleotides.

Column Chromatography.
Gel filtration methods can be used to separate labeled probes from unincorporated nucleotides on the basis of size. Sephadex chromatography is used for this purpose. The separation depends on the abilities of the various sample molecules to enter the pores of the Sephadex beads, which are held in columns. The smaller nucleotide molecules enter the gel pores and are retained for a longer time, while the large probe molecules flow between the beads. To obtain the best separation, the sample is applied to the column in a small volume of buffer and washed into the matrix with small aliquots of buffer. Ready-to-use disposable columns are commercially available. Elution of the radiolabeled DNA can be monitored by a handheld monitor. The fractions collected are pooled and treated with phenol:chloroform:isoamyl alcohol mixture. The aqueous phase can be collected and precipitated with ice-cold ethanol in the presence of 0.1 v of 5 M ammonium acetate (pH 5.3). Probes less than 150 nucleotides in length may be isolated rapidly by chromatography through an alkaline column of Sepharose CL-4B [1].

Centrifugal Filtration.
Labeled nucleic acids may be separated from unincorporated nucleotides by ultrafiltration units. There are several units commercially available for use in centrifuges of many sizes. Because of their large size, labeled probes are retained above a filtration membrane, while small unincorporated molecules pass through the membrane driven by the centrifugation process. The centrifugal filtration process also may be used to concentrate dilute nucleic acid samples because the aqueous solvent also passes through the membrane during centrifugation.

Quantification by Trichloroacetic Acid (TCA) Precipitation.
To ascertain the amount of label incorporated into nucleic acid molecules, a small aliquot from the solution containing the labeled probe can be precipitated with ice-cold TCA (5%). The nucleic acid in the solution will be precipitated by cold TCA, which can be collected over a glass fiber filter and counted in a scintillation counter. The total amount of radioactive label incorporated in probe molecules can be calculated from the percentage of radioactive counts associated with the precipitated sample com-

pared with the total radioactive counts from a similar volume of the reaction. In some instances (e.g., RNA transcription) the total amount of probe synthesized may be inferred from precipitable and total counts. A small amount of radioactive tracer may also be used to confirm and quantify the parallel incorporation of nonradioactive labeled (e.g., biotin, digoxigenin) nucleotides.

REFERENCES

1. Sambrook, J., Fritsch, E.F., Maniatis, T.: Molecular Cloning, A Laboratory Manual. Cold Spring Harbor, NY: Cold Spring Harbor Laboratory, 1989.
2. Keller, G.H., Manak, M.M.: DNA Probes. New York: Stockton Press, 1989.
3. Walker, J.M.: Nucleic Acids: Methods in Molecular Biology, Vol. 2. Clifton, NJ: Humana Press, 1984.
4. Ausubel, F.M., Brent, R., Kingston, R.E., Moore, D.D., Seidman, J.G., Smith, J.A., Struhl, K., Wang-Iverson, P., Bonitz, S.G.: Short Protocols in Molecular Biology. New York: John Wiley and Sons, 1989.
5. Brawerman, G., Mendecki, J., Lee, S.: A procedure for the isolation of mammalian messenger ribonucleic acid. Biochemistry 11:637 (1972).
6. Blobel, G., Potter, V.: Nuclei from rat liver. Isolation method that combines purity with high yield. Science 154:1662 (1966).
7. Hanahan, D., Meselson, M.: Plasmid screening at high colony density. Gene 10:63 (1980).
8. Mass, R.: An improved colony hybridization method with significantly increased sensitivity for detection of single genes. Plasmid 10:296 (1983).
9. Perbal, B.: A Practical Guide to Molecular Cloning, 2nd ed. New York: John Wiley and Sons, 1988.
10. Birnboim, H.C., Doly, J.: A rapid alkaline extraction procedure for screening recombinant plasmid DNA. J. Nucleic Acids Res. &:1513 (1979).
11. Yanisch-Perron, C., Vieira, J., Messing, J.: Improved M-13 phage cloning vectors and host strains: Nucleotide sequences of the M13mp18 and pUC 19 vectors. Gene 33:103 (1985).
12. Green, M.R., Maniatis, T., Melton, D.A.: Human β-globin pre-mRNA synthesized *in vitro* is accurately spliced in *Xenopus oocyte* nuclei. Cell 32:681 (1983).
13. Studier, F.W., Rosenberg, A.H.: Genetic and physical mapping of the late region of bacteriophage T7 DNA by use of cloned fragments of T7 DNA. J. Mol. Biol. 153:503 (1981).
14. Davanloo, P., Rosenberg, A.H., Dunn, J.J., Studier, F.W.: Cloning and expression of the gene for bacteriophage T7 RNA polymerase. Proc. Natl. Acad. Sci. U.S.A. 81:2035 (1984).
15. Tabor, S., Richardson, C.C.: A bacteriophage T7 RNA polymerase/promoter system for controlled exclusive expression of specific genes. Proc. Natl. Acad. Sci. U.S.A. 82:1074 (1985).
16. Casey, J., Davidson, N.: Rates of formation and thermal stabilities of RNA:DNA and DNA:DNA duplexes at high concentrations of formamide. Nucleic Acids Res. 4:1539 (1977).

17. Hanahan, D.: Studies on transformation of *Escherichia coli* with plasmids. J. Mol. Biol. 166:577 (1983).

18. Hames, B.D., Higgins, S.J.: Nucleic Acid Hybridization: A Practical Approach. Oxford: IRL Press, 1985.

19. Rigby, P.W.J., Dieckmann, M., Rhodes, C., Berg, P.: Labeling deoxyribonucleic acid to high specific activity *in vitro* by nick translation with DNA polymerase I. J. Mol. Biol. 113:237 (1977).

20. Maniatis, T., Jeffrey, A., Kleid, D.G.: Nucleotide sequence of the rightward operator of phage λ. Proc. Natl. Acad. Sci. U.S.A. 72:1184 (1975).

21. Feinberg, A.P., Vogelstein, B.: A technique for radiolabeling DNA restriction endonuclease fragments to high specific activity. Anal. Biochem. 132:6 (1983).

22. Feinberg, A.P., Vogelstein, B.: Addendum: A technique for radiolabeling DNA restriction endonuclease fragments to high specific activity. Anal. Biochem. 137:266 (1984).

23. Goulian, M.: Initiation of the replication of single-stranded DNA by *Escherichia coli* DNA polymerase. Cold Spring Harbor Symp. Quant. Biol. 33:11 (1969).

24. Erlich, H.A.: PCR Technology: Principles and Applications of DNA Amplification. New York: Stockton Press, 1989.

25. Innis, M.A., Gelfand, D.H., Sninsky, J.J., White, T.J.: PCR Protocols: A Guide to Methods and Applications. San Diego: Academic Press, 1990.

26. Saiki, R.K., Chang, C.-A., Levenson, C.H., Warren, T.C., Boehm, C.D., Kazazian, H.H. Jr., Erlich, H.A.: Diagnosis of sickle cell anemia and β-thalassemia with enzymatically amplified DNA and nonradioactive allele-specific oligonucleotide probes. N. Engl. J. Med. 319:537 (1985).

27. Mullis, K., Faloona, F., Scharf, S., Saiki, R., Horn, G., Erlich, H.: Specific enzymatic amplification of DNA *in vitro*. The polymerase chain reaction. Cold Spring Harbor Symp. Quant. Biol. 51:263 (1986).

28. Mullis, K.B., Faloona, F.A.: Specific synthesis of DNA *in vitro* via a polymerase catalyzed chain reaction. Methods Enzymol. 155:335 (1987).

29. Saiki, R.K., Gelfand, D.H., Stoffel, S., Scharf, S.J., Higuchi, R., Horn, G.T., Mullis, K.B. Erlich, H.A.: Primer-directed enzymatic amplification of DNA with a thermostable DNA polymerase. Science 239:487 (1988b).

30. Chien, A., Edgar, D.B., Trela, J.M.: Deoxyribonucleic acid polymerase from the extreme thermophile *Thermus aquaticus*. J. Bacteriol. 127:1550 (1976).

31. Wong, C., Dowling, C.E., Saiki, R.K., Higuchi, R.G., Erlich, H.A., Kazazian, H.H. Jr.: Characterization of β-thalassaemia mutations using direct genomic sequencing of amplified single copy DNA. Nature 330:384 (1987).

32. Engelke, D.R., Hoener, P.A., Collins, F.S.: Direct sequencing of enzymatically amplified human genomic DNA. Proc. Natl. Acad. Sci. U.S.A. 85:544 (1988).

33. Kwok, S., Mack, D.H., Mullis, K.B., Poiesz, B., Ehrlich, G., Blair, D., Friedman-Kien, A., Sninsky, J.J.: Identification of human immunodeficiency virus sequences by using *in vitro* enzymatic amplification and oligomer cleavage detection. J. Virol. 61:1690 (1987).

34. Ou, C.-Y., Kwok, S., Mitchell, S.W., Mack, D.H., Sninsky, J.J., Krebs, J.W., Feorino, P., Warfield, D., Schochetman, G.: DNA amplification for direct detection of HIV-1 in DNA of peripheral blood mononuclear cells. Science 239:295 (1988).

35. Almoguera, C., Shibata, D., Forrester, F., Martin, J., Araheim, N., Perucho, M.: Most hu-

man carcinomas of the exocrine pancreas contain mutant c-K-ras genes. Cell 53:549 (1988).

36. Higuchi, R., Von Beroldingen, C.H., Sensabaugh, G.F., Erlich, H.A.: DNA typing from single hairs. Nature 332:543 (1988).

37. Li, H., Gyllensten, U.B., Cui, X., Saiki, R.K., Erlich, H.A., Arnheim, N.: Amplification and analysis of sequences in single human sperm and diploid cells. Nature 335:414 (1988).

38. Bos, J.L., Fearson, E.R., Hamilton, S.R., Verlaan-deVries, M., van Boom, J.H., Van derEb, A.J., Vogelstein, B.: Prevalence of *ras* gene mutations in human colorectal cancers. Nature 327:293 (1987).

39. Farr, C.J., Saiki, R.K., Erlich, H.A., McCormick, F., Marshall, C.J.: Analysis of *RAS* gene mutations in acute myeloid leukemia by polymerase chain reaction and oligonucleotide probes. Proc. Natl. Acad. Sci. U.S.A. 85:1629 (1988).

40. Palazzolo, M., Meyerovitz, E.: A family of lambda phage cDNA cloning vectors, ISWAJ, allowing the amplification of RNA sequences. Gene 52:197 (1987).

41. Johnson, M., Johnson, B.: Efficient synthesis of high specific activity ^{35}S-labeled human β-globin pre-mRNA. Biotechniques 2:156 (1984).

CHAPTER 3

NONISOTOPIC DNA LABELING STRATEGIES

CALVIN P. H. VARY
Maine Medical Center Research Institute,
South Portland, ME 04106

3.1. INTRODUCTION

The chemistries involved in the production and labeling of DNA probes have evolved considerably over the last 10 years. The need for ever more sensitive DNA probe detection systems coupled with the desires of many investigators, institutions, and companies to reduce or eliminate the use of radionucleotides has spurred the refinement of nonisotopic labeling techniques. The purpose of this review is to describe the practical aspects of several commonly used approaches to nonradioisotopic labeling and detection of DNA and RNA probes for use in nucleic acid hybridization assays. No attempt is made to provide an exhaustive survey of all available DNA labeling chemistries. The purpose of this chapter is to provide a summary of those techniques that are of the most practical value to practitioners of hybridization technology. The approaches described fall within the following two broad classes: (1) hapten labeling, typified by the use of affinity labels such as biotin and digoxigenin. These approaches share a reliance on an affinity interaction subsequent to hybridization, necessitating a two-stage recognition format, and are especially useful when labeling nucleic acids derived from biological sources. (2) Direct labeling methods, typified by the covalent DNA–enzyme conjugates. These conjugates are analogous to the much more thoroughly described antibody–enzyme and antibody–antigen conjugates used widely in immunological diagnostics. Finally, several types of systems for signal amplification, as opposed to target am-

Nucleic Acid Analysis: Principles and Bioapplications, pages 47–65
© 1996 Wiley-Liss, Inc.

plification, are described, as well as concerns regarding assay formatting and performance.

3.2. BIOLOGICAL INCORPORATION OF NUCLEOTIDE PRECURSORS

High molecular weight nucleic acids obtained from biological sources, including bacterial plasmids, phage episomal DNAs, or their *in vitro* transcripts can be labeled with nucleotide analogues containing haptenic or antigenic substituents such as biotin [1,2] or digoxigenin [3]. These nucleotide analogues are effectively incorporated using one of several DNA polymerase-based labeling reactions. These regents allow biologically produced DNAs to be labeled to high density. These methods allow incorporation of a range of nucleotide precursors that, in addition to biotinyl nucleotides, include nucleotides carrying derivatizable alkylamine moieties suitable for covalent conjugation to enzymes or other structures.

3.3. INCORPORATION BY NICK TRANSLATION

Escherichia coli DNA polymerase I catalyzes the formation of a phosphodiester bond between the 3'-hydroxyl of a nicked DNA strand [4], produced by limited cleavage with DNAse I, and a label carrying nucleotide such as biotinyl-dUTP [1,2]. DNA pol I also catalyzes the removal of nucleotides from the 5' end of a DNAse I nick. As shown in Figure 3.1, this reaction proceeds with simultaneous incorporation of label and removal of unlabeled DNA from the 5' end of the nicked template in a progressive fashion. In practice, biotinyl-dUTP [5–8] or other affinity labeled nucleotide analogues bearing digoxigenin [9–16] are presented as a mixture with the congeneric nucleotide TTP in most cases. At a commonly used level of 1 part Bio-dUTP to 3 parts TTP, an upper limit of 25% of the T residues will be substituted. Studies have shown that heavy substitutions of T residues with Bio-dUTP can destabilize double helices. At the above levels of substitution, a Bio-dUTP residue will occur, statistically, in about 1 in every 16 base pairs [2]. This level of substitution appears to provide a reasonable compromise between labeling density, and hence signal generating potential, within the stability constraints of double-helix formation.

3.4. INCORPORATION BY RANDOM PRIMED EXTENSION

An alternative procedure to nick translation, shown in Figure 3.2A, involves the use of random hexameric oligonucleotides to prime synthesis, by DNA polymerization, of a labeled complementary strand on the single-stranded DNA template. The Klenow large fragment of *E. coli* DNA polymerase is typically used for this application [17,18]. This enzyme fragment lacks the 5'-exonuclease activity of the DNA pol I holoenzyme. This procedure eliminates the need for optimization of a DNAse

Nick Translation of Double-Stranded DNA

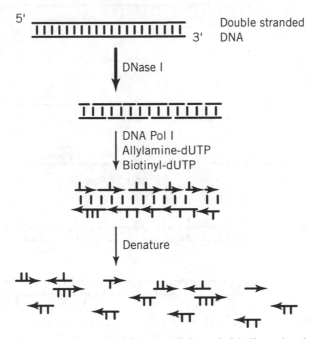

Fig. 3.1. Hapten incorporation method for nonradioisotopic labeling using the "nick translation" reaction.

l digestion reaction prior to labeling and provides labeling efficiencies equivalent to nick translation.

Many other primer-based protocols have been used. Chief among these are reverse transcriptase polymerase chain reaction–based techniques for preparation of nonradioisotopically labeled complementary cRNA probes [11] as well as direct preparation and labeling of probes by the polymerase chain reaction [16].

3.5. TERMINAL LABELING OF DNA

3.5.1. Restriction Fragments

Restriction fragments bearing 3' recessed termini and containing adenosyl residues within the corresponding 5' extension can be end labeled with one or more biotinyl-dUTP residues, in conjunction with the Klenow large fragment of DNA pol I and the other three nucleoside triphosphates as shown in Figure 3.3A [17,19]. This results in a terminally labeled restriction fragment preparation that can be used for nonisotopic detection of restriction fragment patterns or polymorphisms.

Random Primed Elongation

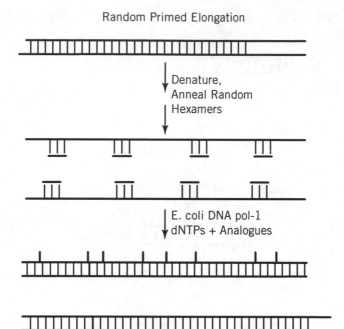

Fig. 3.2. Label incorporation using the "random primering" approach with DNA polymerase.

3.5.2. Linear Duplex DNA

An alternative method for labeling any 3' end of any duplex DNA without prior restriction digestion involves the use of T4 phage DNA polymerase [20]. This enzyme, when presented in the absence of dNTP precursors, catalyzes 3'-directed exonucleolytic hydrolysis of DNA. Shown in Figure 3.3B, this reaction results in 3' recessed DNA ends of extent commensurate with the amounts of enzyme used and the time of incubation. Subsequent addition of all four dNTPs plus the labeled nucleotide analogue of choice allows the polymerase activity to fill in the 3' recessed ends with label. This results in the regeneration of the duplex molecule with label present in the 3' region of each strand.

3.6. INCORPORATION BY TERMINAL TAILING

Perhaps more frequently used as a labeling strategy for oligonucleotides as well as large DNA molecules, 3' terminal tailing reactions, depicted in Figure 3.4A, are useful for incorporation of labels composed of homo- or mixed polynucleotide tracts that reside outside of the region of the probe involved in hybridization.

DNA molecules are commonly "tailed" using the enzyme terminal deoxynu-

Panel A T4 DNA Polymerase Labeling of Double-**Stranded** DNA

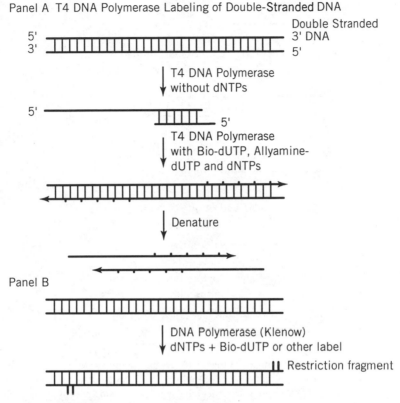

Fig. 3.3. Terminal labeling techniques for duplex DNA: **A:** T4 DNA polymerase. **B:** Restriction gap filling reaction.

cleotidyl transferase (TdT) [21–23]. TdT catalyzes the addition of deoxynucleoside monophosphate residues, derived from biotinylated [7,22] or digoxigenin-labeled deoxynucleoside triphosphates, onto the 3′ terminus of a DNA strand. The reaction proceeds well with mixtures of deoxynucleoside triphosphate substrates, typically dATP, and the labeled nucleotide analogue supplied as 5%–10% of the total deoxynucleotide. Reaction conditions are described for optimized polymerization of labeled homopolymer tracts onto the 3′ terminus of double-stranded or single-stranded DNAs (utilizing Co^{2+} or Mn^{2+} cofactors, respectively). The enzyme can be used to incorporate alkyl- or allylamine groups for subsequent covalent modification as well.

Terminal transferase is also capable of transferring a single ribonucleotidyl residue to the 3′ terminus of a DNA strand [23]. Addition of a second ribonucleotidyl residue is precluded, since a DNA terminus required for phosphodiester formation by this enzyme is no longer present at the ribosyl substituted 3′ terminus. This approach can be used to 3′ label the DNA with a single alkylamine residue (donated by allylamine rUTP) or a single 3′-biotinyl residue derived from biotinyl-

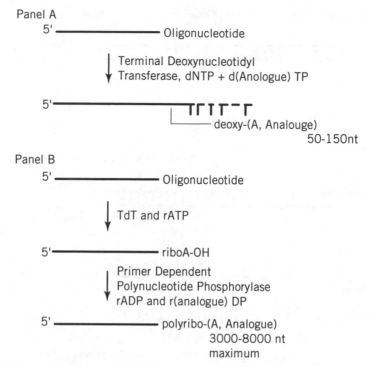

Fig. 3.4. Approaches for terminal tailing DNA oligonucleotides with terminal deoxynucleotidyl transferase. **A:** Oligonucleotide tailing with poly dA and analogues. **B:** Polyriboadenylation of DNA with primer-dependent polynucleotide phosphorylase following monoriboadenylation with TDT.

rUTP. A 3′ ribo-residue also provides a unique site for oxidative coupling to a primary amine through Schiff base formation following oxidation of the 3′ terminal vicinyl diol of the ribose residue. This variation of the TdT reaction also provides an RNA like terminus on the oligonucleotide that can serve as a site for polyriboadenylation by primer-dependent polynucleotide phosphorylase [24]. The latter reaction, depicted in Figure 3.4B, causes polymerization of very long (5–8 kb) polyriboadenylate tracts onto the 3′ terminus of a ribosylated oligonucleotide. These polyA tracts can be transformed into a potent signal generation system involving conversion of the polyA tract to ATP.

3.7. RNA LABELING SCHEMES

Riboprobes can be prepared for nonisotopic detection by incorporation of ribonucleoside triphosphate analogues during DNA-dependent RNA polymerase–directed transcription [13]. Numerous systems are available that allow transcription of DNA sequences into RNA for use as riboprobes. These include SP6, T3, and T7 phage

In Vitro RNA Polymerization Labeling Scheme

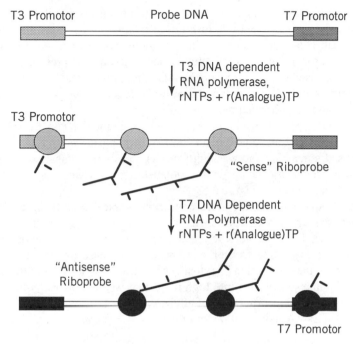

Fig. 3.5. Production of sense and antisense riboprobes from phage RNA polymerase promoters in the presence of hapten-modified ribonucleoside triphosphates.

transcription systems, which provide, as shown in Figure 3.5, considerable flexibility in the production of sense and antisense RNAs from the appropriate transcriptionally competent vector.

Riboprobes prepared from transcription vectors can be labeled for direct covalent modification or indirect affinity labeling using the analogous ribonucleotide substrates. The commonly available ribonucleotide analogues that support transcription include biotinyl rUTP, allylamine rUTP, and fluorescent dye–labeled ribonucleotides.

The advantages of riboprobes over double-strand DNA–derived probes include enhanced sensitivity, which is due primarily to the absence of competing complementary probe strands, and the enhanced stability of RNA–DNA and RNA–RNA hybrids, which can allow greater stringency during hybrid formation. This latter property can also compensate for the destabilizing effects of biotinylation or covalent attachment of enzyme molecules. A disadvantage of riboprobes is that they may require fragmentation, achieved by controlled alkaline hydrolysis, for optimal hybridization especially in applications such as *in situ* hybridization. While biotinyl rUTP is effectively incorporated by the RNA polymerases, digoxigenin-labeled precursors may inhibit these enzymes [13].

3.8. DIRECT CHEMICAL LABELING SCHEMES

A shared disadvantage of the above-described techniques is that they require relatively expensive enzymes and kits to produce relatively small amounts of labeled
probe material. This expense and the limited amounts of mass of probe provided by
the biological production systems limits the concentration that probe can be used in
a hybridization reaction. A consequence of this limitation is that hybridization reactions conducted at these concentrations of probe require long hybridization times,
typically 12–24 hours for maximal sensitivity.

It should be noted that affinity methods, including biotin–avidin, fluorescein–an-
tifluorescein antibody, and digoxygenin–antidigoxygenin antibody, require a second
tier of reagent contacting and washing steps. A schematic comparison of generalized protocols for hybridization with radiolabeled, biotinylated, and direct-labeled
DNA probes is provided in Figure 3.6.

A considerable body of literature concerns chemical modification of nucleic
acids. Chemistries have been developed that achieve O-alkylation of phosphodiester
oxygens, carbonyldiimidazole-based activation and subsequent alkylation of $5'$
phosphates, and bisulfite activation of C residues followed by the introduction of
amino groups at the C5 position of cytosine base. Convenient routes to the introduction of alkylamine substituents into high molecular weight polynucleotides include
polymerase-catalyzed incorporation of allylamine derivatives of rUTP or dUTP and
terminal transferase–catalyzed transfer of hexylamine derivatives of dATP.

Comparison of Generalized Non-Isotopic
Hybridization and Detection Protocols

Affinity Label Protocol	Direct Label Protocol
Southern Membrane Hybridization	Southern Membrane Transfer
Blocking Reaction	Blocking Reaction (15-30 min)
Hybridization Reaction	Hybridization Reaction (15-30 min)
Washing to Remove Unbound Probe	Washing to Remove Unbound Probe (20-30 min)
Incubation in Affinity Label (Avidin-Alkaline Phosphatase)	Enzyme Buffer Conversion (10 min)
Washing to Remove Unbound Reagent	Imaging of Hybrids with Enzyme Substrates (5 min-4 Hhs)
Imaging of Bound Affinity Reagent with Enzyme Substrates	

Fig. 3.6. Schematic summary of the steps involved in the detection of membrane-bound hybrids using direct and indirect labeled probe.

Oligonucleotide probes, however, possess advantages in terms of ease of synthesis, amounts of material available, and target specificity [25–27]. Recognition of the utility synthetic DNA probes and the need for their convenient derivatization has resulted in the commercial availability of several useful reagents for the preparation of synthetic oligonucleotides complete with specifically positioned nucleophilic sites. These reagents, provided as the phosphoramidite derivatives or, in one case, an amino-derivatized column, allow automated synthesis of derivatizable oligonucleotide probes possessing either 5'-3' or single or multiple internal alkylamine groups.

Direct labeling of nucleic acids can be described as involving three stages of synthesis. Native DNA does not possess sites of suitable reactivity for facile covalent modification in an aqueous environment. In addition, most sites found in native DNA will, if modified, alter the affinity or specificity of the base-pairing reactions. Therefore, a site of suitable reactivity must first be incorporated into the polynucleotide. Second, the derivatizable groups must be activated to acceptable levels of electrophilicity or nucleophilicity. This is usually accomplished in a separate step, since the reagents possessing the necessary reactivity for coupling at the relatively low concentrations of macromolecules involved tend to be short lived as a result of competing hydrolysis reactions. The third step involves enzyme coupling and purification. Even relatively efficient coupling schemes leave unacceptable amounts of unreacted precursors. These precursors, especially unbound DNA, must be efficiently removed in order to realize maximum probe sensitivity.

3.9. DERIVATIZATION METHODS

3.9.1. Derivatization of the Oligonucleotide 5' Terminus

Alkylamino phosphoramidite reagents with various capping moieties (N6 TFA, monomethoxytrityl) on the amino group can be added to the oligonucleotide at the last automated coupling step. The structures of the reagent and the resulting linkage in the oligonucleotide are shown in Figure 3.7. An important property of this scheme is that only full length oligonucleotides can bear a derivatizable amino group: Contaminated failure oligonucleotides will not participate in the activation reaction required for enzyme coupling.

3.9.2. Internal Derivatization of the Oligonucleotide

Internal modification of the oligonucleotide can be achieved with either amino derivatives of functional nucleotides [25] or with amine-bearing analogues of the normal synthetic reagents. The latter reagent has the disadvantage that a distorted abasic site results in the DNA molecule, as shown in Figure 3.8. This modification may have an impact on the stability of the resultant hybrids. Both reagents are unique in that they allow the placement of more than one site for coupling in an oligonucleotide. In practice two or more attachment sites may do more to hinder probe function than help it. However, for fluorescent labeling with activated reagents such

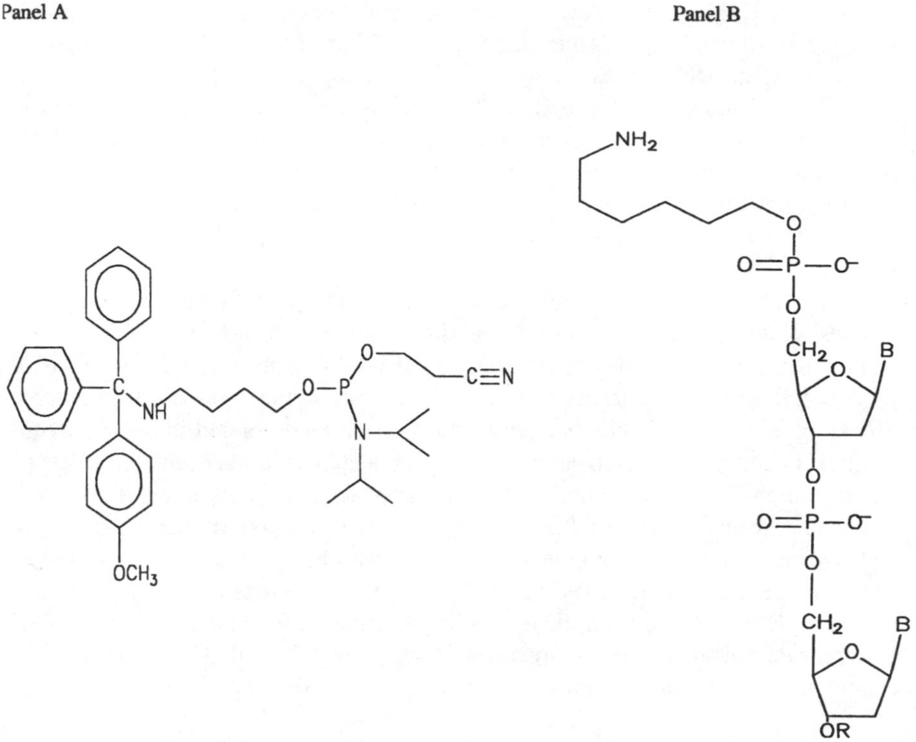

Panel A   Panel B

Fig. 3.7. Reagents for 5′ terminal amino derivatization using automated DNA synthesis. **A:** N6-monomethoxytrityl phosphoramidite, Aminomodifier (Clontech), reagent. **B:** Resulting 5' terminal DNA amine placement.

as the succinimidyl derivative of fluorescein or the sulfonylchloride derivative of Texas Red, multiple dye placement may be of benefit.

3.9.3. Derivatization of the Oligonucleotide 3′ Terminus

A different approach, based on amino-functionalized controlled pore glass supports, is available for automated DNA synthesis. The chemical linkage provided and the resulting oligonucleotide derivative are shown in Figure 3.9. These materials are fundamentally different in that the amino functionality is located at the 3′ terminus of the growing oligonucleotide. A property of this type of derivative is that all oligonucleotides, including failure oligonucleotides, bear a derivatizable amino group [28]. An additional approach involves the initiation of DNA synthesis on a controlled pore glass column bearing the ribo version of one of the four nucleosides. The 3′ terminally directed approaches share the drawback that short failure sequences can compete for activation and coupling reagents and reduce the yield of the desired product. A significant advantage, however, is the elimination of the cost

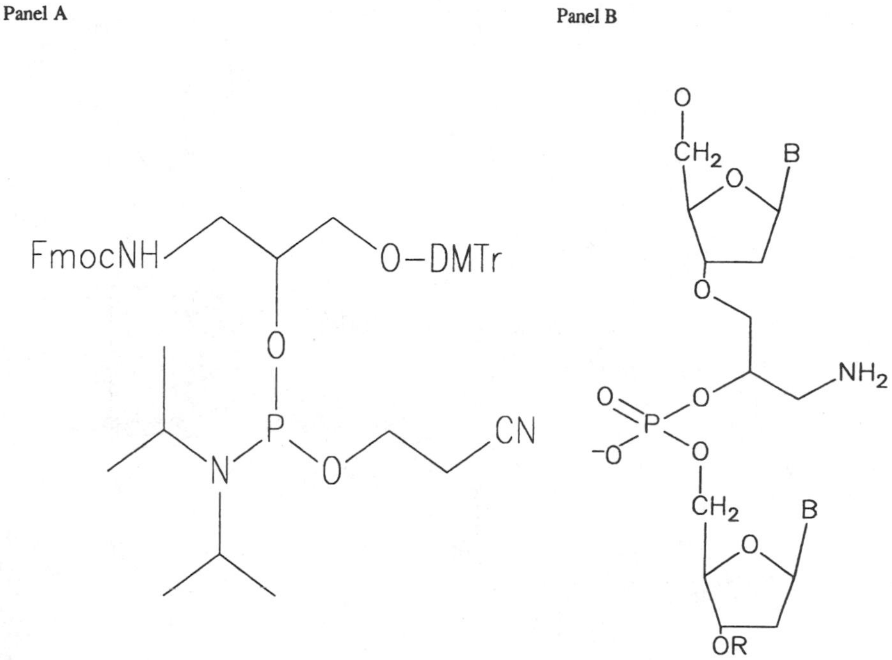

Panel A

Panel B

Fig. 3.8. Reagents for internal amino derivatization using automated DNA synthesis. **A:** Aminomodifier II (Clontech). **B:** Resulting internal DNA amine placement.

and maintenance of an additional set of amidite reagents. In most cases, however, yields of full length oligonucleotides, of length 20–25 bases, are sufficiently high that this is not a problem, though care should be taken to ensure efficient automated DNA synthesis. In addition, relatively straightforward purification schemes will reduce the contribution from these failed sequences.

3.10. PREPARATION AND ACTIVATION OF AMINO LABELED DNA-HETEROBIFUNCTIONAL REAGENTS

Disuccinimidylsuberate (DSS) was one of the first crosslinking agents used to activate an alkylamino group placed in an oligonucleotide during automated DNA synthesis [25]. Homobifunctional crosslinking reagents, such as DSS, must be used in considerable molar excess (80–100-fold) with respect to the derivatizable amino groups in order to avoid dimerization of DNA strands. Efficient removal of this excess reagent must occur before the activated DNA can be subsequently coupled to the enzyme. Since activated DNA represents, at most, 1%–3% of the total DSS reagent on a molar basis, efficient removal of this excess DSS can require rigorous and therefore time-consuming purification steps. Unfortunately, the property that makes the succinimidyl ester of DSS so useful for derivatization of amino groups,

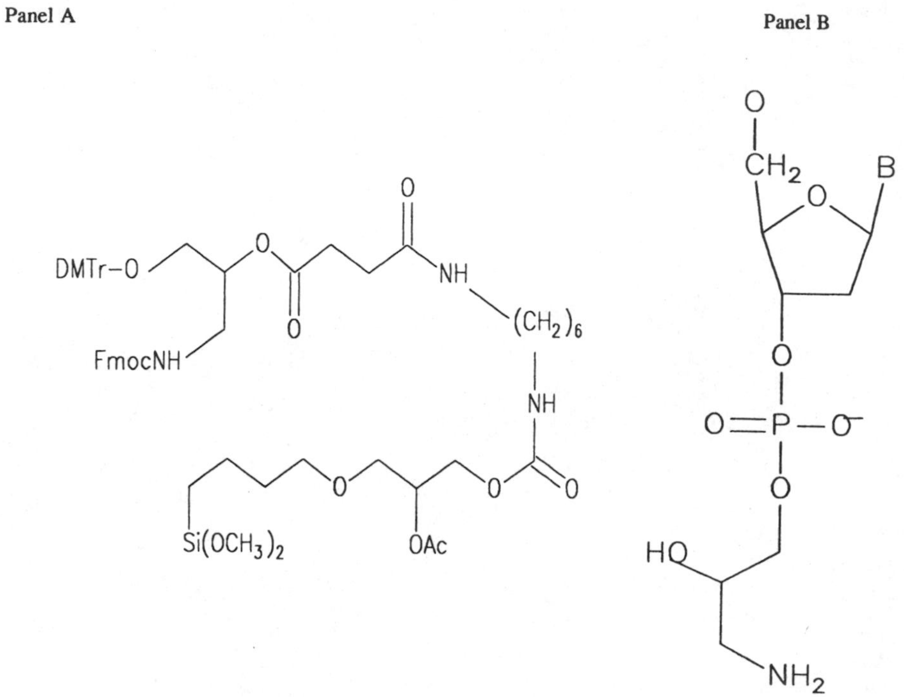

Panel A   Panel B

Fig. 3.9. Reagents for 3′ terminal amino derivatization using automated DNA synthesis. **A:** Structure of the 3′ amine-on (Clontech) reagent. **B:** Resulting 3′ terminal DNA amine placement.

i.e., the relatively high electrophilicity of the reagent, makes DSS relatively prone to hydrolysis during purification. Despite steps to drop the pH and freeze the DNA-succinimidyl intermediate immediately following coupling, adequate yields of final product have been achieved only with difficulty. Given this situation, the extra time and materials involved in the use of heterobifunctional crosslinking reagents (see below) are well worth the investment as yields are high, quality of the product is reproducible, and both the activated A-pase and DNA intermediates can be stored frozen under appropriate conditions. Batches of "activated" reagents can be prepared in larger amounts and hence more efficiently. These properties permit greater quality control of the reagents and better manufacturing practices.

A convenient route to activation of primary amine containing molecules, including but not limited to synthetic amino-labeled DNA or proteins with accessible nonessential lysine residues, involves formation of an amide linkage between the amino function on the DNA or protein and carbonyl group provided as the succinimidyl ester. Commercially available reagents such as succinimidyl 4-(*N*-maleimidomethyl)cyclohexane-1-carboxylate (SMCC) provide rapid and efficient conversion of amino groups to the maleimidyl derivative. An example of an efficient conjugation scheme is portrayed in Figure 3.10. An alkylamino-derivatized synthet-

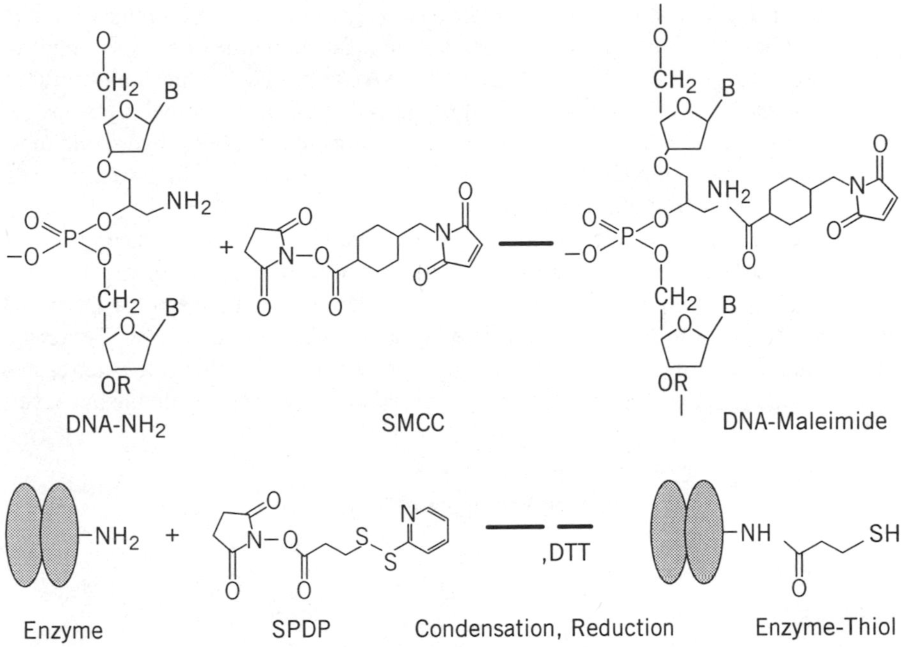

Fig. 3.10. General amine-specific modification scheme illustrated for an aminomodified oligonucleotide and a protein, alkaline phosphatase.

ic oligonucleotide is reacted with SMCC. This reaction efficiently converts the alkylamine to an amide bearing a linker arm terminated with a maleimide group. In contrast to the homobifunctional reagent DSS, the heterobifunctional reagent cannot dimerize amine-bearing molecules. These reagents can, as a result, be used at lower levels of molar excess, i.e., 2–10-fold. This property facilitates rapid and efficient purification of the desired product. Furthermore, the resulting maleimide derivative is much less susceptible to hydrolysis and can be stored under appropriate conditions for months prior to use [28,29].

The maleimide functionalization allows the DNA to couple, rapidly and efficiently, with any sulfhydryl-bearing compound. In a separate reaction, alkaline phosphatase is reacted with either 3-imino-thiolane (2-Imt, Traut's reagent) or *N*-succinimidyl 3-(2-pyridyldithio)propionate (SPDP), as shown in Figure 3.10. Both reagents give good results, and which reagent is used depends on the needs of the investigator. Traut's reagent is somewhat harder to control in terms of determination of optimal reaction conditions but provides a single-step reaction to convert amines to thiol derivatives. SPDP can be used under the same conditions as determined for SMCC but requires a subsequent reduction reaction for release of the thiol prior to coupling. The A-pase–SPDP product can be stored under appropriate conditions. The activation of the SPDP thiol results in the release of stoichiometric amounts of pyridinethiol. The amount of this compound released upon reduction can be determined spectrophotometrically and provides a direct measure of the amount of reactive thiol present. These properties offer advantages if large-scale preparation and quality control measures are to be in force. Thiol functionalities may also be introduced with succinimidyl-*S*-acetylthioacetate (SATA). SATA is especially useful as an alternative to SPDP since the thiol-based reduction and deblocking steps required for SPDP can be replaced with a deblocking step involving hydroxylamine. The latter reagent, unlike the thiol-based reducing reagents, will not react with maleimide-derivatized oligonucleotides.

These reactions ultimately result in the conversion of amino groups on the protein to thiols. Mixture of the maleimide derivative of the DNA and the thiol derivcative of A-pase results in the formation of a covalent bridge between the DNA and the enzyme, as illustrated in Figure 3.11. Blocking steps involving 2-mercaptoethanol, to cap any unreacted maleimide, and *N*-ethylmaleimide to cap unreacted thiol functions are necessary to achieve maximal product stability and the low levels

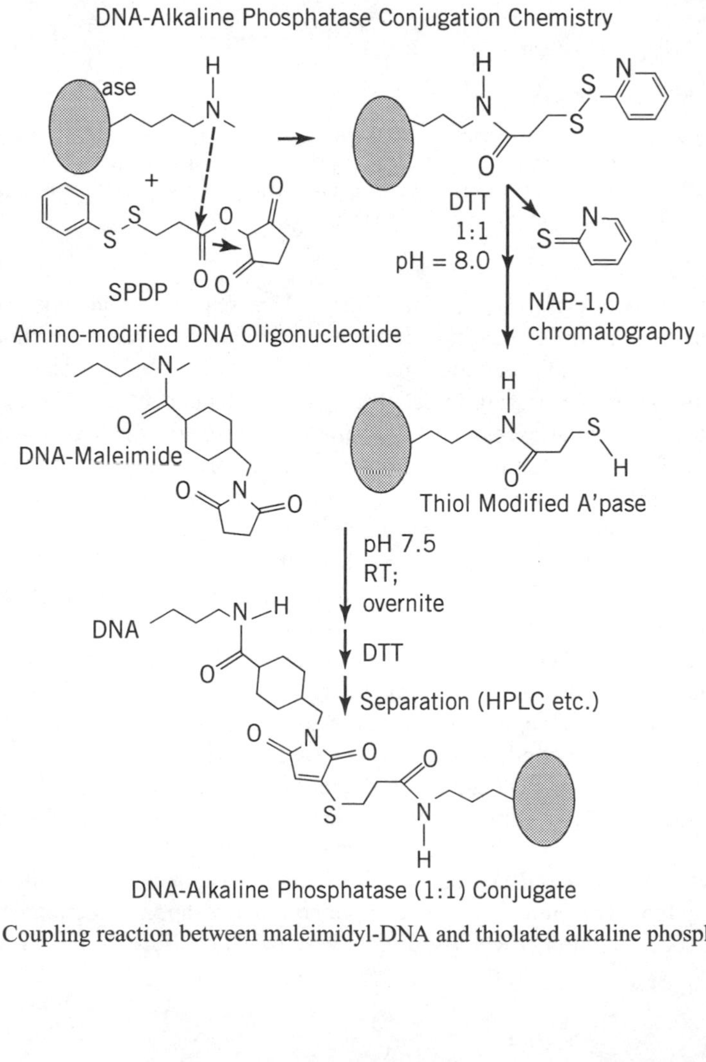

Fig. 3.11. Coupling reaction between maleimidyl-DNA and thiolated alkaline phosphatase.

of background binding typical of these types of conjugate DNA probes. A purification step, such as ion exchange chromatography, is required to remove unreacted DNA strands. The resulting conjugate can be used for hybridizations or stored for an indeterminate amount of time at 4°C. We prepare about 100 nm of conjugate in a single reaction. This is sufficient material for about 30 liters of hybridization mix. One conjugate preparation has been used for at least 1–2 years without loss of potency.

3.11. SIGNAL AMPLIFICATION AND DETECTION

3.11.1. Substrate-Based Amplification

The number of signaling events that occur for every probe-binding event is of critical concern in the design and performance of a hybridization assay. Many approaches have been tried to increase the signal gain realized from each hybridization event. These approaches include direct conversion of nucleic acid to an amplified bioluminescent signal [24,30]. However, the simplest and most widely used systems for signal amplification are luminescent, colorimetric, and fluorescence-based enzyme–substrate systems.

3.11.2. Alternative Enzymes for Direct Conjugation

Alkaline phosphatase appears to possess superior properties for standard types of blot hybridization applications [25]. These properties include excellent stability when exposed simultaneously to combinations of protein denaturants such as sodium laurel sulfate (SDS) and moderately high temperatures (50°–65°C). The enzyme possesses adequate rates of catalysis, insensitivity to product inhibition, and adequate affinities for a variety of substrates. These properties ultimately determine the limiting number of enzyme molecules detectable. When other assay formats such as microtiter plate–based assays are envisaged, assay design can include consideration of potential benefit from the use of an alternative enzyme label. Several candidates are discussed briefly.

Horseradish Peroxidase (HRPO). HRPO, found in plant sources, carries covalently attached carbohydrate residues. Some of these residues bear vicinal hydroxyl groups, which are susceptible to facile oxidative cleavage using sodium periodate. This reaction results in the ultimate conversion of the sugar hydroxyl groups to aldehyde functions. DNA molecules bearing an alkylamine group will form a reversible Schiff base with the sugars. Subsequent treatment with a reducing agent such as sodium cyanoborohydride will reduce the Schiff linkage. This series of reactions results in stable covalent linkage between the alkylamine group and the HRPO molecule. HRPO is detectable with colorimetric and luminescent reaction schemes [31–35]. Enhanced luminescence, discussed in the following chapter, is a particularly sensitive means for the detection of HRPO. Silver enhancement schemes are also available for enhanced detection of *in situ* hybridization applications.

β-Galactosidase. The enzyme β-galactosidase is of bacterial origin and can be used in *in situ* applications without fear of background contributions from endogenous sources. Colorimetric and luminescent substrates are available [31]. Novel fluorescent substrates may be used to enhance detection of DNA sequences in *in situ* hybridization applications and especially in chromosomal analysis.

3.12. SUMMARY

Nonradioisotopic detection of nucleic acids has advantages over radioisotopic techniques. These include personnel safety and elimination of radiation regulatory, containment, and disposal requirements. The disadvantages in the use nonisotopic systems, however, are significant. Biotinylated polynucleotides have not always achieved the sensitivities expected for the level of hapten substitution achieved, and background nonspecific binding can be difficult to eliminate. Direct enzyme-linked oligonucleotide probes achieve the levels of sensitivity expected for a single enzyme label, with good nonspecific background binding properties. A single enzyme label is, however, just barely sensitive enough for detection of single-copy genomic sequences. Hybridization times are, however, extremely short compared with biologically prepared materials. In applications using polymerase chain reaction (PCR) or other amplification schemes, direct labeled oligonucleotide conjugates provide hybridization times of 5–30 minutes, and visualization of bound probe can often be achieved in minutes to hours. These probes avoid the additional reagents, contacting and washing steps, and excessive hybridization times associated with biotinylated DNA probe systems. Particular assay formats need to reflect the requirements of sensitivity, ease of use, and auxiliary equipment requirements [36–38].

In summary, biotinylated high molecular weight cosmid probes would seem to be the choice when high sensitivity is required. These probe constructions have proven utility in *in situ* applications and in unamplified infectious disease detection. Some use of direct enzyme-labeled conjugate probes for imaging of mRNA species using *in situ* techniques are reported. If a detection limit of 10^5–10^6 copies is acceptable, as with repetitive sequence applications, detection of middle to high abundance RNA species or following amplification by PCR or other amplification schemes, the advantages offered by oligonucleotide–enzyme conjugates are considerable. These advantages include automated synthesis, which avoids cloning or PCR-based schemes to obtain the nucleic acids for probe material.

REFERENCES

1. Langer, P.R., Waldrop, A.A., Ward, D.C.: Enzymatic synthesis of biotin-labeled polynucleotides: Novel nucleic acid affinity probes. Proc. Natl. Acad. Sci. U.S.A. 78:6633 (1981).

2. Leary, J.J., Brigati, D.J., Ward, D.C.: Rapid and sensitive colorimetric method for visualizing biotin-labeled DNA probes hybridized to DNA or RNA immobilized on nitrocellulose: Bio-blots. Proc. Natl. Acad. Sci. U.S.A. 80:4045 (1983).

3. Durham, S.K., Goller, N.L., Lynch, J.S., Fisher, S.M., Rose, P.M.: Endothelin receptor B expression in the rat and rabbit lung as determined by *in situ* hybridization using nonisotopic probes. J. Cardiovasc. Pharmacol. 22:S1 (1993).

4. Kelly, R.G., Cozzarelli, N., Deutscher, M.P., Lehrman, I.R., Kornberg, A.: Enzymatic synthesis of deoxyribonucleic acid. J. Biol. Chem. 245:39 (1970).

5. Rose, M., Nagai, T., Niemann, C.: Nonisotopic labelling of a variable number of tandem repeat probe by nick translation with biotinylated nucleotides. Exp. Clin. Immunogenet. 7:200 (1990).

6. Bashir, R., Hochberg, F., Singer, R.H.: Detection of Epstein-Barr virus by *in situ* hybridization. Progress toward development of a nonisotopic diagnostic test. Am. J. Pathol. 135:1035 (1989).

7. Singer, R.H., Byron, K.S., Lawrence, J.B., Sullivan, J.L.: Detection of HIV-1–infected cells from patients using nonisotopic *in situ* hybridization. Blood 74:2295 (1989).

8. Duggan, M.A., Inoue, M., McGregor, S.E., Gabos, S., Nation, J.G., Robertson, D.I., Stuart, G.C.: Nonisotopic human papillomavirus DNA typing of cervical smears obtained at the initial colposcopic examination. Cancer 66:745 (1990).

9. Kim, C.M., Graves, L.M., Swaminathan, B., Mayer, L.W., Weaver, R.E.: Evaluation of hybridization characteristics of a cloned pRF106 probe for *Listeria monocytogenes* detection and development of a nonisotopic colony hybridization assay. Appl. Environ. Microbiol. 57:289 (1991).

10. Miyazaki, M., Nikolic-Paterson, D.J., Endoh, M., Nomoto, Y., Sakai, H., Atkins, R.C., Koji, T.: A sensitive method of non-radioactive *in situ* hybridization for mRNA localization within human renal biopsy specimens: Use of digoxigenin labeled oligonucleotides. Intern. Med. 33:87 (1994).

11. Young, I.D., Stewart, R.J., Ailles, L., Mackie, A., Gore, J.: Synthesis of digoxigenin-labeled cRNA probes for nonisotopic *in situ* hybridization using reverse transcription polymerase chain reaction. Biotech. Histochem. 68:153 (1993).

12. McCreery, T., Helentjaris, T.: Production of hybridization probes by the PCR utilizing digoxigenin-modified nucleotides. Methods Mol. Biol. 28:67 (1994).

13. Heer, A.H., Keyszer, G.M., Gay, R.E., Gay, S.: Inhibition of RNA polymerases by digoxigenin-labeled UTP. Biotechniques 16:54 (1994).

14. Terenghi, G., Polak, J.M.: Detecting mRNA in tissue sections with digoxigenin-labeled probes. Methods Mol. Biol. 28:193 (1994).

15. Davies, E., Hodge, R., Isaac, P.G.: Hybridization and detection of digoxigenin probes on RNA blots. Methods Mol. Biol. 28:121 (1994).

16. Saeki, K., Mishima, K., Horiuchi, K., Hirota, S., Nomura, S., Kitamura, Y., Aozasa, K.: Detection of low copy numbers of Epstein-Barr virus by *in situ* hybridization using non-radioisotopic probes prepared by the polymerase chain reaction. Diagn. Mol. Pathol. 2:108 (1993).

17. Jacobsen, H., Klenow, H., Overgaard-Hansen, K.: The N-terminal amino acid sequences of DNA polymerase I from *Escherichia coli* and of the large and small fragments obtained by a limited proteolysis. Eur. J. Biochem. 45:623 (1974).

18. McCreery, T., Helentjaris, T.: Production of DNA hybridization probes with digoxigenin-

modified nucleotides by random hexanucleotide priming. Methods Mol. Biol. 28:73 (1994).

19. Sanger, F., Nicklen, S., Coulson, A.R.: DNA sequencing with chain terminating inhibitors. Proc. Natl. Acad. Sci. U.S.A. 74:5463 (1977).

20. O'Farrel, P.: Replacement synthesis method of labeling DNA fragments. Bethesda Res. Lab. Focus 3:1 (1981).

21. Roychoudhury, R., Jay, E., Wu, R.: Terminal labeling and addition of homopolymer tracts to duplex DNA fragments by terminal deoxynucleotidyl transferase. Nucleic Acids Res. 3:101 (1976).

22. Fain, J.S., Bryan, R.N., Cheng, L., Lewin, K.J., Porter, D.D., Grody, W.W.: Rapid diagnosis of *Legionella* infection by a nonisotopic *in situ* hybridization method. Am. J. Clin. Pathol. 95:719 (1991).

23. Wu, R., Jay, E., Roychoudhury, R.: Nucleotide sequence analysis of DNA. Methods Cancer Res. 12:87 (1976).

24. Vary, C.P., McMahon, F.J., Barbone, F.P., Diamond, S.E.: Nonisotopic detection methods for strand displacement assays of nucleic acids. Clin. Chem. 32:1696 (1986).

25. Jablonski, E., Moonmaw, E.W., Tullis, R.H., Ruth, J.L.: Preparation of oligodeoxynucleotide–alkaline phosphatase conjugates and their use as hybridization probes. Nucleic Acids Res. 14:6115 (1986).

26. Rotbart, H.A., Eastman, P.S., Ruth, J.L., Hirata, K.K., Levin, M.J.: Nonisotopic oligomeric probes for the human enteroviruses. J. Clin. Microbiol. 26:2669 (1988).

27. Hyman, B.T., Wenniger, J.J., Tanzi, R.E.: Nonisotopic *in situ* hybridization of amyloid beta protein precursor in Alzheimer's disease: Expression in neurofibrillary tangle bearing neurons and in the microenvironment surrounding senile plaques. Brain Res. Mol. Brain Res. 18:253 (1993).

28. Carmody, M.W., Vary, C.P.H.: Inhibition of DNA hybridization following partial dUTP substitution. Biotechniques 15:692 (1993).

29. Ghosh, S.S., Kao, P.M., McCue, A.W., Chappelle, H.L.: Use of maleimide-thiol coupling chemistry for efficient syntheses of oligonucleotide–enzyme conjugate hybridization probes. Bioconjug. Chem. 1:71 (1990).

30. Vary, C.P.H.: A homogeneous nucleic acid hybridization assay based on strand displacement. Nucleic Acids Res. 15:6883 (1987).

31. Bronstein, I., Fortin, J., Stanley, P.E., Stewart, G.S.A.B., Kricka, L.J.: Chemiluminescent and bioluminescent reporter gene assays. Anal. Biochem. 219:169 (1994).

32. van Gijlswijk, R.P., Raap, A.K., Tanke, H.J.: Quantification of sensitive non-isotopic filter hybridizations using the peroxidase catalyzed luminol reaction. Mol. Cell Probes 6:223 (1992).

33. Pollard-Knight, D., Read, C.A., Downes, M.J., Howard, L.A., Leadbetter, M.R., Pheby, S.A., McNaughton, E., Syms, A., Brady, M.A.: Nonradioactive nucleic acid detection by enhanced chemiluminescence using probes directly labeled with horseradish peroxidase. Anal. Biochem. 185:84 (1990).

34. Stone, T., Durrant, I.: Hybridization of horseradish peroxidase–labeled probes and detection by enhanced chemiluminescence. Methods Mol. Biol. 28:127 (1994).

35. Durrant, I., Stone, T.: Preparation of horseradish peroxidase–labeled probes. Methods Mol. Biol. 28:89 (1994).

36. Urdea, M.S., Warner, B.D., Running, J.A., Stempien, M., Clyne, J., Horn, T.: A compari-

son of non-radioisotopic hybridization assay methods using fluorescent, chemiluminescent and enzyme labeled synthetic oligodeoxyribonucleotide probes. Nucleic Acids Res. 16:4937 (1988).

37. Park, J.S., Kurman, R.J., Kessis, T.D., Shah, K.V.: Comparison of peroxidase-labeled DNA probes with radioactive RNA probes for detection of human papillomaviruses by *in situ* hybridization in paraffin sections. Mod. Pathol. 4:81 (1991).

38. Diamandis, E.P., Hassapoglidou, S., Bean, C.C.: Evaluation of nonisotopic labeling and detection techniques for nucleic acid hybridization. J. Clin. Lab. Anal. 7:174 (1993).

CHAPTER 4

APPROACHES TO GENETIC ANALYSIS: GENETIC DISEASE AND SCREENING

LINDA L. McCABE and EDWARD R. B. McCABE
Department of Pediatrics, University of California, Los Angeles, CA 90024-1752

4.1. INTRODUCTION

In this chapter we review genetic analysis using the application of molecular genetic technology to newborn screening as an example of how DNA methods can be utilized to detect individuals with genetic disease within large populations.

Screening is a concept that is integral to all medical evaluations. The traditional history and physical examination are in fact screening tests that cover a broad range of topics and organ systems, with the physician focusing more carefully on those areas that the screening evaluation indicates may be sources of pathology. In the genetic paradigm we discuss, the history and physical examination would be considered a screen targeting the phenotype of the individual.

While the diagnosis of genetic disease can be accomplished utilizing the phenotype and/or genotype, most genetic screening, be it at the clinical or laboratory level, has been phenotypic. As we discuss in this chapter, the modern tools of molecular genetics are beginning to permit screening for genotype.

We will use two terms, *screening* and *diagnosis,* that have very discrete meanings. Screening refers to the evaluation of large populations of individuals with the intent to ascertain a specific disorder so that appropriate intervention can be instituted. Screening tests should be relatively simple and inexpensive and will have the inherent problem of identifying false positives (individuals without disease who are called positive on the screening test) and false negatives (individuals with disease who are called negative on the screening test). Diagnostic testing in general will be

Nucleic Acid Analysis: Principles and Bioapplications, pages 67–77
© 1996 Wiley-Liss, Inc.

more expensive, but also should provide a better discrimination between those with disease and those without disease so that the risk of obtaining false-positive and false-negative results should be much lower than with a screening test.

Genotypic diagnosis is preferred over phenotypic diagnosis when there is a single or a limited number of known genetic aberrations that can be readily identified. There is a decreased risk of missing an affected individual with accurate genotype diagnosis compared with phenotypic diagnosis. An individual's genes do not change with age, diet, or infectious disease. A genotypic diagnosis for sickle cell disease (SCD) would be preferred over a phenotypic diagnosis in the neonatal period because hemoglobin electrophoresis cannot separate all affected, carrier, and normal newborns due to the elevated fetal hemoglobin level in the newborn period. A genotypic diagnosis for phenylketonuria (PKU) would avoid cases missed on tests for serum phenylalanine level when the infant is ingesting low levels of phenylalanine due to poor feeding or illness. In both of these examples, however, phenotypic screening has definite advantages over genotypic diagnosis: Hemoglobin electrophoresis permits detection of a large number of hemoglobinopathy phenotypes in a single analysis, and phenylalanine determination allows phenotypic ascertainment of individuals with classical PKU, as well as variants, each representing a large number of individual genotypes.

The problems with genotypic diagnosis include the labor-intensive nature of the laboratory testing, making the cost relatively high at this time. In addition, some diseases (such as PKU) require testing for multiple markers, further increasing personnel time and costs. Even with multiple markers, not all patients can have a genetic diagnosis because not all mutations have been identified. At the present time genotypic testing for most disorders is very time consuming. Currently genotypic testing for a number of diseases is available in a single research laboratory or a small number of research groups, further increasing the time between obtaining the sample and a definitive diagnosis.

4.2. NEWBORN SCREENING FOR GENETIC DISEASE

A large-scale attempt to identify individuals with genetic disease was begun in the early 1960s. The emphasis of testing for genetic disease centered on newborns. Infants were tested to determine which were likely to have a disease and which were not. Positive newborn screening results were not diagnostic, but suggested that the infant required further evaluation and testing before treatment would be initiated. The reasons for testing newborns for genetic disease include the detection of disease so that treatment can be initiated to prevent irreparable damage; the identification of carriers for a disease that may occur in their children; and the acquisition of information regarding the genetic composition of the tested population.

To provide screening services effectively to large populations, the tests should be rapid, reliable, amenable to automation, and inexpensive [1]. The entire target population should be screened. If this is not the case, affected individuals will be missed. The number of false-negative results should be minimized. A false-negative result occurs when an individual with the disease is classified as an individual with a low likelihood of having the disease on the basis of the newborn screening test. Those with false-negative results will not receive follow-up testing and will not be diagnosed during the newborn period. The number of false-positive results should be minimized. A false-positive result occurs when an individual without the disease is classified as an individual with a high likelihood of having the disease on the basis of the newborn screening test. Those with false-positive results then receive follow-up evaluation to determine the presence or absence of the disease. Of the two, false negatives and false positives, the false-negative results have the most serious consequences, since an individual affected with the disease might be missed. If the child who was determined false negative is subsequently diagnosed, the family might seek legal recourse for the missed diagnosis. Each newborn with a positive screening test result must receive expeditious follow-up testing. Delay could result in mental retardation, neurological deterioration, or death. The benefits of newborn screening should be greater than the costs of the program. The U.S. General Accounting Office suggested that for every $1 spent on newborn screening, $24 was saved in lifetime care costs for untreated patients [2].

Newborn screening test results include not only the false positives and false negatives discussed above, but also true positives and true negatives. True positives occur when an infant with a high likelihood of having a disease is detected and followed up and confirmed with appropriate diagnostic tests; the desired result of expeditious initiation of treatment is then achieved. True negatives occur when an infant without a disease is correctly labeled as having a low likelihood. Screening tests are designed as inexpensive assays for large numbers of individuals and are useful as screens, but are not definitively diagnostic.

4.3. PHENYLKETONURIA AS A MODEL FOR THE DEVELOPMENT OF NEWBORN SCREENING

PKU is an autosomal recessive disease that leads to mental retardation if not treated. The incidence of PKU in the United States is 1 in 10,000 to 1 in 25,000. Individuals with PKU do not metabolize phenylalanine, an amino acid present in protein foods. Treatment for PKU involves restricting the amount of phenylalanine consumed. The newborn screening test measures the level of phenylalanine in the blood.

The initial newborn screening test for PKU, which is no longer used, involved the measurement of phenylpyruvic acid in the urine. Phenylpyruvic acid is made in the body from phenylalanine and therefore is an indirect indicator of the level of this amino acid. The test for phenylpyruvic acid had a high false-negative rate, misclassifying up to one-third to one-half of the infants with the disease as normal.

The most widely used newborn screening test for PKU is the Guthrie bacterial inhibition assay, which is semiquantitative. The assay uses a bacteria that requires phenylalanine for growth. A piece of filter paper impregnated with the infant's blood is placed on media with the bacteria. The higher the phenylalanine in the blood spot, the larger the bacterial growth zone. The use of filter paper containing the newborn's dried blood has a number of advantages for obtaining, handling, and transporting the specimen. In addition, the analytes (such as phenylalanine) are stable in these dried blood specimens.

Another semiquantitative method is paper chromatography, which separates phenylalanine, and other amino acids, as spots on a sheet of paper. One end of this paper is immersed in a solvent. In addition to these semiquantitative methods, there are two tests that provide quantitative diagnostic measures of phenylalanine in the blood. These are the fluorometric method (based on the fluorescence of phenylalanine when reacted with other chemicals) and the high performance liquid chromatographic (HPLC) method (based on the separation and quantitation of each amino acid, including phenylalanine).

PKU is treated with dietary restriction of phenylalanine and supplementation with tyrosine, along with regular monitoring of blood phenylalanine levels. Treatment should begin as early in the newborn period as possible, definitely before 4 weeks of age, in order to prevent mental retardation. This diet is maintained throughout life. Dietary restriction of phenylalanine is especially important during pregnancy of women with PKU. Those not on a diet may produce an infant with microcephaly, mental retardation, congenital heart disease, or other disorders.

Given the large number of mutations leading to PKU, PKU is not currently amenable to DNA analysis for either newborn screening or routine diagnosis. However, there are a number of other diseases that lend themselves to DNA diagnosis as part of newborn screening or later diagnosis.

4.4. DNA ANALYSIS USING DRIED BLOOD SPECIMENS

We had originally demonstrated that dried blood spots on filter paper blotters could provide a source of DNA for diagnostic purposes [3]. We could microextract approximately 300 ng of DNA from a single 0.5 inch diameter circle. This represented the dried equivalent of about 50 μl of whole blood. This was enough DNA for one Southern blot analysis. As these original investigations were completed, the technique of polymerase chain reaction (PCR) amplification was described [4]. We utilized PCR technology to amplify DNA obtained from dried blood spots [5]. Improvements in the microextraction technique allowed us to obtain DNA from as little as a one-eighth inch diameter punched disc. This represents the dried equivalent of about 3 μl of whole blood [5]. The development of the heat stable *Thermus aquaticus* (*Taq*) polymerase [6], and a programmable heating block [7], led to automated temperature cycling, which was a great improvement over the previous tedious manual PCR procedures.

We have applied this technology to demonstrate the feasibility of DNA confirmation of the hemoglobinopathies, Duchenne muscular dystrophy, and cystic fibrosis using the original newborn screening specimen. The technology has implications for the use of dried blood specimens for applications in addition to newborn screening [8]. Some of these additional applications await the development of automation of the technology in order to obtain reasonable cost effectiveness.

4.4.1. The Hemoglobinopathies

A randomized trial of penicillin prophylaxis for patients with sickle cell disease [9] clearly demonstrated the need for newborn screening to provide early identification and to facilitate treatment of patients with sickle cell disease. This study showed an 84% decrease in infection in the treated group. There were no deaths from pneumococcal sepsis among the treated children, but there were three deaths in the placebo group. The researchers concluded that children should be screened for sickle cell disease and that penicillin prophylaxis should begin by 4 months of age in infants with sickle cell disease. An NIH consensus development conference recommended universal neonatal screening for hemoglobinopathies to prevent early death from septicemia in children with sickle cell disease [10].

Newborn screening for sickle cell disease and other hemoglobinopathies depends on the detection of hemoglobin phenotype [11,12]. Many hemoglobinopathy screening programs recommend repeat analysis at 2–4 months of age when the proportion of hemoglobin F would be diminished so that the adult β-globin phenotype would be more readily determined. Molecular genetic methods are not influenced by the developmental expression of the hemoglobins, thereby providing diagnostic confirmation from the original newborn screening filter paper blotter. This facilitates more efficient diagnosis and more efficacious management [5]. Our group and others showed that DNA could be obtained from newborn screening specimens by microextraction [5,13] or boiling [5] and that this DNA could be analyzed for the A and S alleles after PCR amplification [5,13,14].

We have completed a blinded analysis of 75 specimens from the Texas Department of Health Newborn Screening Laboratory, comparing the results from our DNA analysis with the hemoglobin electrophoresis phenotype from the state laboratory [15]. DNA was extracted from a one-half inch semicircle representing the dried equivalent of approximately 25 μl of whole blood. The microextraction procedure was similar to that used previously [5], except that the proteinase K digestion proceeded overnight. This change produced improved yields of DNA. From 75 specimens of approximately 25 μl dried whole blood equivalent, the range of DNA recovered by microextraction was 900–2,000 ng, with a median of 1,500 ng. The PCR amplified products from each specimen were analyzed using four molecular genetic techniques: hybridization with allele-specific oligonucleotide (ASO) probes labeled with ^{32}P or horseradish peroxidase (HRP), direct digestion with the restriction enzyme *Dde*I, and Southern blot analysis using ^{32}P-labeled ASO probes.

On 5 of the 75 specimens, there were disagreements between the DNA results and the hemoglobin phenotypes. Two specimens in the same run disagreed on all four DNA methods, but agreed after extraction of the second semicircle of the original spot received in the DNA Laboratory. This discrepancy was probably due to contamination during microextraction or PCR. Contamination during PCR was also suspected with a third sample, which disagreed on only one of the four initial analyses and agreed with the state laboratory on all four of the repeat analyses. Two of the 75 samples showed consistent disagreement between the state laboratory and the DNA laboratory on the initial and repeat extractions. A repeat specimen was required for these two samples. One of these samples had shown poor amplification initially, and the repeat specimen showed no amplification. This sample was of inadequate quality for amplification. The second sample showing consistent disagreement between the two laboratories had a hemoglobin phenotype of FS and a genotype of AS on four different sets of analyses using two different spots from the original newborn screening system. This discrepancy was explained when the clinical diagnosis was determined to be S/β-thalassemia.

This molecular genetic approach is now being used to confirm positive newborn screening tests and for the clarification of those that are inconclusive by hemoglobin electrophoresis at the state laboratory. We have developed methods for evaluating the A, S, C, and E alleles. We are also adapting existing screening methods for β-thalassemia using microextracted RNA [16]. We are developing ASO tests for the most common African-American β-thalassemia point mutations [17].

4.4.2. Multiplex PCR for the Diagnosis of Duchenne Muscular Dystrophy

Duchenne muscular dystrophy (DMD) is an X-linked recessive disease with an estimated incidence of 1:3,000 to 1:5,000 male live births [18]. With progressive muscle deterioration, these patients become wheelchair bound by 12–15 years of age, and death occurs in the second or third decade. Newborn screening programs have measured blood creatine kinase (CK) activity in dried blood specimens.

Probes derived from 14 kb DMD cDNA can be used for Southern blot analysis of DNA from patients with this disorder [19]. These DMD cDNA probes detect deletions in approximately 56% of patients with DMD [20]. The patients' deletions are not distributed uniformly throughout the DMD genomic region, but rather there are definite concentrations of deletions [19,20]. Multiplex PCR can be used to target the regions with the highest frequencies of deletions. The original 9-plex reaction developed by Chamberlain and colleagues detects up to 80%–90% of the deletions [21,22]. Nearly 50% of all patients with DMD can be detected by this method [20]. The multiplex PCR can replace the histological evaluation of the muscle biopsy for confirmation of the DMD diagnosis in many patients.

We have been able to perform multiplex PCR on DNA obtained from dried blood specimens with excellent results [23]. Using this method for newborn screening for DMD would eliminate the need for a repeat specimen for about 50% of the true positives identified in the initial screening. Naylor's group in Pittsburgh is using multiplex PCR as part of their DMD newborn screening [24,25].

4.4.3. DNA Testing for Cystic Fibrosis Alleles

CF is an autosomal recessive disorder with a frequency among whites of 1:2,000 live births [18]. The gene for CF, which maps to 7q31, has been cloned [26]. The most common mutant allele among whites is ΔF508 [27]. This allele is present in 75.8% of CF chromosomes, leading to the conclusion that population-based screening would identify approximately 57% of non-Ashkenazic white couples at risk for

CF [28]. A recent NIH-sponsored workshop recommended that mutation analysis for carrier detection should be completed in families with a history of CF but that population-based carrier screening was not yet appropriate [29]. In the interim the NIH has issued grants for pilot studies on population screening for CF carriers.

The use of newborn screening specimens for CF haplotype analysis had been demonstrated even before the CF gene was cloned and the ΔF508 mutation was characterized [30,31]. One of these reports demonstrated the remarkable stability over time of DNA in dried blood specimens [31]. A Guthrie spot that had been obtained in 1971 was analyzed in 1988 in order to permit informed genetic counseling of a sibling of a deceased CF patient. No other source of DNA was available.

The analysis of the ΔF508 allele for use with newborn screening for CF has been described by Naylor and his group in Pittsburgh [32], and similar efforts by our group in collaboration with Seltzer in Colorado have also produced very favorable results [33]. Both groups have used PCR amplification followed by polyacrylamide gel electrophoresis (PAGE) for separation of homoduplexes and heteroduplexes [34]. Assuming that the ΔF508 allele is present in 75.8% of CF chromosomes, 57% of true-positive individuals in a CF newborn screening program would be homozygous for ΔF508, 37% would be mixed heteroxygotes with one ΔF508 allele, and 6% would not be detected with a test targeted solely for the ΔF508 mutation [33].

Pilot studies are in progress to determine the efficacy of newborn screening for CF based on the measurement of immunoreactive trypsinogen (IRT) in Guthrie spots [18]. The use of direct genotypic analysis for the ΔF508 mutation on samples from the original blotters with initial elevated IRT could eliminate the need for a follow-up specimen for repeat IRT analysis in over 50% of true-positive CF individuals ascertained by the initial IRT screen. Genotypic analysis on the initial specimen would represent a significant reduction in the cost of a newborn screening program for collection of repeat specimens and the performance of confirmatory sweat electrolyte analyses. The savings would increase as additional CF alleles were added to such a program. At an allele detection rate of 95%, direct genotyping would confirm approximately 90% of true positives on the initial IRT.

4.4.4. Allele Frequency Ascertainment Using DNA from Dried Blood Specimens

Newborn screening programs have provided substantial data on the frequencies of various diseases in different populations [18]. Molecular genetic analysis of Guthrie spots will allow for ascertainment of allele frequencies in any study population. This approach does not require the evaluation of large numbers of homozygous individuals to have an accurate assessment of disease frequency. A much smaller sample can be used since the identification of the frequency of a specific allele among the heterozygotes will permit calculation of the frequency of that allele in the population. The stability of DNA in the filter paper sample, and the ease of shipment of dried blood specimens to a centralized laboratory, will allow the investigation of allele frequencies in populations that may be geographically or culturally isolated [3].

This technology has been utilized to determine the frequency of specific alleles

for a variety of disorders, including those for several alleles for the hemoglobinopathies in China [35,36] and one PKU allele in Southern Europe and Northern
Africa [37,38]. Other investigators have established the feasibility of determining
allele frequencies among affected individuals and/or heterozygotes in populations
for medium-chain acyl CoA dehydrogenase deficiency [39,40] and hereditary fructose intolerance [41]. The value of this information to newborn screening programs
and other health care planning activities will undoubtedly lead to many more of
these studies in the future.

4.4.5. Automation of DNA Diagnosis from Dried Blood

Using direct amplification without microextraction, we currently estimate the cost
of DNA analysis in our hemoglobinopathy follow-up program to be $5–$10 per
specimen, with approximately $2 or less representing the reagent costs and the bal-
ance due to personnel time [15]. Before these molecular genetic techniques can be
used for primary screening, they will require automation.

Automated direct sequencing is one possible approach that may become economically feasible in the near future. We have demonstrated that microextracts of
DNA can be utilized for direct sequencing of β-globin alleles using the ABI automated DNA sequencer [15]. As a consequence of the activities supported by the
Human Genome Initiative, this technology will become increasingly more efficient
and less expensive, increasing its viability. The use of fluorescent primers and a
competitive oligonucleotide priming strategy permits automated analysis of the
products and has been demonstrated to be yet another potential approach to decrease personnel time and increase sample throughput [42].

A major contributor to personnel expense in the past was the requirement for microextraction in order to perform these analyses reliably [5]. Direct amplification
from aliquots of dried blood spots [15,43] has significantly decreased personnel
costs. Innovation in developing automated approaches to DNA analysis from the
Guthrie spots will be required for any large-scale, population-based screening using
DNA technology. We have successfully demonstrated an automated method that
will significantly reduce labor intensity and costs while increasing sample throughput [44].

4.5. SUMMARY

Molecular genetic techniques for analysis of dried blood specimens in a filter paper
matrix rely on PCR amplification of the small amounts of DNA available from
these specimens. Experience is showing that a wide variety of methods, including
ASO hybridization, restriction enzyme digestion, competitive oligonucleotide priming, and direct automated sequencing, can be applied to DNA obtained from the
Guthrie spots. We have shown how these DNA-based techniques are beginning to
be incorporated into newborn screening programs for the hemoglobinopathies,

DMD, and CF in order to facilitate confirmation of positive specimens. The cost-effective implementation of molecular genetic technology as an adjunct to newborn screening will require improved automation.

REFERENCES

1. McCabe, E.R.B.: Principles of newborn screening for metabolic disease. Perinatol. Neonatol. 6:63 (1982).

2. McCabe, L., McCabe, E.R.B.: Newborn screening. In Encyclopedia of Human Biology. New York: Academic Press, 1991.

3. McCabe, E.R.B., Huang, S.-Z., Seltzer, W.K., Law, M.L.: DNA microextraction from dried blood spots on filter paper blotters: Potential applications to newborn screening. Hum. Genet. 75:213 (1987).

4. Saiki, R.K., Scharf, S., Faloona, F., Mullis, K.B., Horn, G.T., Erlich, H.A., Arnheim, N.: Enzymatic amplification of β-globin genomic sequences and restriction site analysis for diagnosis of sickle cell anemia. Science 230:1350 (1985).

5. Jinks, D.C., Minter, M., Tarver, D.A., Vanderford, M., Hejtmancik, J.F., McCabe, E.R.B.: Molecular genetic diagnosis of sickle cell disease using dried blood specimens from newborn screening blotters. Hum. Genet. 81:363 (1989).

6. Saiki, R.K., Gelfand, D.H., Stoffel, S., Scharf, S.J., Higuchi, R., Horn, G.T., Mullis, K.B., Erlich, H.A.: Primer-directed enzymatic amplification of DNA with a thermostable DNA polymerase. Science 239:487 (1988).

7. Oste, C.: PCR automation. In H.A. Erlich (ed.): PCR Technology—Principles and Applications for DNA Amplification. New York: Stockton Press, 1989.

8. McCabe, E.R.B.: Utility of PCR for DNA analysis from dried blood spots on filter paper blotter. PCR Methods Applications 1:99 (1991).

9. Gaston, M.H., Verter, J.I., Woods, G., Pegelow, C., Kelleher, J., Presbury, G., Zarkowsky, H., Vichinsky, E., Iyer, R., Lobel, J.S., Diamond, S., Holbrook, C.T., Gill, F.M., Ritchey, K., Falletta, J.M.: Prophylaxis with oral penicillin in children with sickle cell anemia. A randomized trial. N. Engl. J. Med. 314:1593 (1986).

10. Wethers, D.L.: Panel, Newborn screening for sickle cell disease and other hemoglobinopathies. National Institutes of Health Consensus Development Conference Statement 6:1–22 (1987).

11. Schneider, R.G.: Laboratory identification of hemoglobin variants in the newborn. In T.P. Carter, A.M. Willey (eds.): Genetic Disease—Screening and Management. New York: Alan R. Liss, 1986.

12. Garrick, M.D.: Technical options for screening newborns for hemoglobinopathies. In B.L. Therrell (ed.): Advances in Neonatal Screening. Amsterdam: Elsevier, 1987.

13. Rubin, E.M., Andrews, K.A., Kan, Y.W.: Newborn screening by DNA analysis of dried blood spots. Hum. Genet. 82:134 (1989).

14. McCabe, E.R.B., Zhang, Y.-H., Descartes, M., Therrell, B.L., Erlich, H.A.: Rapid detection of β-S DNA from Guthrie cards using chromogenic probes. Lancet ii:741 (1989).

15. Descartes, M., Huang, Y., Zhang, Y.-H., McCabe, L., Gibbs, R., Therrell, B.L., Jr., McCabe, E.R.B.: Genotypic confirmation from the original dried blood specimens in a neonatal hemoglobinopathy screening program. Pediatr. Res. 31:217 (1992).

16. Zhang, Y.-H., McCabe, E.R.B.: RNA analysis from newborn screening dried blood specimens. Hum. Genet. 89:311 (1992).

17. Sylvester-Jackson, D.A., Page, S.L., White, J.M., McCabe, H.M., Zhang, Y.-H., Therrell, B.L., Jr., McCabe, E.R.B.: Unbiased analysis of the frequency of β-thalassemia point mutations in a population of African-American newborns. Arch. Pathol. Lab. Med. 117:1110 (1993).

18. Committee on Genetics, American Academy of Pediatrics: Newborn screening fact sheets. Pediatrics 83:449 (1989).

19. Koenig, M., Hoffman, E.P., Bertelson, C.J., Monaco, A.P., Feener, C., Kunkel, L.M.: Complete cloning of the Duchenne muscular dystrophy (DMD) cDNA and preliminary genomic organization of the DMD gene in normal and affected individuals. Cell 50:509 (1987).

20. Baumbach, L.L., Chamberlain, J.S., Ward, P.A., Farwell, N.J., Caskey, C.T.: Molecular and clinical correlations of deletions leading to Duchenne and Becker muscular dystrophies. Neurology 39:465 (1989).

21. Chamberlain, J.S., Gibbs, R.A., Ranier, J.E., Nguyen, P.N., Caskey, C.T.: Deletion screening of the Duchenne muscular dystrophy locus via multiplex DNA amplification. Nucleic Acids Res. 16:11141 (1988).

22. Chamberlain, J.S., Gibbs, R.A., Ranier, J.E., Caskey, C.T.: Multiplex PCR for the diagnosis of Duchenne muscular dystrophy. In M.A. Innis, B.H. Gelfand, J.J. Sninsky, T.J. White (eds.): PCR Protocols—A Guide to Methods and Applications. New York: Academic Press, 1990.

23. McCabe, E.R.B., Huang, Y., Descartes, M., Zhang, Y.H., Fenwick, R.G.: DNA from Guthrie spots for diagnosis of DMD by multiplex PCR. Biochem. Med. Metab. Biol. 44:294 (1990).

24. Naylor, E., Paulus-Thomas, J., Reid, K., Mitchell, B., Wessel, H., Schmidt, B.: Newborn screening for Duchenne muscular dystrophy: Description of screening program. Abstracts of Vth International Congress of Inborn Errors of Metabolism, W15.9 (1990).

25. Paulus-Thomas, J., Wessel, H., Johns, M., Zahorchak, A., Naylor, E.: Deletion analysis and dystrophin studies aid presymptomatic diagnosis of Duchenne-Becker muscular dystrophy in Neonates. Abstracts of Vth International Congress of Inborn Errors of Metabolism, OC 5.6 (1990).

26. Rommens, J.M., Iannuzzi, M.C., Kerem, B., Drumm, M.L., Melmer, G., Dean, M., Rozmahel, R., Cole, J.L., Kennedy, D., Hidaka, N., Zsiga, M., Buchwald, M., Riordan, J.R., Tsui, L.-C., Collins, F.S.: Identification of the cystic fibrosis gene: Chromosome walking and jumping. Science 245:1059 (1989).

27. Kerem, B., Rommens, J.M., Buchanan, J.A., Markiewics, D., Cox, T.K., Chakravarti, A., Buchwald, M., Tsui, L.-C.: Identification of the cystic fibrosis gene: Genetic analysis. Science 245:1073 (1989).

28. Lemna, W.K., Feldman, G.L., Kerem, B., Fernbach, S.D., Zevkovich, E.P., O'Brien, W.E., Riordan, J.R., Collins, F.S., Tsui, L.-C., Beaudet, A.L.: Mutation analysis for heterozygote detection and the prenatal diagnosis of cystic fibrosis. N. Engl. J. Med. 322:291 (1990).

29. Beaudet, A.L., Kazazian, H.H.: Statement from the NIH workshop on population screening for the cystic fibrosis genes. N. Engl. J. Med. 323:70 (1990).

30. Williams, C., Weber, L., Williamson, R., Hjelm, M.: Guthrie spots for DNA-based carrier testing in cystic fibrosis. Lancet ii:693 (1988).

31. McIntosh, I., Strain, L., Brock, D.J.H.: Prenatal diagnosis of cystic fibrosis where a single affected child has died. Guthrie spots and microvillar enzyme testing. Lancet ii:1085 (1988).

32. Spence, C., Paulus-Thomas, J., Zahorchak, A., Naylor, E.: Molecular analysis of the cystic fibrosis gene using dried filter paper blood samples as part of a neonatal screening program. Abstracts of the Vth International Congress on Inborn Errors of Metabolism, OC 7.8 (1990).

33. Seltzer, W.K., Accurso, F., Fall, M.Z., Descartes, M., Huang, Y., McCabe, E.R.B.: Screening for cystic fibrosis: Feasibility of molecular genetic analysis of dried blood specimens. Biochem. Med. Metab. Biol. 46:105 (1991).

34. Rommens, J., Kerem, B.-S., Greer, W., Chang, P., Tsui, L.C-C., Ray, P.: Rapid radiation detection of the major cystic fibrosis mutation. Am. J. Hum. Genet. 46:395 (1990).

35. Zeng, Y.-T., Huang, S.-Z., Ren, Z.-R., Li, H.-J.: Identification of Hb D-Punjab gene: Application of DNA amplification in the study of abnormal hemoglobins. Am. J. Hum. Gen. 44: 886 (1989).

36. Huang, S.-Z., Zhou, X., Ren, Z.-R., Zeng, Y.-T.: Detection of β-thalassemia mutations in the Chinese using amplified DNA from dried blood specimens. Hum. Genet. 84:129 (1990).

37. Lyonnet, S., Caillaud, C., Rey, F., Berthelon, M., Frezal, J., Rey, J., Munnich, A.: Guthrie cards for detection of point mutations in phenylketonuria. Lancet ii:507 (1988).

38. Lyonnet, S., Caillaud, C., Rey, F., Berthelon, M., Frezal, J., Rey, J., Munnich, A.: Molecular genetics of phenylketonuria in Mediterranean countries: A mutation associated with partial phenylalanine hydroxylase deficiency. Am. J. Hum. Genet. 44:511 (1989).

39. Matsubara, Y., Narisawa, K., Miyabayashi, S., Tada, K., Coates, P.M.: Molecular lesion in patients with medium-chain acyl-CoA dehydrogenase deficiency. Lancet i:1589 (1990).

40. Matsubara, Y., Narisawa, K., Tada, K., Ikeda, H., Ye-Qi, Y., Danks, D.M., Green, A., McCabe, E.R.B.: Prevelence of K329E mutation in medium-chain acyl-CoA dehydrogenase gene determined from Guthrie cards. Lancet ii:552 (1991).

41. Tolan, D.R.: Screening and detection of hereditary fructose intolerance alleles in newborn blood samples using polymerase chain reaction. Abstracts of Vth International Congress of Inborn Errors of Metabolism, W15.10 (1990).

42. Chehab, F.F., Kan, Y.W.: Detection of sickle cell anemia mutation by colour DNA amplification. Lancet i:15 (1990).

43. Schwartz, E.I., Khalchitsky, S.E., Eisensmith, R.C., Woo, S.L.C.: Polymerase chain reaction amplification from dried blood spots on Guthrie cards. Lancet ii:639–640 (1990).

44. Zhang, Y.-H., McCabe, L., Wilborn, M., Therrell, B.L., Jr., McCabe, E.R.B.: Application of molecular genetics in public health: Improved follow-up in a neonatal hemoglobinopathy screening program. Biochem. Med. Metab. Biol. 52:27 (1994).

CHAPTER 5

APPROACHES TO GENETIC ANALYSES: FORENSICS

CATHERINE THEISEN COMEY and BRUCE BUDOWLE
Forensic Science Research and Training Center,
FBI Academy,
Quantico, VA 22135

5.1. INTRODUCTION

It is the goal of the forensic serologist to determine whether or not evidentiary items can be associated with victims or suspects in a criminal investigation. This determination is made through the characterization of genetic markers in body fluids or tissue. Traditionally, serologists attempted to identify various polymorphic protein markers present in biological evidence [1]. It can be difficult at times, though, to characterize protein markers in tissues other than blood, and, even in blood or bloodstains protein markers can be degraded or present in insufficient quantity to characterize. However, DNA has the potential of yielding more genetic information than do classic serological markers. Furthermore, DNA is much more stable than protein markers, making possible the analysis of old or degraded samples or samples exposed to a variety of environmental insults. The presence of DNA in all nucleated cells enables genetic analysis of a variety of tissues and body fluids commonly encountered in forensic cases. Notably, rape cases represent a significant portion of a crime laboratory's caseload, and therefore semen represents the predominant form of evidence originating from a suspect. The only polymorphic protein, other than the ABO blood group proteins, present in semen in sufficient quantity to be analyzed routinely is phosphoglucomutase-1, and often even that is in too low a quantity to be characterized [2]. Thus, the use of DNA analysis in rape cases allows a much higher degree of genetic characterization of semen and a much high-

Nucleic Acid Analysis: Principles and Bioapplications, pages 79–104
© 1996 Wiley-Liss, Inc.

er success rate than had previously been possible. DNA analysis of bloodstains as well enables a higher degree of characterization than formerly attainable. Therefore, the use of molecular biological procedures for the analysis of DNA polymorphisms has given forensic scientists an extremely powerful means of analyzing biological evidence.

5.2. RESTRICTION FRAGMENT LENGTH POLYMORPHISM ANALYSIS OF VARIABLE NUMBERS OF TANDEM REPEAT LOCI

Highly polymorphic genetic loci offer great potential for characterization of biological material. Wyman and White [3] described the first locus that demonstrated a highly variable length polymorphism in DNA. Subsequently, other loci were shown to have length polymorphisms. These length polymorphisms are characterized by short sequences repeated in tandem a varying number of times. Loci that exhibit this form of polymorphism are referred to as *variable number of tandem repeat* (VNTR) loci [4]. The use by Jeffreys et al. [5] of a DNA probe that recognizes multiple hypervariable tandem repeat loci to detect DNA "fingerprints" and the use of this approach by Gill and Werrett [6] to identify a rape and murder suspect in Leicestershire, England, ushered in a new era in forensic serology characterized by the use of variations in DNA sequences to include or exclude crime victims and suspects as contributors of biological evidence.

The primary means of forensic DNA analysis in the United States makes use of single-locus VNTR probes. This analysis method has proven to be an extremely powerful tool for excluding and associating evidentiary material with suspects and victims. The single-locus probe approach, which uses procedures based on restriction fragment length polymorphism (RFLP) analysis of VNTR loci, has been used in thousands of criminal cases in the United States. As of 1995, more than 50 state and local crime laboratories in North America, as well as private laboratories and crime laboratories in countries throughout Europe, Australia, and Asia, use single-locus VNTR probes for DNA analysis of evidentiary samples.

Even though RFLP VNTR analysis is a generally accepted technique in the research community, specific validation is required for forensic analysis. This validation is designed to (1) test whether or not DNA subjected to a variety of stresses shows altered VNTR patterns, (2) establish matching criteria for VNTR patterns, and (3) estimate the frequency of occurrence of VNTR patterns by analyzing VNTR patterns from various general population groups. Any analytical procedure also must be tailored to accommodate the needs of an application-oriented laboratory, which may have a high throughput and many people performing the analyses. For these reasons, many laboratories in North America have chosen to use a protocol that was developed to streamline procedures, reduce cost, increase speed of analysis, decrease handling, and obtain enhanced sensitivity of detection [7].

Forensic RFLP VNTR analysis consists of several steps: (1) extraction of DNA from a substrate, (2) digestion of DNA with a restriction endonuclease, (3) size fractionation of the digested DNA by electrophoresis, (4) blotting of DNA to a membrane support, (5) hybridization of membrane-bound DNA to a ^{32}P-labeled VNTR probe, (6) autoradiography, and (7) interpretation of resulting band patterns. Since these methods are described in detail elsewhere, general comments regarding only those aspects of the protocol of special interest, as well as descriptions of validation experiments and profile matching procedures for evaluating the weight of DNA evidence, are described here.

5.2.1. DNA Extraction

Procedures suitable for extraction of DNA from small liquid blood samples and bloodstains are based on standard DNA extraction procedures and have been described elsewhere [7,8]. However, since the majority of forensic cases analyzed in a forensic serology laboratory are rape cases (60%–70%), and the evidence from these types of cases contain nucleated cells from both the suspect (consisting primarily of sperm) and the victim, it could be difficult at times to elucidate the DNA profiles of these mixtures. Therefore, a procedure to separate the DNA originating from sperm from mostly female, epithelial cells or blood contained in the evidentiary material was required. For semen-containing evidentiary material, a two-part extraction procedure, which separates a significant portion of sperm DNA from nonsperm DNA, can be used [9,10].

In this procedure, nonsperm cells are lysed first using SDS and proteinase K. The sperm heads, which are resistant to lysis by SDS and proteinase K alone due to the presence of thiol-rich proteins in their membranes, can be separated from that lysate by centrifugation. Subsequently, the sperm are lysed in the presence of SDS, proteinase K, and dithiothreitol. The extracted DNA from each of these sources is purified by phenol/chloroform extraction and ethanol precipitation [9,10]. The end result is that enriched fractions can be obtained, allowing separate DNA profiles from the sperm and nonsperm fractions of the evidentiary material.

5.2.2. Restriction Endonuclease Digestion of DNA

In RFLP analysis, DNA must be digested with a restriction endonuclease following extraction. A variety of restriction enzymes were tested for robustness and effectiveness with the probes used and with the sample types that would be encountered in a forensic laboratory. The restriction endonuclease *Hae*III was chosen because it met the following criteria suitable for forensic applications [8]. First, the restriction enzyme must be compatible with the locus of interest, i.e., it should not digest within the VNTR sequence. Second, *Hae*III, a restriction enzyme that recognizes and digests at a four-base site (5′-GGCC-3′), generally yields fragments that are smaller in size than restriction enzymes that recognize longer sites, such as *Pst*I [11]. Third, since different tissues may exhibit different methylation patterns, use of a methylation-sensitive enzyme for forensic analysis could lead to difficulties when comparing evidence from an unknown origin, for example, DNA from sperm, with a known DNA sample from, for example, blood. However, *Hae*III is generally insensitive to mammalian methylation patterns. The methylation pattern observed in mammals can occur at C residues generally only when followed by a G. If the 5′ C in the HaeIII recognition site is methylated, digestion is inhibited. However, this 5′ C is followed by another C, and thus the 5′ C would not be expected to be methylated in humans. HaeIII digestion is unaffected by methylation of the 3′ C residue even though it could be methylated if a G follows the recognition site. Fourth, an additional advantage of using HaeIII is that it is robust and thus active in the presence of a wide variety of salt, pH, and temperature conditions. This is an important consideration when dealing with forensic samples that may not be nearly as clean as DNA from drawn liquid blood [8].

5.2.3. Membrane Hybridization

In the procedures employed by the FBI Laboratory, DNA is subjected to agarose gel electrophoresis following endonuclease digestion, then denatured and transferred in 0.4 N NaOH to a nylon membrane. Following fixation of the DNA to the membrane, hybridization is carried out with VNTR probes. A simplified hybridization method was developed that offered certain advantages. By the use of a hybridization solution containing only 10% polyethylene glycol, 7% SDS, and phosphate buffer, the prehybridization step was eliminated, and yet a high signal intensity with reduced background was achieved. This hybridization buffer has additional advantages in that it is simple and inexpensive to prepare. Furthermore, it contains relatively few components, which makes it easier to troubleshoot in a situation where a reagent may have gone bad. Following hybridization, membranes are subjected to stringency washes to remove unbound probe [7]. Washed membranes are then wrapped in plastic wrap, placed between two sheets of Kodak X-OMAT x-ray film, and this complex is sandwiched between two Dupont Cronex Lightning Plus intensifying screens. Film is exposed at −80°C for an appropriate time period and developed.

5.2.4. Single-Locus VNTR Probes

In 1985, Jeffreys et al. [5] reported that DNA probes containing core sequences of tandemly repeated minisatellite regions of the genome could be used to identify complex RFLP patterns in Southern blots of human DNA. These probes, which simultaneously detected multiple loci in Southern blots of genomic DNA, produced highly individual specific DNA patterns [12]. Multilocus analysis was shown to be effective on laboratory-prepared forensic-type samples, such as bloodstains, vaginal swabs, semen, semen stains, hair roots, and buccal swabs [9,13], and was used in some of the earliest forensic cases in which DNA typing was employed [6,14].

Relatively early in the investigation of the use of DNA typing for forensic application, however, it became apparent that probes that detected single VNTR loci, rather than multilocus probes, would be more suitable for forensic analysis. Single-locus probes, like multilocus probes, detect VNTR sequences, but the high stringency hybridization conditions are such that only the VNTR sequence at one locus is detected. The use of single-locus VNTR probes offers a number of advantages for forensic analysis. First, single-locus probes are more sensitive and allow banding patterns to be detected with as little as 50 ng of genomic DNA using single-locus probes, while minisatellite analysis requires an amount of DNA (0.5–1.0 μg, at a minimum) that may not be present in some forensic specimens. Second, hybridization with single-locus probes generally gives one- or two-banded patterns, while the resulting band pattern using multilocus probes is complex and may be difficult to interpret for some complicated or unknown situations encountered in forensic science. For example, if a specimen contains more than one source of DNA, it could be difficult to determine which bands arose from which contributors. Third, the application of population genetics principles is much more straightforward using single-locus probes than multilocus probes. With multilocus probes, the determination of pattern frequency cannot use traditional approaches to assess estimates of genotype or phenotype frequencies. Fourth, the probes used for multilocus analysis cross-hybridize with DNA from a number of different species [15–18], while single-locus probes cross-hybridize only with DNA from other higher primates.

Several characteristics of VNTR loci need to be considered when choosing probes for forensic use. It is desired that they recognize highly polymorphic loci, be sensitive, cross-react only with higher primates, be compatible with the restriction enzyme used, and recognize alleles within the range of resolution of the gel system used. A variety of single-locus VNTR probes have been examined. Probes for the following loci are routinely used in the FBI Laboratory: D2S44 [19], D17S79 [20], D1S7 [21], D4S139 [22], D10S28 [23], and D5S110 [24]

Typically, four to six single-locus probes are sequentially hybridized to DNA immobilized on Southern blots. This enables the user to increase the amount of genetic information that can be obtained from small quantities of DNA (50 ng or greater).

5.2.5. Interpretation of RFLP VNTR Profiles

In a forensic case employing DNA analysis, DNA patterns from unknown, evidentiary sources are compared with DNA patterns from a known source, such as a suspect or victim. The result of a comparison of DNA profiles can fall into one of three categories: (1) a match, where the profiles are operationally similar such that the known and unknown samples cannot be excluded as being from the same source; (2) an exclusion, where the profiles differ such that the known and unknown samples could not have originated from the same source; and (3) an inconclusive result, where insufficient data are present to make an interpretation.

To determine whether or not a DNA profile from an unknown sample matches that of a known sample, a visual comparison of the banding patterns is made first. Figure 5.1 shows DNA patterns resulting from evidence in two cases. Figure 5.1a represents RFLP VNTR typing results from a rape case, in which the evidence consisted of a semen stain on an item of clothing. The pattern in lane 6, arising from DNA from the sperm fraction of the differentially extracted stain, appears similar to the pattern in lane 4, arising from DNA from blood from the suspect. The DNA pattern from the DNA from the nonsperm fraction in lane 7 is similar to that arising from blood from the victim (lane 3), as expected. Figure 5.1b also represents typing

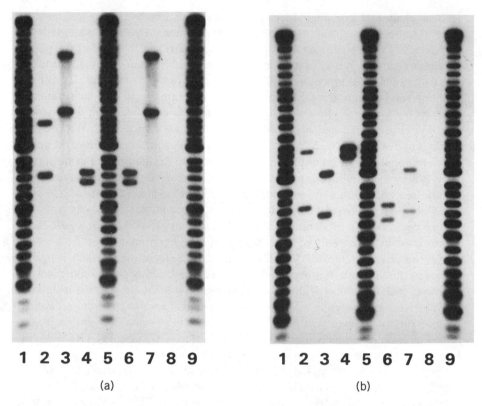

1 2 3 4 5 6 7 8 9   1 2 3 4 5 6 7 8 9

(a)   (b)

Fig. 5.1. Representative autoradiograms of RFLP VNTR patterns from two cases. The blots have been hybridized with the probe D2S44. **a:** Lanes 1, 5, and 9, size markers consisting of digested lambda DNA. Lane 2, DNA from K562 cell line. Lane 3, DNA from victim's blood. Lane 4, DNA from suspect's blood. Lanes 6 and 7, the sperm and nonsperm fractions, respectively, from the evidence, a differentially extracted semen stain. **b:** Lanes 1, 5, and 9, lambda DNA size markers. Lane 2, DNA from K562 cell line. Lane 3, DNA from victim's blood. Lane 4, DNA from suspect's blood. Lanes 6 and 7, the sperm and nonsperm fractions, respectively, from the evidence, a differentially extracted semen stain.

results from a rape case. In this case, the evidence consisted of a semen stain on a bedsheet. The pattern in lane 6, from the sperm fraction of the differentially extracted stain, appears dissimilar to that in lane 4, arising from DNA from blood from the suspect. Again, the typing pattern from the DNA from the nonsperm fraction, lane 7, is similar to that arising from blood from the victim (lane 3).

A visual match in a case is confirmed or refuted by computer-assisted image analysis [25]. The sizes of the bands are determined by comparison with size ladders consisting of viral DNA fragments of known size. Repeated measurements of bands derived from ideal samples as well as measurements from bands arising from DNA from the female fraction of vaginal swabs from rape cases with DNA from the victim's blood [26] were used to determine the amount of measurement variation that might be expected in a case and, thus, the amount of variation tolerated when confirming or refuting a visual match.

Once a visual match has been confirmed by computer-assisted image analysis, the significance of the match can be established to determine the proportion of the relevant population(s) that could contribute a sample giving that DNA typing pattern. To accomplish this, laboratories have compiled data bases from various population groups [26–28]. Genotype frequencies are calculated for each locus, and an estimate of the likelihood of occurrence of a DNA profile in a general population is determined [26].

Some have raised concerns about the frequency estimates used for determining the likelihood of occurrence of DNA profiles [29–32]. Critics of current methods for estimating profile frequencies contend that large differences in profile estimates could be obtained when general population group data bases (e.g., Caucasian, African-American) are used rather than data bases from population subgroups. To this end, VNTR profiles from over 100 groups around the world at several loci were compiled [33] and compared [34,35]. The data show that differences in allele frequencies within major groups have little effect on profile estimates. Rather, the largest differences are seen across major population groups. These data support the use of major population group data bases for forensic estimates.

5.2.6. Effects of Environmental Insults and Contaminants

Because forensic evidence can be subjected to a variety of adventitious substances and environmental effects prior to its analysis by a laboratory, it is important to evaluate the effects of these various influences on DNA typing patterns. This enables analysts to understand the effectiveness and limitations of DNA typing. It is difficult to imagine how a DNA type could be changed or altered in ways other than loss of banding patterns due to degradation. However, it is incumbent upon the forensic scientist to show that DNA stressed in a variety of ways yields reliable results.

A number of studies have been conducted by the FBI and others to examine the effects of a variety of factors on biological evidence. DNA analysis was performed on specimens derived from different body fluids and tissues, such as semen [10,36], bloodstains [8,36,37], and bone [38]. The results were compared with control DNA extracted from liquid blood and demonstrated that the DNA source does not affect typing results. Mixtures of body fluids from different donors gave RFLP patterns that were consistent with the DNA types of the contributing individuals [36]. Bloodstains were deposited on a variety of substrata, such as synthetic fabrics, denim, cotton, glass, wood, aluminum, plastic, and carpet [11,36], in order to establish that under the less-than-ideal conditions to which evidentiary samples may be exposed only two effects are observed: either the expected DNA type is not altered or no result is obtained (i.e., an inconclusive outcome).

Additionally, bloodstains were exposed to sunlight, in a greenhouse in the springtime (March–May), and both in a greenhouse and unprotected outdoors in the summer (July) [36]. RFLP patterns, consistent with those of control bloodstains, were obtained from springtime exposure through the 10 week period tested, as well as from the entire 10 day outdoor summer testing period and the first 11 days of the

12 day greenhouse testing period. DNA obtained from the day 12 sample was too degraded to produce a pattern. Again, in all cases, either the correct result or no result was obtained. McNally et al. [39] reported similar results from DNA extracted from bloodstains exposed directly to an ultraviolet light source.

To test the effects of various substrates and storage conditions [36], bloodstains were made on six substrates (cotton, nylon, denim, glass, wood, and aluminum) and stored at 4°, 22°, and 37°C and at ambient outdoor temperatures for 2 days and 1, 3, and 5 months. Additionally, some stains, placed on cotton, were stored at 37°C for 4 years. Some samples (those on cotton and denim exposed at higher temperatures, and prepared bloodstains maintained at 37°C for 4 years) failed to give RFLP results. All samples that gave RFLP tying results produced patterns consistent with those from untreated DNA from the same donors.

Bloodstains were exposed to a variety of biological contaminants *(Escherichia coli, Bacillus subtilis, Candida albicans, Candida valida,* and *Staphylococcus epidermidis)* and to chemical contaminants (gasoline, motor oil, detergent, bleach, salt, acid, base, and soil) [36]. All samples produced typing patterns consistent with control (untreated) bloodstains except those exposed to soil. The samples exposed to soil produced no typing results. McNally et al. [39] observed similar results. DNA from the biological contaminants alone did not produce typing patterns arising from VNTR probe hybridization. However, human samples contaminated with *E. coli,* as well as the *E. coli* control DNA, showed a pattern of seven fragments, smaller than 1,600 bp, when the blots were hybridized simultaneously with the higher primate-specific VNTR probe and probes specific for lambda phage and phi X 174 phage DNA. The viral DNA probes detect a series of lambda and phi X DNA fragments that are run adjacent to sample lanes. These viral DNAs serve as a size standard (i.e., a ladder) to extrapolate sizes of bands from human DNA in sample. When the blots were hybridized with the VNTR probe alone the expected typing pattern appeared. However, when the *E. coli* blots were hybridized with the ladder probe alone, the seven-fragment pattern appeared. This cross-hybridization would not be mistaken for a human DNA profile. These fragments, consistent in subsequent hybridizations with any typing probe used in combination with the ladder probe, are due to cross-hybridization with the ladder DNA and *E. coli* DNA.

Hochmeister et al. [40,41] investigated effects on RFLP typing of various presumptive test reagents commonly used by laboratories to identify blood or semen and of nonoxinol-9, the active ingredient of many spermicides. As with other contaminants, either RFLP results consistent with controls were obtained from bloodstains or semen stains exposed to reagents (as seen with luminol, benzidine, phenolphthalein, *o*-tolidine, leucomalachite green, and nonoxinol-9) or no RFLP results (with benzidine dissolved in glacial acetic acid, leucomalachite green, and *o*-tolidine) were obtained.

The conclusion that can be drawn from all of the validation experiments described above is that in situations where the DNA is adversely affected by a contaminant or environmental influence, no interpretable typing pattern, rather than an altered typing pattern, was obtained. Although not every possible forensic situation can be tested in the laboratory, these studies, which examined several extreme conditions, demonstrated the robustness and reproducibility of the single locus probe RFLP methodology.

5.3. POLYMERASE CHAIN REACTION-BASED FORENSIC ANALYSES

While RFLP VNTR typing is a reliable and valid means of analyzing forensic biological evidence, it has certain limitations. These include the lengthy time period needed for analysis (6–8 weeks if four VNTR probes are used in succession), cost of analysis, sensitivity limits, and need for isotopic assays. To alleviate these limitations for forensic DNA analysis, DNA typing methodologies based on the polymerase chain reaction (PCR) [42,43] have been and are being examined for their applicability to forensic casework. (However, the advent of chemiluminescent detection of VNTR loci [43a,b] has decreased the time necessary for RFLP typing and obviated the need for radioisotopes.) PCR amplifies specific DNA sequences by repetitively denaturing template DNA, hybridizing specific DNA primers flanking the region to be amplified to the template DNA, and extending the primers. Thus, the region of interest can be replicated. PCR is a means of obtaining relatively large amounts of specific DNA sequences from minute quantities of genomic DNA and obviates the need for isotopic DNA detection methods, because enough specific DNA is generated that less sensitive, nonisotopic detection methods can be used. Additionally, degraded DNA samples can be amplified by PCR and subsequently typed because alleles are much smaller in size than alleles detected by RFLP analysis. Thus, PCR-based analyses offer potential benefits for forensic work.

5.3.1. Analysis of the HLA-DQA1 Gene

The first PCR-based typing method that has undergone an extensive validation process to date for forensic application involves the amplification of the HLA-DQA1 gene and typing by dot blot analysis with allele-specific oligonucleotide probes [44]. The HLA-DQA1 locus is polymorphic, exhibiting 6 common alleles and 21 genotypes. The variability of the HLA-DQA1 gene is contained within a region (242 or 239 base pairs long, depending on the allele) in its second exon. A kit is commercially available that contains the components needed for amplifying and typing the HLA-DQA1 gene.

Initially, HLA-DQA1 typing was done using a conventional dot blot format in which amplified HLA-DQA1 DNA was applied to a nylon membrane, and then the immobilized DNA was hybridized with an allele-specific probe [44]. However, HLA-DQA1 typing can be accomplished as well using a "reverse dot blot" typing procedure in which amplified HLA-DQA1 DNA is hybridized to nylon membranes to which are affixed allele-specific oligonucleotide probes in a dot pattern [45]. The reverse dot blot system allows for analysis of one sample on a single membrane strip (as opposed to strips for each probe required for the conventional dot blot ap-

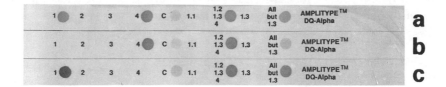

Fig. 5.2. HLA-DQA1 a typing strips showing 1.2 and 4 alleles **(a)**, the 4 allele **(b)**, and the 1.2 allele **(c)**.

proach). Also, the user does not have to be concerned with balancing probe concentrations for a number of hybridizations with different probes in order to obtain similar dot intensities; the probes fixed on the strip are relatively balanced. Because the primers are biotinylated, hybridization can be detected via a horseradish peroxidase–streptavidin complex, which oxidizes tetramethylbenzidine, converting it from a colorless solute to a blue precipitate. Therefore, the presence of a blue dot indicates hybridization of amplified sample DNA to a particular allele-specific probe. Figure 5.2a–c shows three HLA-DQA1 typing strips displaying patterns resulting from amplifying DNA samples containing the 1.2 and 4 alleles (a), the 4 allele (b), and the 1.2 allele (c), respectively.

Many reports indicate that HLA-DQA1 typing of amplified DNA will prove useful for analyzing forensic evidence. In fact, this procedure has been used in hundreds of criminal cases since 1986. Westwood and Werrett [46] and Comey and Budowle [47] have conducted extensive validation studies of the HLA-DQA1 typing system and have demonstrated that the system can reliably type a variety of contaminated or environmentally stressed samples (see below). Others have reported on the success of the test of forensic-type specimens: Higuchi, et al. [48] reported that HLA-DQA1 results could be obtained from hair roots, Hochmeister et al. [49,50] demonstrated HLA-DQA1 typing results on cigarette butts and bones from putrefied bodies, Sajantila et al. [51] reported HLA-DQA1 typing results on tissue obtained from fire victims, and Westwood and Werrett [46] described HLA-DQA1 typing of hair roots, bloodstains, semen stains, and vaginal swabs. Blake et al. [52] and Comey et al. [53] described HLA-DQA1 typing on forensic casework material. Potsch et al. [54] reported HLA-DQA1 results from DNA extracted from dental pulp. Tahir and Watson [55] reported HLA-DQA1 results from nail material. Additionally, population studies of HLA-DQA1 allele and genotype frequencies have been conducted [47,56–59] that allow the assignment of statistical weight to an HLA-DQA1 match. The following section summarizes experiments performed by the FBI Laboratory designed to evaluate this typing system.

Validation of HLA-DQA1 Typing. There are many important issues regarding the validity of PCR-based typing procedures. Some of these are addressed here. One of the most important issues regarding the forensic use of PCR is contamination of evidentiary samples with other sources of DNA, especially from individuals within the laboratory performing the analyses. One potential source of contamination is the inadvertent mixing of extracted DNA samples (yet to be subjected to PCR) with amplified DNA. This kind of contamination is minimized by performing pre- and post-PCR procedures in different rooms and using separate reagents and equipment for these procedures. Another potential source of contamination is from the analyst. A study has shown [47] that DNA from a stain containing 1 μl of saliva from one individual and 10 μl of blood from another individual yields HLA-DQA1 profiles showing alleles from each individual with similar intensities. Thus, the wearing of masks by analysts during DNA extraction and sample preparation will minimize this type of contamination. Negative controls, which check all reagents coming into contact with samples, and positive controls, containing DNA of a known type, enable analysts to monitor laboratory-induced contamination. Duplicate analyses of unknown samples, when feasible, also aid in monitoring contamination. These measures, which can be addressed by implementing proper protocols in the laboratory, limit and monitor contamination effects, allowing a forensic laboratory to be confident that interpretable results are both valid and reliable.

Experiments similar to those performed to evaluate RFLP VNTR analysis were performed using the HLA-DQA1 typing system. These are summarized below.

DNA Extraction. The organic DNA extraction procedure for bloodstains used in RFLP analysis [8] often yielded DNA that could not be amplified. It has been suggested that hematin, a breakdown product of heme, and a potent inhibitor of *Taq* polymerase, may copurify with bloodstain DNA [60] and inhibit amplification. Thus, it was necessary to modify the bloodstain DNA extraction procedure used for RFLP VNTR analysis for PCR-based systems. Various methods were tried [47], including presoaking the bloodstain in water for 2 hours at 56°C prior to extraction, a nonorganic extraction procedure [61], and an organic/dialysis extraction method that makes use of Centricon microconcentrator devices [47,62]. The organic/dialysis method entails extraction of DNA from bloodstains as described through the organic extraction steps [8], but, instead of concentrating the DNA by ethanol precipitation, the samples are dialyzed and concentrated in Centricon 100 (Amicon) microconcentrator tubes. Initially, the organic/dialysis method was chosen for its ease and reduced chance of sample-to-sample contamination as compared with the water presoak method, which entails placing the stain substrate in the punctured lid of a microcentrifuge tube and removing the water from the stain by centrifugation. The organic/dialysis method succeeded in yielding amplifiable DNA from most samples. Presumably dialysis removes the PCR inhibitor.

A simple DNA extraction method was described which uses Chelex 100 ion exchange resin (BioRad, Richmond, CA) to extract DNA from a variety of sources [63,64]. This extraction method consists of placing a small cutting of the evidentiary substrate (for example, a 3 × 3 mm bloodstain) in a 5% (w/v) suspension of Chelex in water, heating the sample at 56°C for 30 minutes, and then boiling the sample for 8 minutes. A portion of the extract can then be amplified without any further purification. This method succeeds in yielding amplifiable DNA (for the HLA-DQA1 system) from bloodstains on many substrate types. This extraction method has been used for the approximately 1,000 cases analyzed by the FBI Laboratory for HLA-DQA1 since 1992 and has been used by many other laboratories as well. Recently, however, it has been found that Chelex-extracted DNA was not suitable for amplified fragment length polymorphism analysis (see section 5.3.3) [65]. Thus, use of the organic/dialysis method would allow a laboratory to extract DNA suitable for a variety of PCR-based analyses methods, as well as for RFLP analysis, determine the quantity and quality of the extracted DNA, and then decide which analytical methods would be best suited to that sample.

Environmental Insults. Bloodstains were subjected to a variety of environmental influences to determine the effects of these influences on amplification and typing [47]. The influences included exposure to a variety of microorganisms; sunlight (up to 20 weeks); to acid, base, detergent, salt, bleach, gasoline, and motor oil; and to a variety of substrates, including many different fabric types, drywall, carpet, upholstery, linoleum, metal, leaves, and soil. These varied influences either had no effect on amplification and typing or led to amplification failure. Importantly, though, none of these caused an incorrect HLA-DQA1 type to be obtained. DNA exposed to leaves or soil could not be amplified; either the DNA was completely degraded or it could not be extracted from these substrates. Because envelope flaps, stamps, and cigarette butts are often processed for latent fingerprints but also often contain biological material that can be typed for HLA—DQA1, Presley et al. [66] looked at the effects of a variety of latent fingerprint processes on HLA-DQA1 typing. This work showed that samples could be processed for fingerprints using a variety of methods and still yield HLA-DQA1 typing results. One method, physical developer, diluted or degraded DNA to an extent that it was not typeable.

Induced DNA Contamination. The sensitivity of PCR raises the concern of contamination of samples by DNA from sources other than that of the evidentiary material. To test the possibility that various ways of handling or treating stains could introduce contamination that could be detected in HLA-DQA1 typing, various scenarios were constructed [47]. Bloodstains (both wet and dry) with different HLA-DQA1 types were stored in close contact with one another. Scissors that were used to cut a bloodstain with a particular HLA-DQA1 type were then used to cut a bloodstain with a different HLA-DQA1 type without being cleaned between cuttings. Stains were handled extensively, coughed upon, exposed to shed scalp cells, or made on a perspiration-soaked shirt. The only one of these treatments that introduced contamination was the storage of two wet stains together, as might be expected.

Mixed stains, comprised of saliva and blood, were made, and the DNA was extracted, amplified, and typed. As little as 1 μl saliva could be detected in a stain containing 10 μl of blood, with the equivalent dot intensities. The wearing of masks by analysts will limit this level of contamination arising in the laboratory.

To determine the effects of sample handling in a forensic setting, evidence from a mock crime scene was analyzed. The evidence consisted of four items of clothing stained with pig blood. Two items of clothing had been worn extensively by a laboratory worker but washed prior to their use in the mock crime scene, while the other two items were purchased new and then washed. Since the DNA from the pig blood would not amplify or type with the primers and probes in the AmpliType kit, any HLA-DQA1 types obtained would have arisen from either the wearer of the clothing or from the persons handling the evidence. Twenty cuttings were taken from various areas of the clothing; no HLA-DQA1 type was obtained from any of them. Thus, at the level of sensitivity recommended for the HLA-DQA1 test (2 ng genomic DNA), contamination introduced by sources other than the evidentiary stain does not appear to be significant. However, it is always desirable for a laboratory performing PCR analyses to take precautions against laboratory-induced contamination, such as using separate pre- and post-PCR work areas, equipment, and reagents and wearing of face masks.

Population Studies. Population studies of HLA-DQA1 genotype frequencies were done: 1,055 individuals (298 Caucasians, 338 African-Americans, and 429 Hispanics) were typed [47,67]. The population samples meet Hardy-Weinberg expectations, and the genotype frequencies are generally similar to other population groups [47]. These data bases can be used to assess the significance of HLA-DQA1 matches in a casework setting. The rarest genotype (1.3, 1.3 in these populations) is present at less than 1% in these populations, while the most common types may be present in 11%–20% of the populations.

Analysis of Old Case Samples. Old case samples, previously analyzed by RFLP typing, were typed at the HLA-DQA1 locus [53,68]. All casework sample comparisons that were included by RFLP analysis (i.e., cannot be excluded as arising from the same source) also were included by HLA-DQA1 typing (67 cases in the two studies), and all samples that were excluded as arising from the same source by HLA-DQA1 typing were excluded by RFLP typing (13 cases in the two studies). As expected, based on the lower discrimination potential of the HLA-DQA1 approach, some RFLP exclusions were HLA-DQA1 inclusions (5/18 RFLP exclusions in the two studies).

For analyses of old case samples, fixed volumes of Chelex extracts were subjected to amplification. Thus, the quantity of genomic DNA used as a PCR template varied. Some samples gave weak HLA-DQA1 typing results, and, had more DNA been subjected to amplification, a stronger typing result may have been obtained. Quantification of the amount of human DNA would ensure that sufficient DNA is present to attempt to produce a typing result. Thus, for case analysis, DNA extracted from evidentiary material is subjected to quantification using a slot blot analysis procedure with hybridization of a probe that detects alphoid repeat DNA and is specific to higher primates [69]. For HLA-DQA1 analysis, 2–10 ng genomic DNA is subjected routinely to amplification.

It became apparent from these studies (as well as from the induced contamination studies) that the possibility that an evidentiary sample is composed of tissues or body fluids from more than one source should be considered when interpreting dot blots. Because of the high level of sensitivity afforded by PCR-based methods, the possibility of two or more contributors to a stain should be considered if no additional information regarding the origin of a stain is available. In most situations, the presence of more than one contributor will be evident in that three or more alleles are present in a sample and/or dots are present in different intensities, signalling a mixture.

Allele Dropout. During the course of the validation studies of HLA-DQA1 typing, a problem referred to as allele dropout was encountered. Samples that were previously determined to be heterozygotes containing one of the 1 alleles (i.e., 1.1, 1.2, or 1.3), in combination with a 2, 3, or 4 allele, showed only a single allele. The allele missing from each sample was the 1 allele. The cause of this problem was determined to be insufficient heating of the samples during the denaturation step of the PCR process. Each of the 1 alleles has three more GC pairs contained within the amplified sequence than do alleles 2, 3, and 4 and, thus, if there is insufficient heat available, could be more difficult to denature [70]. In samples that exhibited allele dropout, the denaturation temperature was high enough to denature alleles 2, 3, or 4, but not high enough to denature the 1 alleles. This problem can be alleviated by the addition of formamide to the PCR reaction mix to a final concentration of 5% [68], or by avoidance of the front two wells of the model of the thermal cycler used (the original DNA Thermal Cycler model, Perkin-Elmer Corp.; these wells were slower to heat than the wells in the back 4 rows), as well as by ensuring the thermal cycler is performing to specifications.

5.3.2. Multiplex Amplification and Typing of Six Polymorphic Loci

A kit is commercially available (the AmpliType PM PCR Amplification and Typing Kit, Perkin-Elmer, Norwalk, CT) that enables the simultaneous amplification of six polymorphic loci. These loci are HLA-DQA1 low-density lipoprotein receptor, glycophorin A, hemoglobin G gammaglobin, D7S8, and group-specific component. A portion of the amplified product can be used for typing the HLA-DQA1 locus and another portion for typing the other five loci (PM loci), also using a reverse dot blot approach where allele-specific oligonucleotide probes are immobilized on a nylon membrane strip. Fildes and Reynolds demonstrated the consistency and reproducibility of the PM system through a field study [70a]. This typing system was shown to be reliable for forensic use by conducting a number of experiments [67]. These included mixed body fluid studies; chemical contaminant effects on the DNA in body fluid samples; the effect of typing DNA from body fluid samples deposited on various substrates; the effect of microorganism contamination on typing DNA derived from blood and semen; the effect of sunlight and storage conditions on DNA typing; determination of the sensitivity of detection of the PM test kit; determination of cross-reactivity of DNA from species other than human; typing DNA derived from various tissues from an individual; and an evaluation of the hybridization temperature of the assay. The data demonstrate that DNA exposed to a variety of environmental insults yields reliable PM typing results. Roy and Reynolds [70b] reported PM typing results from pap smear, semen smear, and postcoital slides and Hochmeister et al. [70c] reported PM results from human skeletal remains.

Allele and genotype frequencies for PM loci, as well as for HLA-DQA1, were determined in African-Americans, Caucasians, southeastern Hispanics, southwestern Hispanics [67], and Koreans [70d]. All loci meet Hardy-Weinberg expectations, and there is little evidence for association of alleles between the loci. The frequency data can be used in forensic analyses and paternity tests to estimate the frequency of a multiple locus DNA profile in various general U.S. populations. Because the PM system is multiplexed with DQA1 and the typing procedure is essentially the same as that for HLA-DQA1, the forensic laboratory can type six loci with no more DNA than required to type HLA-DQA1 alone and no additional effort. Hochmeister et al. [70e] showed that genomic DNA previously subjected to HLA-DQA1 amplification can be amplified and typed using the PM system.

5.3.3. Analysis of Amplified Fragment Length Polymorphisms

An alternative PCR-based method of DNA analysis is currently being investigated by the FBI Laboratory for the purpose of identification of biological evidence. This method, amplified fragment length polymorphism (AMP-FLP) analysis, combines the speed, sensitivity, and specificity of PCR with the information content of VNTR loci. In this procedure, VNTR sequences are amplified by PCR using primers that are specific to sequences that flank a particular VNTR locus. Subsequently, the alleles are separated by gel electrophoresis and visualized by ethidium bromide, silver staining, autoradiography, or fluorescence detection. The loci being investigated for AMP-FLP analysis, like the VNTR loci used for RFLP analysis, consist of blocks of sequences repeated in tandem, but the range of allele sizes is smaller than that for the loci used in RFLP analysis.

AMP-FLP analysis offers several advantages over RFLP VNTR analysis. In addition to obviating the need for isotopic detection, the size of the alleles and the resolution of the gels are such that alleles differing by one repeat unit can be distinguished (generally not possible with the loci used in RFLP VNTR analysis). Size standard ladders consisting of the different alleles of a locus can be run in lanes adjacent to sample lanes, and samples can be typed by comparison to these allelic ladders. The alleles are designated by the number of repeat units they contain. This obviates the need for actually sizing fragments, reduces measurement error, and facilitates interlaboratory comparisons of DNA profiles.

The first AMP-FLP locus that was studied in the FBI Laboratory was D1S80 (pMCT118) [71,72]. D1S80 alleles can be separated by electrophoresis on horizontal ultrathin-layer polyacrylamide gels [71,73,74] poured by the flap technique [75]. The ultrathin-layer polyacrylamide gels are run using a discontinuous buffer system. Resolution can be altered by changing concentration of the leading ion, substituting different trailing ions, using additives such as glycerol, and changing the percent total acrylamide and percent cross-linker. To visualize the AMP-FLP alleles, the gels are silver-stained [71]. More recently, D1S80 fragments have been separated on vertical polyacrylamide gels [76] also using a discontinuous buffer system.

Several studies indicate that D1S80 typing is a reliable means of forensic DNA analysis. D1S80 analysis has been used on forensic samples, such as tissue from fire victims [51], bone from badly decomposed bodies [38,49], and cigarette butts [50], as well as from casework [56,76a]. Cosso and Reynolds [76b] conducted validation studies that included typing of liquid and stained blood, semen, and saliva, typing of mixed DNA samples, typing of degraded DNA samples, as well as population studies. A large number of experiments have been conducted by the FBI Laboratory that further validate the system for use in forensic case analysis [76]. These experiments were similar to those conducted during the evaluation of RFLP, HLA-DQA1, and PM typing and included evaluation of the performance of the system on samples exposed to sunlight, a variety of substrates and chemicals including spermicides, and microorganism contamination. Different body fluid mixtures were amplified and typed (combinations of blood/blood, blood/saliva, blood/semen, and saliva/semen), as well as DNA samples from a variety of animals. As seen with all other DNA validation studies to date, exposure to adverse conditions either does not affect the typing results or causes no result to be produced. Mixtures are detected when both components are present in sufficient quantity. Samples from 50 old RFLP cases were amplified and typed for D1S80; no discordances were found com-

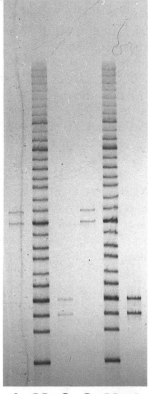

1 M 2 3 M 4

Fig. 5.3. D1S80 typing patterns from victim's DNA (lane 1, type 25-24), suspect's DNA (lane 2, type 18-17), the nonsperm fraction DNA (lane 3, type 25-24), and the sperm fraction DNA (lane 4, type 18-17) from a differentially extracted vaginal swab. Lanes designated M contain D1S80 allelic ladders (containing alleles 14 and 16-41).

pared with the RFLP typing results. Figure 5.3 shows an example of a D1S80 typing gel containing samples from one of these old cases. Samples have been typed from various population groups [77] in order to determine allele frequency estimates and, hence, DNA profile frequency estimates. Alleles are classified based on their migration relative to a standard allelic ladder [78] and based on the number of core repeat units they contain. Frequency distributions meet Hardy-Weinberg expectations in the four groups examined (African-American, Caucasian, southeastern Hispanic, and southwestern Hispanic).

Other VNTRs that show promise for forensic use are short tandem repeats (STRs). Trimeric and tetrameric repeats are abundant in the human genome [79], are polymorphic, and can easily be amplified by PCR [80,81]. Their overall short allele size (generally less than 400 bp) makes the typing of these loci less susceptible to degradation than longer VNTR sequences. Three STR markers, CSF1PO [82], TPOX [83], and HUMTH01 [80], can be amplified in a multiplex system (Promega Corporation, Madison, Wisconsin) and typed on vertical denaturing polyacrylamide gels (essentially DNA sequencing gels). As with D1S80, alleles in samples can be identified by their position relative to allelic ladders for each locus. Population studies on allele frequencies for these loci have been conducted [82–86]. Data show that for the Caucasians, African-Americans, southeastern Hispanics, and southwestern Hispanics examined, frequencies meet Hardy-Weinberg expectations and there is little evidence for association of alleles between loci [86]. Forensic validation studies are under way to establish the reliability of STR typing. By using a panel of two or three STR multiplex systems, a high degree of discrimination can be obtained from a limited amount of DNA (1 ng or less DNA can be typed with this system).

5.4. FUTURE OF FORENSIC DNA ANALYSIS

The primary goal of biological forensic evidence analyses is to obtain genetic information with the highest probability of distinguishing between two samples that are not from the same individual. Additional goals are to have sensitive procedures in order to type minute specimens and the ability to accomplish the typing rapidly. The development of new procedures will help forensic scientists achieve these goals more effectively.

5.4.1. Automated Detection and Analysis of AMP-FLPs

Automated detection and analysis of AMP-FLPs is an alternative approach to polyacrylamide gel electrophoresis and silver staining, described above. An example of this method relies on detection of DNA molecules labeled with fluorescent dyes. Specifically, the PCR primers are labeled with fluorescein derivatives. Amplified DNA can be subjected to electrophoresis in an agarose gel in an Applied Biosystems 362 Gene Scanner or on polyacrylamide gels in a 373A Sequencer (ABI, Foster City, CA) [87,88]. The labeled DNA fragments are detected during electrophoresis by argon laser excitation of the fluorescein-labeled primers at a designated window. After the emissions from the dyes are filtered and collected by a photomultiplier tube, the signal is digitized and analyzed by a software program designed for this purpose. Allele sizes are assigned based on comparisons with DNA fragments of known sizes run in the same lane as the unknown sample. Because four different dyes are available, this approach allows for the electrophoresis of a size standard or allelic ladder (labeled with one dye) to be run in the same lane as a sample (labeled with a second dye). This can minimize lane-to-lane electrophoretic variation. This system also offers the possibility of running multiple samples, amplified at the same locus but labeled with different dyes, in the same lane. Additionally, the system can accommodate samples subjected to multiplexed PCR, or amplify several loci in one sample, each labeled with a different dye. Fregeau et al. [89] and Kimpton et al. [90] have described systems in which several loci can be typed simultaneously.

An example of an alternative automated detection system, which does not require the electrophoretic apparatus to be coupled with the detection apparatus, is one in which AMP-FLP or STR gels can be run manually and then stained with a fluorescent dye such as SYBR Green (Molecular Probes). The DNA can then be detected on an instrument such as the FluorImager (Molecular Dynamics, Sunnyvale, CA). Alleles in a sample can be determined with reference to allelic ladders, and these data can be stored in a computer.

5.4.2. Sequencing of Mitochondrial DNA

The use of PCR followed by direct sequencing of amplified DNA [70] and automated sequencing [91,92] can allow the forensic scientist to generate sequence data easily and rapidly from small samples. Sequencing of nuclear DNA (with the possible exception of Y chromosome DNA), however, may not be practical for forensic use because the presence of two copies of each gene complicates interpretation of sequence data. However, analysis of mitochondrial DNA may prove useful for forensic analysis.

Mitochondrial DNA (mtDNA) is well characterized, highly polymorphic, maternally inherited and thus monoclonal in individuals, and present in high copy number. Since mtDNA presents many more potential targets for analysis than nuclear genes, the analysis of smaller samples should be feasible (see Budowle et al. [93] for a review). The greatest amount of variability in mtDNA occurs in two hypervariable regions in the D-loop region of the mtDNA. Although validation experiments remain to be completed prior to use of mtDNA analysis for forensic use, there is evidence that mtDNA analysis could prove useful, especially for hairs or other material that may contain too little nuclear DNA for routine analysis. Sullivan et al. [94] and Hopgood et al. [95] described extraction, amplification, and sequencing of D-loop DNA from semen and hair shafts. To accomplish successful sequencing, this group amplified mtDNA in two stages, first amplifying the entire D-loop region using primers that bind to conserved regions of mtDNA. Asymmetric amplification of the products of the first amplification was then performed using nested primers.

This second amplification generated single-stranded DNA, which was then sequenced using an automated sequencing system [91]. More recently, the availability of cycle sequencing [96] has improved the efficiency in which mtDNA can be typed. There are several reports of successful extraction and characterization of mtDNA from very old tissue specimens [62,97–99].

Before mtDNA analysis can be routinely used for forensic analyses, population studies are necessary to characterize variations in D-loop sequences and to assess the statistical significance of mtDNA sequence variations. These studies are currently being conducted by the FBI Laboratory, as well as by others.

5.5. CONCLUSION

DNA typing has proved to be a powerful method for associating or excluding biological evidence with individuals involved in a violent crime. Advances in technology are leading to continual evolution of techniques used in forensic DNA analysis. The development of new DNA analytical procedures, especially PCR-based procedures, with their increased sensitivity and decreased analysis time, and their potential for direct assessment of genotypes and automated detection, will increase the ability of the forensic scientist to exclude individuals falsely associated with evidence and possibly, ultimately, to determine the source of biological evidence with an extremely high degree of certainty.

ACKNOWLEDGMENTS

The authors thank Dr. F. Samuel Baechtel and Ms. Jill B. Smerick for the gel that appears in Figure 5.3.

REFERENCES

1. Sensabaugh, G.F.: Biochemical markers of individuality. In Saferstein, R. (ed.): Forensic Science Handbook. Englewood Cliffs, NJ: Prentice-Hall, 1982.
2. Budowle, B., Murch, R.S., Davidson, L.C., Gambel, A.M., Kearney, J.J.: Subtyping phosphoglucomutase-1 in semen stains and bloodstains: A report on the method. J. For. Sci. 31:1341–1348 (1986).
3. Wyman, A., White, R.: A highly polymorphic locus in human DNA. Proc. Natl. Acad. Sci. U.S.A. 77:6754–6758 (1980).
4. Nakamura, Y., Leppert, M., O'Connell, P., Wolff, R., Holm, T., Culver, M., Martin, C., Fujimoto, E., Hoff, M., Kumlin, E., White, R.: Variable number of tandem repeat (VNTR) markers for human gene mapping. Science 235:1616–1622 (1987).
5. Jeffreys, A.J., Wilson, V., Thein, S.L.: Hypervariable "minisatellite" regions in human DNA. Nature 314:67–73 (1985).

6. Gill, P., Werrett, D.J.: Exclusion of a man charged with murder by DNA fingerprinting. For. Sci. Int. 35:145–148 (1987).

7. Budowle, B., Baechtel F.S.: Modifications to improve the effectiveness of restriction fragment length polymorphism typing. Appl. Theor. Electrophoresis 1:181–187 (1990).

8. Budowle, B., Waye, J.S., Shutler, G.G., Baechtel, F.S.: Hae III—A suitable restriction endonuclease for restriction fragment length polymorphism analysis of biological evidence samples. J. For. Sci. 35:530–536 (1990).

9. Gill, P., Jeffreys, A. J., Werrett, D.J.: Forensic application of DNA "fingerprints." Nature 318:577–579 (1985).

10. Giusti, A., Baird, M., Pasquale, S., Balazs, I., Glassberg, J.: Application of deoxyribonucleic acid (DNA) polymorphisms to the analysis of DNA recovered from sperm. J. For. Sci. 31:409–417 (1986).

11. McNally, L., Shaler, R.C., Baird, M., Balazs, I., Kobilinsky, L., De Forest, P.: The effects of environment and substrata on deoxyribonucleic acid (DNA): The use of casework samples from New York City. J. For. Sci. 34, 1070–1077 (1989).

12. Jeffreys, A.J., Wilson, V., Thein, S.L.: Individual-specific "fingerprints" of human DNA. Nature 316:76–79 (1985).

13. Gill, P., Lygo, J.E., Fowler, S.J., Werrett, D.J.: An evaluation of DNA fingerprinting for forensic purposes. Electrophoresis 8:38–44 (1987).

14. Honma, M., Yoshii, T., Ishiyama, I., Mitani, K., Kominami, R., Muramatsu, M.: Individual identification from semen by the deoxyribonucleic acid (DNA) fingerprint technique. J. For. Sci. 34:222–227 (1989).

15. Hill, W.G.: DNA fingerprinting applied to animal and bird populations. Nature 327:98–99 (1987).

16. Wetton, J.H., Carter, R.E., Parkin, D.T., Walter, D.: Demographic study of a wild house sparrow population by DNA fingerprinting. Nature 327:147–152 (1987).

17. Dallas, J.F.: Detection of DNA fingerprints of cultivated rice by hybridization with a human minisatellite DNA probe. Proc. Natl. Acad. Sci. U.S.A. 85:6831–6835 (1988).

18. Weiss, M.L., Wilson, V., Chan, C., Turner, T., Jeffreys, A.J.: Application of DNA fingerprinting probes to old world monkeys. Am. J. Primatol. 16:73–79 (1988).

19. Nakamura, Y., Gillilan, S., O'Connell, P., Leppert, M., Lathrop, G.M., White, R.: Isolation and mapping of a polymorphic DNA sequence pYNH 24 on chromosome 1 (D2S44). Nucleic Acids Res. 15:10073 (1987).

20. Balazs, I., Baird, M., Clyne, M., Meade, E.: Human population genetic studies of five hypervariable DNA loci. Am. J. Hum. Genet. 44:182–190 (1989).

21. Wong, Z., Wilson, V., Patel, I., Povey, S., Jeffreys, A.J.: Characterization of a panel of highly variable minisatellites cloned from human DNA. Ann. Hum. Genet. 51:269–288 (1987).

22. Milner, E.C.B., Latshaw, C.L., Willems van Dijk, K., Charmley, P., Concannon, P., Schroeder, H.W.: Isolation and mapping of a polymorphic DNA sequence pH30 on chromosome 4 (HGM provisional no. D4S139). Nucleic Acids Res. 17:4002 (1989).

23. Bragg, T., Nakamura, Y., Jones, C., White, R.: Isolation and mapping of a polymorphic DNA sequence (cTBQ7) on chromosome 10 (D10S28). Nucleic Acids Res. 16:11395 (1988).

24. Armour, J.A., Povey, S., Jeremiah, S., Jeffreys, A.J.: Systematic cloning of human minisatellites from ordered charomid libraries. Genomics 8:501–512 (1990).

25. Monson, K.L., Budowle, B.: A system for semi-automated analysis of DNA autoradi-

ograms. Proceedings of the International Symposium on the Forensic Aspects of DNA Analysis. Federal Bureau of Investigation, Washington, DC., 1989.

26. Budowle, B., Giusti, A.M., Waye, J.S., Baechtel, F.S., Fourney, R.M., Adams, D.E., Presley, L.A., Deadman, H.A., Monson, K.L.: Fixed-bin analysis for statistical evaluation of continuous distributions of allelic data from VNTR loci, for use in forensic comparisons. Am. J. Hum. Genet. 48:841–855 (1991).

27. Baird, M., Balazs, I., Giusti, A.M., Miyazaki, L., Nicholas, L., Wexler, K., Kanter, E., Glassberg, J., Allen, F., Rubinstein, P., Sussman. L.: Allele frequency distribution of two highly polymorphic DNA sequences in three ethnic groups and its applications to the determination of paternity. Am. J. Hum. Genet. 39:489–501 (1986).

28. Budowle, B., Monson, K.L., et al.: A preliminary report on binned general population data on six VNTR loci in Caucasians, Blacks, and Hispanics from the United States. Crime Lab. Digest 18:9–26 (1991).

29. Lander, E.S.: Invited editorial: Research on DNA catching up with courtroom application. Am. J. Hum. Genet. 48:819–823 (1991).

30. Lander, E.S.: Population genetic consideratioins in the forensic use of DNA typing. In Sensabaugh, G., Witkowski, J. (eds.): DNA Technology and Forensic Science. Cold Spring Harbor, NY: Cold Spring Harbor Laboratory Press, 1989.

31. Lewontin, R.C. Hartl, D.L.: Population genetics in forensic DNA typing. Science 254:1745–1750 (1991).

32. National Research Council: DNA Typing: Statistical bases for interpretation. In DNA Technology in Forensic Science. Washington, D.C.: National Academy Press, 1992.

33. FBI Laboratory: VNTR Population Data: A Worldwide Study, Vols. I–IV. Washington, D.C.: Federal Bureau of Investigation, 1993.

34. Budowle, B., Monson, K.L., Giusti, A.M., Brown, B.L.: The assessment of frequency estimates of Hae III–generated VNTR profiles in various reference databases. J. For. Sci. 39:319–352 (1994).

35. Budowle, B., Monson, K.L., Giusti, A.M., Brown, B.L.: Evaluation of Hinf I-generated VNTR profile frequencies determined using various ethnic databases. J. For. Sci. 39:998–1008 (1994).

36. Adams, D.E., Presley, L.A., Baumstark, A.L., Hensley, K.W., Hill, A.L., Anoe, K.S., Campbell, P.A., McLaughlin, C.M., Budowle, B., Giusti, A.M., Smerick, J.B., Baechtel F.S.: Deoxyribonucleic acid (DNA) analysis by restriction fragment length polymorphisms of blood and other body fluid stains subjected to contamination and environmental insults. J. For. Sci. 36:1284–1298 (1991).

37. Kanter, E., Baird, M., Shaler, R., Balazs, I. Analysis of restriction fragment length polymorphisms in deoxyribonucleic acid (DNA) recovered from dried bloodstains. J. For. Sci. 31:403–408 (1986).

38. Lee, H.C., Pagliaro, E.M., Berka, K.M., Folk, N.L., Anderson, D.T., Ruano, G., Keith, T.P., Phipps, P., Herrin, G.L., Jr., Garner, D.D., Gaensslen, R.E.: Genetic markers in human bone I: Deoxyribonucleic acid (DNA) analysis. J. For. Sci. 36:320–330 (1991).

39. McNally, L., Shaler, R.C., Baird, M., Balazs, I., De Forest, P., Kobilinsky, L.: Evaluation of deoxyribonucleic acid (DNA) isolated from human bloodstains exposed to ultraviolet light, light, heat, humidity, and soil contamination. J. For. Sci. 34:1059–1069 (1989).

40. Hochmeister, M.N., Budowle, B., Borer, U.V., Dirnhofer, R.: Effects of nonoxinol-9 on the ability to obtain DNA profiles from postcoital vaginal swabs. J. For. Sci. 38:442–447 (1993).

41. Hochmeister, M.N., Budowle, B., Baechtel, F.S.: Effects of presumptive test reagents on

the ability to obtain restriction fragment length polymorphism (RFLP) patterns from human blood and semen stains. J. For. Sci. 36:656–661 (1991).

42. Saiki, R.K., Scharf, S., Faloona, F., Mullis, K.B., Horn, G.T., Erlich, H.A., Arnheim, N.: Enzymatic amplification of beta-globin genomic sequences and restriction site analysis for diagnosis of sickle cell anemia. Science 230:1350–1354 (1985).

43. Saiki, R.K., Gelfand, D.H., Stoffel, S., Scharf, S.J., Higuchi, R., Horn, G.T., Mullis, K.B., Erlich, H.A.: Primer-directed enzymatic amplification of DNA with a thermostable DNA polymerase. Science 239:487–491 (1988).

43a. Budowle, B., Baechtel, F. S., Comey, C. T., Giusti, A.M., Klevan, L.: Simple protocols for typing forensic biological evidence: Chemiluminescent detection for human DNA quantitation and RFLP analyses and manual typing of PCR amplified polymorphisms. Electrophoresis, in press (1996).

43b. Giusti, A.M., Budowle, B.: A chemiluminescent based detection system for human DNA quntitation and restriction fragment length polymorphism [RFLP] analysis. Appl. Theor. Electrophoresis, in press (1996).

44. Saiki, R.K., Bugawan, T.L., Horn, G.T., Mullis, K.B., Erlich, H.A.: Analysis of enzymatically amplified β-globin and HLA-DQα DNA with allele-specific oligonucleotide probes. Nature 324:163–166 (1986).

45. Saiki, R.K., Walsh, P.S., Levenson, C.H., Erlich, H.A.: Genetic analysis of amplified DNA with immobilized sequence-specific oligonucleotide probes. Proc. Natl. Acad. Sci. U.S.A. 86:6230–6234 (1989).

46. Westwood, S.A., Werrett, D.J.: An evaluation of the polymerase chain reaction method for forensic applications. For. Sci Int. 45:201–215 (1990).

47. Comey, C T., Budowle, B.: Validation studies on the analysis of the HLA-DQα loucs using the polymerase chain reaction. J. For. Sci. 36:1633–1648 (1991).

48. Higuchi, R., von Beroldingen, C.H., Sensabaugh, G.F., Erlich, H.A.: DNA typing from single hairs. Nature 332:543–546 (1988).

49. Hochmeister, M.N., Budowle, B., Borer, U.V., Eggmann, U., Comey, C.T., Dirnhofer, R.: Typing of DNA extracted from compact bone from human remains. J. For. Sci. 36:1649–1661 (1991).

50. Hochmeister, M.N., Budowle, B., Jung, J., Borer, U.V., Comey, C.T., Dirnhofer, R.: PCR-based typing of DNA extracted from cigarette butts. Int. J. Leg. Med. 104:229–233 (1991).

51. Sajantila, A., Strom, M., Budowle, B., Karhunen, P.J., Peltonen, L.: The polymerase chain reaction and post-mortem forensic identity testing: Application of amplified D1S80 and HLA-DQα loci to the identification of fire victims. For. Sci. Int. 51:23–34 (1991).

52. Blake, E., Mihalovich, J., Higuchi, R., Walsh, P.S., Erlich, H.A.: Polymerase chain reaction (PCR) amplification and human leukocyte antigen (HLA)-DQα oligonucleotide typing on biological evidence samples: Casework experience. J. For. Sci. 37:700–726 (1992).

53. Comey, C.T., Budowle, B., Adams, D.E., Baumstark, A.L., Lindsey, J.A., Presley, L.A.: PCR amplification and typing of the HLA DQα gene in forensic samples. J. For. Sci. 38:239–249 (1993).

54. Potsch, L., Meyer, U., Rothschild, S., Schneider, P.M., Rittner, C.: Application of DNA techniques for identification using human dental pulp as source of DNA. Int. J. Legal Med. 105:139–143 (1992).

55. Tahir, M., Watson, N.: Typing of DNA HLA-DQAlpha alleles extracted from human nail material using polymerase chain reaction. J. For. Sci. 40:634–635 (1995).

56. Helmuth, R., Fildes, N., Blake, E., Luce, M.C., Chimera, J., Madej, R., Gorodezky, C., Stoneking, M., Schmill, N., Klitz, W., Higuchi, R., Ehrlich, H.A.: HLA-DQα allele and genotype frequencies in various human populations, determined by using enzymatic amplification and oligonucleotide probes. Am. J. Hum. Genet. 47:515–523 (1990).

57. Sajantila, A., Budowle, B., Strom, M., Johnson, V., Lukka, M., Peltonen, L., Enholm, C.: Amplification of alleles at the D1S80 locus by the polymerase chain reaction: Comparison of a Finnish and a North American Caucasian sample, and forensic case-work evaluation. Am. J. Hum. Genet. 50:816–825 (1992).

58. Harrington, C.S., Dunaiski, V., Williams, K.E., Fowler, C.: HLA DQα typing of foren-sic specimens by amplification restriction fragment polymorphism (ARFP) analysis. For. Sci. Int. 51:147–157 (1991).

59. Kloosterman, A.D., Budowle, B., Riley, E.L.: Population data of the HLA DQα locus in Dutch Caucasians. Comparison with seven other population studies. Int. J. Legal Med. 105:233–238 (1993).

60. Walsh, S., Higuchi, R.: PCR inhibition and bloodstains. In Proceedings of the Inter-national Symposium on the Forensic Aspects of DNA Analysis. Washington, D.C.: Federal Bureau of Investigation, 1989.

61. Grimberg, J., Nawoschik, S., Belluscio, L., McKee, R., Turch, A., Eisenberg, A.: A sim-ple and efficient non-organic procedure for the isolation of genomic DNA from blood. Nucleic Acids Res. 17:8390 (1989).

62. Paabo, S., Gifford, J.A., Wilson, A.C.: Mitochondrial DNA sequences from a 7000-year old brain. Nucleic Acids Res. 16:9775–9787 (1988).

63. Singer-Sam, J., Tanguay, R.L., Riggs, A.: Use of Chelex to improve the PCR signal from a small number of cells. Amplifications 3:11 (1989).

64. Walsh, P.S., Metzger, D.A., Higuchi, R.: Chelex 100 as a medium for simple extraction of DNA for PCR-based typing from forensic material. BioTechniques 10:506–513 (1991).

65. Comey, C.T., Koons, B.W., Presley, K.W., Smerick, J.B., Sobieralski, C.A., Stanley, D.M., Baechtel, F.S.: DNA extraction strategies for amplified fragment length polymor-phism analysis. J. For. Sci. 39:1254–1269 (1994).

66. Presley, L.A., Baumstark, A.L., Dixon, A.: The effects of specific latent fingerprint and questioned document examinations on the amplification and typing of the HLA DQ al-pha gene region in forensic casework. J. For. Sci. 38:1028–1036 (1993).

67. Budowle, B., Lindsey, J.A., DeCou, J.A., Koons, B.W., Giusti, A.M., Comey, C.T.: Validation and population studies of the loci LDLR, GYPA, HBGG, D7S8, and Gc (PM loci), and HLA-DQα using a multiple amplification and typing procedure. J. For. Sci. 40:45–54 (1995).

68. Comey, C.T., Jung, J.M., Budowle, B.: Use of formamide to improve amplification of HLA-DQα sequences. BioTechniques 10:60–61 (1991).

69. Waye, J.S., Presley, L.A., Budowle, B., Shutler, G.G., Fourney, R.M: A simple and sen-sitive method for quantifying human genomic DNA in forensic specimen extracts. BioTechniques 7:852–855 (1989).

70. Gyllensten, T.B., Erlich, H.A.: Generation of single-stranded DNA by the polymerase chain reaction and its application to direct sequencing of the HLA-DQA locus. Proc. Natl. Acad. Sci. U.S.A. 85:7652–7656 (1988).

70a. Fildes, N., Reynolds, R.: Consistency and reproducibility of AmpliType PM results between seven laboratories: Field trial results. J. For. Sci. 40:279–286 (1995).

70b. Roy, R., Reynolds, R.: AmpliType PM and HLA DQAlpha typing from Pap smear, semen smear, and postcoital slides. J. For. Sci. 40:266–269 (1995).

70c. Hochmeister, M.N., Budowle, B., Borer, U.V., Rudin, O., Bohnert, M., Dirnhofer, R.: Confirmation of the identity of human skeletal remains using multiplex PCR amplification and typing kits. J. For. Sci. 40:701–705 (1995).

70d. Woo, K.M., Budowle, B.: Korean population data on the PCR-based loci LDLR, GYPA, HBGG, D7S8, Gc, HLA-DQA1, and D1S80. J. For. Sci. 40:645–648 (1995).

70e. Hochmeister, M.N., Budowle, B., Borer, U.V., Dirnhofer, R.: A method for the purification and recovery of genomic DNA from an HLA-DQA1 amplification product and its subsequent amplification and typing with the AmpliType PM PCR amplification and typing kit. J. For. Sci. 40:649–653 (1995).

71. Budowle, B., Chakraborty, R., Giusti, A.M., Eisenberg, A.J., Allen, R.C.: Analysis of the VNTR locus D1S80 by the PCR followed by high-resolution PAGE. Am. J. Hum. Genet. 48:137–144 (1991).

72. Kasai, K., Nakamura, Y., White, R.: Amplification of a variable number of tandem repeats (VNTR) locus by the polymerase chain reaction (PCR) and its application to forensic science. J. For. Sci. 35:1196–1200 (1990).

73. Allen, R.C., Graves, G., Budowle, B.: Polymerase chain reaction amplification products separated on rehydratable polyacrylamide gels and stained with silver. BioTechniques 7:736–744 (1989).

74. Budowle, B., Allen, R.C.: Discontinuous polyacrylamide gel electrophoresis of DNA fragments. In Mathew, G.C. (ed.): Protocols in Human Molecular Genetics—Methods in Molecular Biology. Vol. 9. Human Press, Clifton, New Jersey, 1991.

75. Allen, R.C.: Rapid isoelectric focusing and detection of nanogram amounts of proteins from body tissues and fluids. Electrophoresis 1:32–37 (1980).

76. Baechtel, F.A., Presley, K.W., Smerick, J.B.: D1S80 typing of DNA from simulated forensic specimens. J. For. Sci. (1995). Vol. 40, 536–545.

76a. Kloosterman, A.D., Budowle, B., Daselaar, P.: PCR-amplification and detection of the human D1S80 VNTR locus: Amplification conditions, population genetics, and application in forensic analysis. Int. J. Legal Med. 105:257–264 (1993).

76b. Cosso, S., Reynolds, R.: Validation of the AmpliFLP D1S80 PCR amplification kit for forensic casework analysis according to TWGDAM guidelines. J. For. Sci. 40:424–434 (1995).

77. Budowle, B., Baechtel, F.S., Smerick, J.B., Presley, K.W., Giusti, A.M., Parsons, G., Alevy, M.C., Chakraborty, R.: D1S80 population data in African Americans, Causasians, southeastern Hispanics, southwestern Hispanics, and Orientals. J. For. Sci. 40:38–44 (1995).

78. Baechtel, F.S., Smerick, J.B., Presley, K.W., Budowle, B. Multigenerational amplification of a reference ladder for alleles at locus D1S80. J. For. Sci. Vol. 38:1176–1182 (1993).

79. Weber, J.L., May, P.E.: Abundant class of human DNA polymorphisms which can be typed using the polymerase chain reaction. Am. J. Hum. Genet. 44:388–396 (1989).

80. Edwards, A., Civitello, A., Hammond, H.A., Caskey, C.T.: DNA typing and genetic

mapping with trimeric and tetrameric tandem repeats. Am. J. Hum. Genet. 49:746–756 (1991).

81. Edwards, A., Gibbs, R.A., Nguyen, P.N., Ansorge, W., Caskey, C.T.: Automated DNA sequencing methods for detection and analysis of mutations: Applications to the Lesch-Nyhan syndrome. Trans. Assoc. Am. Physicians 102:185–194 (1989).

82. Hammond, H.A., Jin, L., Zhong, Y., Caskey, C.T., Chakraborty, R.: Evaluation of 13 short tandem repeat loci for use in personal identification applications. Am. J. Hum. Genet. 55:175–189 (1994).

83. Anker, R., Steinbrueck, T., Donis-Keller, H.: Tetranucleotide repeat polymorphism at the human thyroid peroxydase (hTPO) locus. Hum. Mol. Genet. 1:137 (1992).

84. Puers, C., Hammond, H.A., Jin, L., Caskey, C.T., Schumm, J.W.: Identification of repeat sequence heterogeneity at the polymorphic short tandem repeat locus HUMTH01 [AATG]$_n$ and reassignment of alleles in population analysis by using a locus-specific ladder. Am. J. Hum. Genet. 53:953–958 (1993).

85. Huang, N., Budowle, B.: Chinese population data on three tetrameric short tandem repeat loci—HUMTH01, TPOX, and CSF1PO—derived using multiplex PCR and manual typing. J. For. Sci. (submitted) (1995).

86. Comey, C.T., Koons, B.W., Budowle, B.: Analysis of four populations at the tetrameric short tandem repeat (STR) loci CSF1PO, TPOX, and TH01. Proceedings of the Fifth International Symposium on Human Identification, Promega Corporation, 1995.

87. Robertson, J.M., Kronick, M.: Automating DNA fingerprinting—a multifluorophore approach to reduce chance in match calling. Crime Lab. Digest 18:179–182 (1991).

88. Robertson, J., Schaefer, T., Kronick, M., Budowle, B.: Automated analysis of fluorescent amplified fragment length polymorphism for DNA typing. Adv. For. Haemogenet. 4:35–37 (1991).

89. Fregeau, C.J., Fourney, R.M.: DNA typing with fluorescently tagged short tandem repeats: A sensitive and accurate approach to human identification. BioTechniques 15:100–119 (1993).

90. Kimpton, C.P., Gill, P., Walton, A., Urquhart, A., Millican, E.S., Adams, M.: Automated DNA profiling employing multiplex amplification of short tandem repeat loci. PCR Methods Applications 3:13–22 (1993).

91. Smith, L.M., Sanders, J.Z., Kaiser, R.J., Hughes, P., Dodd, C., Connell, C.R., Heiner, C., Kent, S.B.H., Hood, L.E.: Fluorescence detection in automated DNA sequence analysis. Nature 321:674–679 (1986).

92. Prober, J.M., Trainor, G.L., Dam, R.J., Hobbs, F.W., Robertson, C.W., Zagursky, R.J., Cocuzza, A.J., Jensen, M.A., Baumeister, K.: A system for rapid DNA sequencing with fluorescent chain-terminating dideoxynucleotides. Science 238:336–341 (1987).

93. Budowle, B., Adams, D.E., Comey, C.T., Merril, C.R.: Mitochondrial DNA: A possible genetic material suitable for forensic analysis. In Lee, H.C., Gaensslen, R.E. (eds.): Advances in Forensic Sciences: DNA and Other Polymorphisms in Forensic Science. Chicago: Year Book Medical Publishers, 1990.

94. Sullivan, K.M., Hopgood, R., Lang, B., Gill, P.: Automated amplification and sequencing of human mitochondrial DNA. Electrophoresis 12:17–21 (1991).

95. Hopgood, R., Sullivan, K.M., Gill, P.: Strategies for automated sequencing of human mitochondrial DNA directly from PCR products. BioTechniques 13:82–92 (1992).

96. Carothers, A.M., Urlaub, G., Mucha, J., Grunberger, D., Chasin, L.A.: Point mutation analysis in a mammalian gene: Rapid preparation of total RNA, PCR amplification of cDNA and *Taq* sequencing by a novel method. BioTechniques 7:494–499 (1989).

97. Paabo, S.: Ancient DNA: Extraction, characterization, molecular cloning, and enzymatic amplification. Proc. Natl. Acad. Sci. U.S.A. 86:1939–1943 (1989).

98. Smith, M.F., Patton, J.L.: PCR on dried skin and liver extracts from the same individual gives identical products. Trends Genet. 7:4 (1991).

99. Hagelberg, E., Sykes, B., Hedges, R.: Ancient bone DNA amplified. Nature 342:485 (1989).

CHAPTER 6

APPROACHES TO INFECTIOUS DISEASE DIAGNOSIS

WILLIAM C. WILSON
Arthropod-borne Animal Diseases Research Laboratory,
USDA, Agricultural Research Service,
Laramie, WY 82071-3965

6.1. INTRODUCTION

The use of nucleic acid technology has become increasingly routine for disease diagnosis. Many reviews, workshops, and symposia on this technology have therefore resulted [1–4]. The quantity of literature on the approaches to nucleic acid diagnostic tests is staggering. Although nucleic acid technology is being used in diagnostic laboratories, there is a lack of commercially produced nucleic acid diagnostic tests. The biotechnology industry has failed either to recognize or to address the potential markets. Therefore, the majority of the nucleic acid tests being used have been developed or adapted by the individual diagnostic user, which has resulted in a lack of standardization. Nucleic acid tests, however, are very versatile and can be utilized in a variety of situations. In fact, many of these tests could be used in moderately equipped on-site medical or veterinary clinics. Advances in this technology may even promote the use of these techniques in home or farm test kits. Many large-scale animal production units already have on-site laboratories that could utilize this technology. With the increased use of such tests and their potential movement out of the diagnostic laboratories, it is important to understand the various approaches used in the development of these new tests. This information is necessary for users to appreciate the versatility of nucleic acid–based tests and to select an appropriate test for their needs.

When choosing an approach to disease diagnosis via nucleic acid detection tech-

Nucleic Acid Analysis: Principles and Bioapplications, pages 105–129
© 1996 Wiley-Liss, Inc.

nology, the developer must examine several issues. The issues described in this chapter include the advantages and disadvantages of detection methods, specificity requirements, sample type, genome target, and methodology. The following section characterizes and illustrates approaches to address these issues.

6.1.1. Advantages and Disadvantages of Detection Methods

If there is an acceptable immunochemical test available, then the necessity for using a nucleic acid test may be less apparent. Diagnosticians often use immunochemical techniques to detect foreign antigens; usually these are proteins produced by the pathogen. The use of immunofluorescence, immunocytochemistry, or enzyme-linked immunosorbent assays (ELISA) will continue to be important for the diagnostician. These immunochemical tools are especially useful with the availability of monoclonal antibodies specific for a given pathogen. However, development and selection of appropriate monoclonal antibodies can be difficult and require a long time. Furthermore, the incorporation of monoclonal antibodies into an immunological approach greatly enhances specificity of the assay, but may reduce sensitivity of the assay in comparison to a test using polyclonal antisera.

Nucleic acid tests provide an alternative diagnostic approach in situations where immunochemical tests are ineffective or not available. Nucleic acid tests can also be used to confirm or complement existing immunological tests. In some cases, a nucleic acid test is needed because the pathogen's antigens are degraded and/or disrupted by sample collection, fixation, or storage procedures, whereas the nucleic acid can be more stable. The intrinsic function of nucleic acids to serve as blueprints allows cells to generate copies of selected sequences as required for mitosis and gene transcription. In contrast, other macromolecules such as proteins cannot directly serve as templates for their own replication. Through various *in vitro* polymerase-driven reactions, molecular biologists have exploited the template function of nucleic acids to sequence genes, to increase the sensitivity of nucleic acid detection assays through amplification schemes (e.g., PCR), and to incorporate labeled molecules into nucleic acid probes (e.g., nick translation). Beyond providing alternative detection technology, nucleic acid technology provides an advantage over immunochemical technology in that they are relatively easier to develop and modify and can offer the user a greater degree of control over assay sensitivity and specificity.

There are many other advantages to the use of nucleic acid technology for detection of infectious pathogens [2]. An important advantage is the ability to differentiate virulent from avirulent organisms and vaccine from wild-type strains. This discrimination is based on the pathogen's genotype and not phenotype like other assays. This is important in detecting pathogens in which expression of a particular phenotype or protein is critical to its identification, but this expression may be variable or time dependent. Detection of the genotype is independent of phenotype expression, thus providing a more reliable target test system. For example, nucleic acid probes have been used to distinguish pathogenic *Treponema hyodysenteriae* from nonpathogenic treponemes [5]. The potential pathogenicity of *Escherichia coli* has also been screened with nucleic acid probes [6–8]. Another advantage of nucleic acid probes is that the homology shared by nucleic acid probes and related pathogens is a powerful tool in determining genetic relatedness of different pathogens. Nucleic acid technology has been used extensively in studies of pathogenic mechanisms and epidemiology and to ensure the absence of pathogenic microorganisms in our food and water supply [for review, see ref. 2].

Several disadvantages are associated with the use of nucleic acid tests. New technology requires retraining of technical staff and acquiring new equipment and reagents. Lack of experience with this technology and the requirement for the use of radioisotopes in many early tests contributed to the slow acceptance of nucleic acid technology in infectious disease diagnostic laboratories. This is less of a disadvantage now because there are several commercially available alternatives to radioisotopic procedures, but these procedures are generally thought to be less sensitive. An important disadvantage that is common to immunochemical procedures is that high background signals in some tests may make it difficult to detect weak positive samples.

Another disadvantage that may be more apparent with nucleic acid–based detection than with immunochemical schemes is the inconsistencies in correlating test results with the results of diagnostic isolation procedures to assess the presence and concentration of infectious pathogens. Because of the marked sensitivity of some nucleic acid detection systems (e.g., PCR) and the relative hardiness of nucleic acids, residual nucleic acid sequences from pathogens may be detected long after the pathogenic agents can be cultivated from clinical samples. This is complicated, particularly in viral infections, by a high ratio (e.g., 10:1 to 103:1) of noninfectious, defective pathogen particles relative to the number of infectious particles. In some cases, such as when the pathogen is not easily cultured or reliably recovered from clinical samples, positive nucleic acid detection may be deemed more sensitive and desirable. Alternatively, in the case when reliable and sensitive isolation procedures exist for a pathogen, the additional sensitivity offered by nucleic acid detection may only complicate interpretation of test results by identifying the pathogen or its components at concentrations that are epidemiologically irrelevant. This latter scenario is illustrated by the PCR detection of bluetongue viral RNA in blood from infected calves for at least 12 weeks longer than the virus could be recovered by inoculation of embryonated chicken eggs [9]. Blood specimens from previously infected calves that test positive by PCR but negative by virus isolation may not contain a sufficient infectious virus concentration to contaminate the minute blood meal drawn by the gnat, which serves as a vector of the virus. Consequently, the sensitivity of PCR detection may in some instances yield results that appear inconsistent with respect to the epidemiology of virus transmission.

6.1.2. Specificity

The required specificity of a diagnostic test is important. It may be of diagnostic value to discriminate among pathogens belonging to a common taxonomic group. To use the example of human herpesviruses, members of the same group may or may not be pathogenic and, if pathogenic, may not cause similar clinical signs. Several common herpesviruses may infect humans, including varicella-zoster virus, herpes simplex type 1, and Epstein-Barr virus. Although all are classified as herpesviral pathogens of humans, each causes distinct disease manifestations: chicken pox/shingles, cold sores, and infectious mononucleosis, respectively.

Alternatively, closely related pathogens may be associated with similar disease syndromes. The importance of discriminating between specific pathogens in this case may be based more on epidemiologic considerations rather than on concerns of clinical outcome. Bluetongue virus, a pathogen of ruminants, exists as 24 serotypes worldwide. Only five bluetongue serotypes have been identified within the United States. In some instances it may be useful to distinguish bluetongue virus infection from infections with other viral pathogens of ruminants. Accordingly, an appropriate bluetongue group assay would identify all serotypes of bluetongue virus. In other instances, however, the ability to identify specific bluetongue virus serotypes as common to the United States versus serotypes of foreign origin is important. The ability to distinguish between pathogen strains within a taxonomic group might be applied within national borders, in some instances even at the community level, to follow the movement of a pathogenic strain through a population or to determine disease management strategies such as vaccination.

6.1.3. Sample Type

Another issue is the type of samples to be tested. The sample should have sufficient amount of the pathogen of interest with a reasonable volume of sample. Excessive sample volumes can contribute to sample handling problems, poor recovery of pathogen, and high background signals. When a sample should be or will be taken must also be considered, because the concentration of the agents in tissues varies with the status or time course of the infection. For example, some pathogens like rotaviruses are shed in feces at certain times during their infectious cycle. Feces provide a sample obtained by noninvasive means for excreted pathogens, but the abundance of mixed populations of organisms found in feces complicates the detection design. Another problem with fecal samples is that they are rich in inhibitors of enzymatic amplification procedures, requiring more extensive nucleic acid extraction procedures. Saliva or nasal swabs are other noninvasive samples useful for pathogens that can be detected in those secretions. Because of their proximity and exposure to the external environment, they are also susceptible to contamination by mixed opportunistic and commensal microbes.

Probably the most common sample type is blood, because many pathogens circulate in the bloodstream. Pathogens are often found associated with the cellular fractions of blood. White blood cells can be separated from whole blood, thereby greatly reducing the sample volume to be extracted and easing the extraction of microbial nucleic acid. In contrast, extraction of pathogen-derived nucleic acid from association with red blood cells is more difficult because of the larger volume and high quantities of hemoglobin The common extraction solvent phenol denatures hemoglobin into a proteinaceous mass that may obstruct the extraction procedure.

RNases or enzymes that degrade RNA are in abundance in red blood cells; thus precautions such as including inhibitors of RNases need to be taken when the target nucleic acid is RNA. Also, common anticoagulants (heparin, EDTA) inhibit enzymatic amplification procedures, and their presence may necessitate more detailed nucleic acid purification procedures prior to the analysis.

Other pathogens such as some retroviruses are not readily found in excreta or body fluids but are only readily found in certain tissues or cells. Thus, a more invasive procedure such as tissue biopsy may be required for these pathogens. The pathogen's nucleic acid can be extracted from these biopsy samples after they have been mechanically dispersed using a tissue grinder into a cell-lysis solution. The technical difficulty in extracting the pathogen's nucleic acid will vary with the tissue. As an alternative to extracting pathogen nucleic acid from the tissue, the pathogen can be detected within the tissue sections by an *in situ* hybridization procedure, which may or may not include an enzymatic amplification step (e.g., PCR *in situ*). This may be the only effective detection procedure for pathogens that are found in very low amounts in certain cell types. This procedure requires the microscopic analysis of the tissue sections by an experienced technician and is very labor intensive.

6.1.4. Genome Target

The complementary binding of the probe and target sequences defines the specificity of the test; thus care should be taken in choosing the specific target nucleic acid sequence. This genomic region should be conserved among strains of the pathogen, but not present in related organisms. The type of pathogen affects the selection of the target genomic sequences. The knowledge of conserved and variable genes may also aid in the selection of target sequences. Genes that are conserved within a species or type of pathogen provide good group-specific targets. For example, the gene encoding a protein involved in forming a virus particle may be conserved among the various virus types within that group of viruses. This gene could provide a good target for a group-specific detection protocol. In bacteria and protozoa the ribosomal RNA sequences contain regions that are conserved within a species, thus providing good target sequences. Variable genes are also useful for detection of the specific type of a pathogen. For example, a gene encoding for a protein on the cell wall of a bacterial species may be quite variable. Infectious agents may have specific proteins that are on their outer surface, which may be important in their pathogenesis. The genes or messenger RNA for these proteins may provide good targets for detection. Another important target sequence in bacteria are the genes that encode toxins, which can be important from distinguishing pathogenic organisms from nonpathogenic organisms.

Sometimes it is difficult to determine a specific target sequence because gene sequences of the organism are not well characterized and a random selection procedure is required. A random selection procedure simply uses arbitrarily selected pieces of the pathogen's nucleic acid as probes against a battery of target sequences taken from known positive and negative samples. Specificity of the probes is con-

·firmed by cross-hybridization experiments using target nucleic acids from closely and distantly related sources. It is important that nonspecific repetitive sequences are excluded and that the kinetics or rates of hybridization are considered. The length of the probe, probe concentration, homology between the probe and target, and temperature and composition of the hybridization medium are all factors that may affect hybridization kinetics.

6.1.5. Methodology

The experience and requirements of the user are important considerations in the test system selection and design. The time required to perform the test may be an important consideration. Hybridization procedures can take 2–3 days, whereas amplification procedures can be done in only 1 day. The type of facility for which the test is being designed also needs to be considered. If a test is to be performed in on-site clinics, then the use of hazardous materials should be limited if not eliminated. Amplification procedures need to be done in controlled environments because of the potential for aerosol contamination. The methodology is available and versatile enough to meet the requirements of a variety of users. Although viruses, bacteria, and parasites are quite different organisms, the basic approaches to developing a nucleic acid test strategy are similar. The type of sample and the concentration of the pathogen can determine the most appropriate procedure for detection of the pathogen's nucleic acid.

The standard procedures used are (1) *in situ* hybridization, (2) solution hybridization, (3) solid phase hybridization, and (4) *in vitro* amplification. In *in situ* hybridization, small specific pieces of nucleic acid or probes are used to detect complementary nucleic acid sequences in a tissue section on a glass slide (Fig. 6.1). In solution hybridization, the probes are labeled with a detectable marker and bind or hybridize to the complementary target nucleic acid sequences while in solution in an aqueous medium (Fig. 6.2). The bound probe and target sequences are then collected, usually by filtration or affinity binding, and the labeled probe detected. In solid phase hybridization, the target sequences are immobilized on a solid support, usually a nitrocellulose or nylon membrane (Fig. 6.3). The membrane is then immersed in a solution containing a labeled nucleic acid probe that binds to the target sequence. The presence of the bound labeled probe is then determined typically by autoradiography for radiolabeled probes or by chromogenic substrates for enzyme-conjugated probes. Currently, the most popular nucleic acid detection procedure uses *in vitro* amplification, which was discussed in greater detail in a previous chapter, and an enzymatic reaction to replicate the nucleic acid to a detectable level (Fig. 6.4).

In Situ *Hybridization*. *In situ* hybridization is the logical choice if the test being developed is for a histopathology laboratory, where available samples are typically tissue sections. When a pathogen is present in low amounts centered in specific cells, *in situ* hybridization may be the most reliable procedure. This procedure is quite useful because it provides information regarding the relative amount and location of the targeted nucleic acid, which is important in pathogenesis studies to asso-

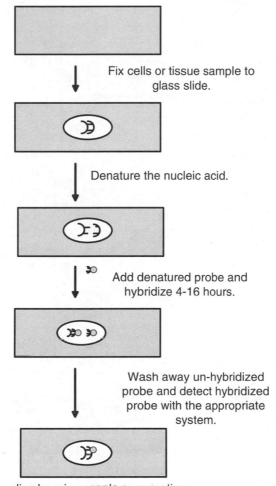

Fig. 6.1. Schematic representation of *in situ* hybridization.

ciate pathogens with specific cells or tissues. This approach is probably most applicable to intracellular agents such as latent viruses. Although the *in situ* hybridization procedures have been automated [10], they still require the microscopic reading of slides. This is a tedious procedure, especially when only determination of the presence or absence of a pathogen is required. Recently, computer-based image analyzing systems with specifically designed software were developed to count objects of specific color and size, such as stained and unstained cells within a given dimension (microscope field), thus making the procedures less time consuming.

Solution Hybridization. Biological samples can be dispersed in quanidine isothiocyanate and the hybridization done in the resulting solution [11]. Solution hy-

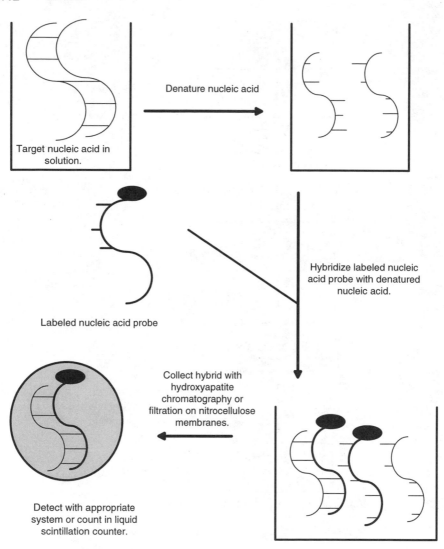

Fig. 6.2. Schematic representation of solution hybridization.

bridization does not require lengthy nucleic acid purification procedures and can be very useful in determining the presence and indicating the titer of a pathogen. Since these tests use RNA transcription vectors to generate the RNA hybridization probes, the development of a solution hybridization protocol requires the developer to be capable of performing molecular biological manipulations. This strategy was employed for the detection of the agriculturally important double-stranded RNA virus, bluetongue virus [11]. The use of the strong protein denaturant quanidine isothiocyanate, which inhibits Rnases, was important because bluetongue virus is associat-

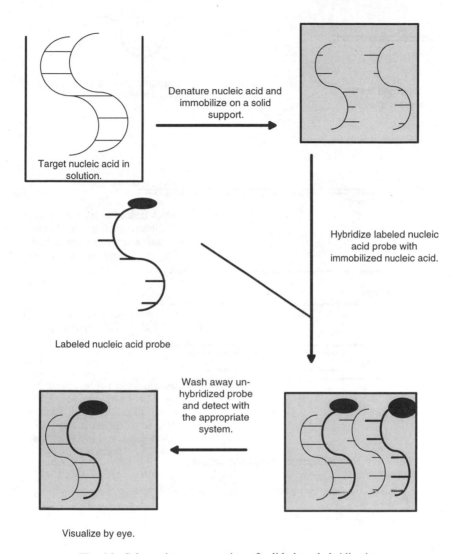

Fig. 6.3. Schematic representation of solid phase hybridization.

ed with red blood cells, which contains abundant RNases. Guanidine isothiocyanate–based extraction solutions are also used in solid phase tests.

Solid Phase Hybridization. Solid phase hybridization is used more frequently because it is easier to develop and can be used to examine many samples. The ability to remove an initial probe from a solid phase immobilized target and rehybridize with another probe is another important advantage of solid phase over solution hybridization.

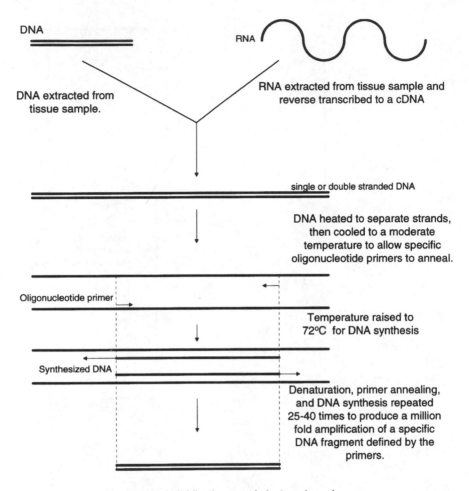

Fig. 6.4. Schematic representation of *in vitro* amplification.

For solid phase hybridization, the pathogen's nucleic acid must be isolated for detection. This may be done by any one of a variety of procedures. The standard isolation procedure usually concentrates the cellular fraction of the sample that contains the pathogen followed by a cell lysis procedure. Detergents such as sodium dodecyl sulfate are often used to lyse the cells. Proteinase K digestion can be included to facilitate the release of the nucleic acid. After lysis, phenol extraction is used to remove the protein from the sample. The nucleic acid is then ethanol precipitated or further purified by some form of chromatography. These standard procedures are very labor intensive, and loss of target nucleic acid probably accounts for some of the lack of sensitivity of solid phase hybridization. A rapid extraction procedure was developed for the hybridization detection of bluetongue virus [12]. The procedure, however, uses a toxic compound, diethyl pyrrocarbonate, and requires cell culture amplification of the virus. Biotechnology companies have developed kits for RNA or DNA isolations using a variety of protocols.

There are two basic types of solid phase hybridization protocols that can be used. The first method is a spot, slot, or dot hybridization in which the isolated nucleic acid is immobilized directly onto the hybridization membrane. This is done by simply spotting the solution on the membrane or by vacuum filtration using a slot or dot manifold. This type of analysis is very rapid to set up but requires a large amount of target sequence. Electrophoresis is used in the second method to separate the pathogen's nucleic acid based on molecular size. This can be done on intact nucleic acid or after it has be digested to smaller fragments using appropriate enzymes. After electrophoresis, the nucleic acid can be transferred to the hybridization membrane using a capillary, electrical, or vaccum procedure. Electrophoresis blot procedures analyze at the pathogen's genotype but take more time to prepare and perform than the dot blot methods.

The main disadvantage to solid phase hybridization, especially for RNA, is low sensitivity. One approach to increase sensitivity is sandwich hybridization as described by Dunn and Hassel [13] and modified by Ranki and coworkers [14] for crude samples. This procedure utilizes oligonucleotides or small DNA fragments bound to a solid support to capture the target nucleic acid. The target is then detected by hybridization with detector or reporter probe. The initial procedures utilized hybridization membranes as a solid support and were successfully used to increase the sensitivity of bluetongue virus detection from the infected insect vector [15]. An interesting variation of sandwich hybridization detected the β-lactamase gene in uropathogenic *E. coli*, using a capture probe immobilized on microtiter strips by UV cross-linking [16]. This intriguing procedure has not been repeated for other pathogens. Another modification used magnetic beads as the solid support for the detection of Dengue-2 viral RNA in a reversible target capture hybridization procedure (RTC-hyb) [17]. RTC-hyb procedures are attractive since they bypass the lengthy phenol extraction/nucleic acid purification procedures. Disadvantages of RTC-hyb are that they are more difficult to set up and are less versatile than standard solid phase hybridization.

Amplification Procedures. Methods to amplify the target or the signal have been sought to enhance the sensitivity of hybridization procedures. The most popular nucleic acid detection procedure has been *in vitro* amplification by PCR. This procedure has been used to identify genetic defects, to detect pathogens, and for a variety of other uses. In some cases, PCR can be used with crude samples, thus bypassing lengthy extraction procedures. The high sensitivity of PCR is both an advantage and a disadvantage, because precautions must be taken to avoid sample contamination. Although PCR is expensive, the development of more diagnostic PCR tests may lower costs.

Product detection must be considered in the development design. Gel electrophoresis and ethidium bromide staining are sufficient to detect the product but do not provide sufficient confirmation of the product. Endonuclease restriction analysis of the PCR product is the simplest method for product confirmation. To increase the sensitivity and confirm the PCR product, hybridization with an internal probe can be used. These three methods can be used separately or in combination. Some procedures use a marker on one primer that can be detected with a colorimetric assay [18,19]. Colorimetric assays are very useful for screening numerous samples, but care must be used in designing the primer so that no nonspecific products are generated.

PCR amplification of viral genes inside of cells combined with *in situ* hybridization increased the sensitivity for detection of visna virus DNA by more than two orders of magnitude [20]. Visna virus is the prototype animal lentivirus of which the human prototype lentivirus is the most widely known virus, the human immunodeficiency virus (HIV). When this technique was applied to a biopsy sample from an HIV-infected individual, HIV DNA was detected in lymphocytes and lymphocytes infiltrating a tumor only by PCR *in situ* [21]. Other studies have since shown the presence of HIV-1 provirus in a latent or defective form in peripheral blood monocytes [22,23]. Further modification allowed the determination of how many CD4$^+$ lymphocytes are latently infected by HIV using PCR *in situ* double-label methods [21]. Even the "Hot-Start" technique, which has been used to decrease nonspecific products with standard PCR, has been applied to PCR *in situ* [24].

6.2. APPROACHES

6.2.1. Approaches to Viral Infection Diagnosis

The determination of the virus template sequence is aided by the relative genetic simplicity of the viruses. Another advantage, especially for important viral pathogens, is that the viral genes are often well defined. The development of a hybridization approach to detect and serotype epizootic hemorrhagic disease (EHD) virus, for example, was simplified by previous studies on the molecular biology of the virus [25]. Gene segment 3 of EHD viruses encodes a conserved inner core protein and therefore affords a genetic probe specific for EHD viruses. The outer core protein of EHD virus encoded by gene segment 2 contains the serotype-specific epitopes, providing a serotype specific probe from gene 2 [25]. Others have used early and late expressed antigens for detection of cytomegalovirus by slot blot or *in situ* hybridization irrespective of the stage of infection [26].

In general, a conserved gene from a given virus pathogen is selected as a probe region because it should detect various strains of the pathogen. The ability of probes from the selected genomic region to bind to other pathogens causing similar clinical signs or the host nucleic acid is then examined. Because of potential similarities of cloning vector sequences with nucleic acids from the host and other contaminating organisms, it is usually recommended that only the insert from a recombinant plasmid be used as a probe [27,28]. The use of insert probes, however, can result in a loss in signal intensity (sensitivity) presumably due to increased labeled nucleic acid and binding or hybridization of vector sequences [29]. Gene regions that contain tandem repeats should be avoided due to potential background hybridization with common tandem repeat regions. With Epstein-Barr virus, however, a specific tandem repeat region was found to be advantageous as a target sequence for an oligomer probe in a nonradioactive automated *in situ* hybridization procedure [30].

The majority of virus detection methods use the standard extraction procedures described earlier to obtain virus nucleic acid. Another method that has been used is to lyse the infected cell; then the lysate is filtered directly onto a hybridization membrane [12,31]. This bypasses the extraction methods and relies on the specificity of the hybridization and the high titer of the test sample. With some viruses, the virus titer is not sufficient for detection by solid phase hybridization and must rely on cell culture amplification, e.g., bluetongue virus [12]. To avoid cell culture procedures, signal amplification such as with PCR is used.

PCR has been used to detect viruses from biological samples ranging from blood to formalin-fixed tissues. Sample preparation can affect PCR performance as formalin fixation of brain tissue samples did for JC virus [32]. Formalin fixation also inhibited the hybridization detection of bluetongue virus from infected insects compared with unfixed or alcohol-fixed infected insects [33]. In many cases, phenol extraction of the test sample is required prior to PCR, but phenol extraction of the test sample may be avoided in other cases. Pseudorabies virus was detected in cells collected from nasal swabs by heating to 115°C for 10 minutes prior to PCR [34]. Heat treatment apparently removes any inhibitors of PCR, such as those found in urine samples in the PCR detection of cytomegalovirus [35]. The presence of inhibitors in rotaviral RNA samples extracted from feces, however, required differential chromatography prior to PCR [36]. Thus, the sample preparation required varies with the pathogen and the biological sample.

Once an appropriate sample preparation is established, the power of PCR can be exploited to detect a variety of viruses. By using a reverse transcriptase reaction prior to PCR amplification, a semiquantitative PCR technique with a 100% sensitivity for symptomatic patients was correlated to HIV-1 infectious virus titer in plasma [37]. Reverse transcription PCR (rt/PCR) has been used to detect other viruses such as picornaviruses [38] and Norwalk virus, a calicivirus [39]. Double-stranded RNA viruses such as African horse sickness viruses [40,41], bluetongue viruses [42–45], EHD viruses [46–48], and rotaviruses [49,50] have also been detected by rt/PCR. Combining the reverse transcriptase reaction and PCR in a single tube for a noninterrupted test has been reported for rotaviruses [51]. Ross River virus [52] and infectious bursal disease virus [53].

Verification of the PCR product is often done by the use of a restriction endonuclease digestion that cuts the product nucleic acid in a specific position. Addition of restriction endonuclease analysis to an rt/PCR for infectious bursal disease virus allowed the detection of genetic variants [54]. Restriction endonuclease analysis for restriction site mapping of PCR products has also been used with human papillomavirus [55] and dengue viruses [56].

The determination of the presence of a specific virus group may not be sufficient to determine the proper treatment or control measures. In many cases viral vaccines only provide effective immunity to the same virus type. Virus typing can also be done using multiplex PCR that uses multiple primer pairs in a single reaction. This is advantageous because of the cost of the *Taq* polymerase and the reduction in labor involved in running multiple reactions. Multiplex PCR has been used in the simultaneous detection of several types of human papilloma virus from boiled clinical samples [57]. It has also been used to determine the serotypes of rotaviruses [49,50] and bluetongue viruses [45]. One group used gene-specific primer pairs, restriction endonuclease digestion, and hybridization for confirmation and to type equine herpes viruses [58].

The development of multiplex PCR procedures is even more dependent on the design of the primer pairs because of potential interactions between primers. Larzul and coworkers [59] compared 16 primer pairs for amplification of hepatitis virus and suggested optimal primer parameters. Primer designs should contain 50%–70% GC and range from 20 to 30 bp in length; the 3' end sequence should have a high GC content, if possible, and not allow for primer dimerization.

Sensitivity can be increased using nested or internal primers in a second amplification. Using a nested PCR for added sensitivity in combination with restriction endonuclease analysis allowed the detection and typing of clinical human picornavirus isolates [38]. Nested PCR accentuates the biggest drawback to PCR detection: sample contamination. The concern of PCR contamination artifacts is illustrated by the circumstantial evidence in one case that fragmented DNA resulting from the autoclave disposal of cytomegalovirus, one floor below the laboratory running PCR, was a source of contamination [60]. False-positive results can also result from the *Taq* polymerase "jumping" from one DNA fragment to another [61]. Procedures to avoid sample crossover and/or false-positive results have been discussed previously [62].

6.2.2. Approaches to Bacterial Infection Diagnosis

One method used to obtain probes in bacteria for which there is little genetic information is to generate a random genomic recombinant library. Recombinant plasmids are chosen at random and evaluated as pathogen-specific probes. This random selection approach from genomic clones of the causative agent of Lyme disease, *Borrelia burgdorferi,* was used to find a probe that is distinct for *B. burgdorferi* but not for the most closely related member of the genus, *B. hermsii* [63]. A sequence of this clone was then used to develop a PCR detection assay [63]. Selection of clones that detect the causative agent of Potomac horse fever, *Ehrlichia risticii,* was done using random clones [64]. This approach is quite useful and is likely to be successful.

A more direct approach would be to select a genetic region that encodes a protein unique to a given pathogen. This approach was used in the detection of *Chlamydia trachomatis.* The gene encoding a major outer membrane protein (MOMP) was used in a novel detection scheme. This gene was amplified with PCR, complexed with a biotinylated RNA probe in a solution hybridization, and detected by an antibiotin immunoassay [65]. The macrophage infectivity potentiator gene was used to develop a PCR assay to detect the causative agent of Legionnaires' disease, *Legionella pneumophila* [66]. Richettisial disease is commonly undiagnosed in endemic areas and, when diagnosed, is only detected with serologic assays. To detect the Karp, Gilliam, and Kato strains of *Richettisa tsutsugamushi* (Scub typhus), but not other *Rickettsia* species, a 1,477 base pair coding region for a 56 kDa antigen was used to develop a diagnostic test [67]. A great deal of basic research is required to develop a nucleic acid test using the direct approach.

Another approach requiring less genetic characterization is the use of probes based on rRNA genes. This approach is useful because these sequences are present in each ribosome of an organism; therefore there are many copies per organism. Although the rRNA sequences are conserved among organisms, there are sequence regions that are specific for a given pathogen. A PCR scheme based on 5S rRNA has been used to detect *Legionella* species [66]. Gen-Probe Corp. was the first to get FDA approval for clinical diagnostic tests for mycoplasma, *Legionella* spp., and mycobacteria [68]. These cDNA probes were cDNA that target the 16S and 23S rRNA.

PCR-based assays using 16S rRNA genes as targets have been developed for *Actinobacillus actinomycetemcomitans, Ehrlichia canis, Erysipelothrix rhusiopathiae,* and *Helicobacter pylori* [69–72]. One of the more comprehensive 16S rRNA PCR tests was developed to detect bacteria found in cerebrospinal fluid. In this study, three series of oligonucleotide probes were developed to detect the PCR products from a set of broad-range PCR primers. The first series contained a universal bacterial probe, a gram-positive probe, and probes for three gram-negative species. The second series was designed to detect PCR products for seven major bacterial species or groups causing meningitis. The final series was designed to detect bacteria known to contaminate clinical samples. This test provided an accurate identification of bacterial species in cerebrospinal fluid and will aid the medical treatment of patients with bacterial meningitis [73].

A 16S rRNA gene has been used to detect the causative agent of Q fever, *Coxiella burnelii* [74]. The Coxiella study also described an alternative approach for detection of bacterial pathogens. The authors noted that all *Coxiella* strains contained 25 kb of plasmid-related sequences. Those strains that contained QpRS plasmid or that incorporated related sequences into their chromosomes were associated with the occasional chronic endocarditis type of the disease, which is often fatal. Using this information, a PCR method that differentiates between the chronic and acute strains of Q fever was developed [74].

The use of naturally occurring bacterial plasmids as diagnostic probes is generally not feasible because they are not usually species specific. When the plasmids are species specific, they are very useful diagnostic probes. A problem in the medical treatment of some bacterial pathogens is penicillin resistance. Penicillin-resistant bacteria are usually detected by agar dilution susceptibility, but this test requires a pure primary culture of the pathogen. A 7.5 or 5.4 kb plasmid is commonly associated with penicillin resistance in the most widely spread sexually transmitted disease, gonorrhea *(Neisseria gonorrhea)*. These plasmids contain a TEM-1 β-lactamase gene, which was used in a solid-support capture hybridization test for the detection of penicillin-resistant *N. gonorrhea* and some *Hemophilus* species [75].

This type of resistance could be differentiated from that of *E. coli, Shigella sonnei,* and *Salmonella typhii* [75]. Bacterial dysentery is often caused by *Shigella* species and by enteroinvasive *Escherichia coli.* The pathogenesis of these bacteria is mediated by an invasion plasmid that can be lost or reduced by culture. Therefore, a direct detection of the invasion-associated plasmid in stool specimens was developed [76].

The yopA structural gene encoded by a virulence plasmid was used to distinguish nonpathogenic species of *Yerisinia* from the food-borne pathogen *Yerisinia enterocolitica* [77]. This virulence plasmid is also found in *Y. pseudotuberculosis* and *Y. pestis* [77]. In the detection of pathogenic *Yerisinia* species, both oligomer and polynucleotide probes were investigated. The advantage of oligomeric probes is that they are very specific and can be produced inexpensively in large quantities containing a high degree of label. The disadvantage is that they are more sensitive to base mutations and therefore may not detect some strain variants. Polynucleotide probes are less sensitive to base changes; however, they can have more background hybridization and are difficult to produce in industrial quantities. This is an important consideration when developing an assay for food- and water-borne pathogens.

Coliform and enterobacteria (*Escherichia, Enterobacter, Citrobacter, Klebsiella,* etc.) are used for monitoring the safety of water supplies. The standard method of detecting these pathogens is by culture procedures. Problems in storage viability and nonculturable bacteria led to the development of a PCR test for *Coliform* bacteria [78] that is a multiplex PCR combined with capture probes to provide a colorimetric detection that is effective, efficient, and rapid.

The utility of nucleic acid tests for bacterial pathogens is especially evident for *Mycobacterium.* This genus contains notoriously slow-growing, acid-fast, nonmotile, rod-shaped bacteria, including the causative agents of leprosy and tuberculosis. Although usually associated with ancient history, leprosy is still a major health problem in many parts of the world. Leprosy is a very disabling disease in terms of morbidity and social stigma. Because the causative agent, *Mycobacterium leprae,* does not grow in culture, it is usually cultured in armadillos. Several laboratories have developed nucleic acid tests for *M. leprae* because conventional diagnosis is time consuming and requires technical expertise. The random genomic library approach provided a 2.2 kb clone that is repeated at least 19 times in the chromosomal DNA yet is specific for *M. leprae* by dot blot hybridization [79]. Another laboratory selected a 360 bp fragment from a 18 kDa gene that, by hybridization analysis, was specific for *M. leprae* [80]. A PCR test was developed and determined not to detect other bacteria common to clinical samples tested for *M. leprae. M. leprae* strains from various geographical areas were detected to show the efficacy of the test [80].

Most PCR tests incorporate negative and positive tests as quality control measures. Hartskeerl and coworkers [81], investigating a PCR test for *M. leprae,* took the positive control a step further by developing a PCR product from a 36 kDa antigen whose gene weakly hybridizes with other mycobacteria. By using optimizing conditions, this product was specific for *M. leprae.* As a positive internal control, primers were included that amplified a smaller PCR product from a 65 kDa heat shock protein that is common in most bacteria. This test then determines if bacteria are present in a given sample and if the bacteria present is *M. leprae* [81]. If the test samples potentially contain inhibitors of the *Taq* polymerase, a control target and primers could be incorporated into a test to ensure that the PCR reaction runs properly.

Another pathogen that is a significant health problem, both globally and nationally, especially in underprivileged areas, is *Mycobacterium tuberculosis*. As with *M. leprae,* a nucleic acid test for *M. tuberculosis* has been sought because of limited sensitivity with nonculture type procedures and difficulties using culture procedures. A classic differential screening approach was used to generate recombinants specific for *M. tuberculosis*. In this screen, replicate *M. tuberculosis* genomic clones were hybridized with *M. bovis* (a distant member of the tuberculosis complex) and with *M. kansasis* (not a member of the tuberculosis complex). The recombinants that hybridized to *M. bovis* and not to *M. kansasis* were selected for further characterization, and three were found to be specific for *M. tuberculosis* [82]. A PCR test was later developed from the nucleic acid sequence of one of these clones, which is present in multiple copies in the *M. tuberculosis* genome and contains an internal Sal 1 site [83]. The authors suggested that the digestion of the PCR product with Sal 1 could be used as confirmation of the PCR product. In this case, endonuclease digestion could be used as a confirmation since a base mutation would only affect one copy of the gene. Caution should be used when the target sequence is derived from a single copy gene.

A similar approach was used to develop a nucleic acid test for *Mycobacterium avium,* the most prevalent among opportunistic mycobacterial pathogens in immunosuppressed patients [84]. Genomic clones from *M. avium, M. tuberculosis,* and *M. intracellulare* were screened independently by hybridization. One clone was identified that was specific for mycobacteria, and two clones were specific for *M. avium*. Using the nucleic acid sequences of these clones, PCR primers that were specific for mycobacteria and for *M. avium* were developed [84].

Screening of a recombinant expression library with specific antibodies is another approach used to find a specific nucleic acid target. This approach is especially useful when well-characterized monoclonal antibodies are available. A lambda GT11 gene library from *M. tuberculosis* was screened with a monoclonal antibody [85]. A recombinant was selected that hybridized specifically to DNA from the *M. tuberculosis* complex. A PCR test was then developed from sequences contained within this clone that was specific for this complex of bacterial strains in clinical samples [85].

Often not only the detection of a given bacterial species is needed; the detection of specific strains or types is also required for a sufficient diagnostic assay. Atrophic rhinitis in pigs is caused by strains of *Pasteurella multocida* containing a *tox*A gene encoding a 143 kDa dermonecrotic toxin [86]. A PCR assay was developed based on this *tox*A gene that differentiated toxigenic from nontoxigenic *P. multocida* strains [86]. A similar assay was developed to detect the presence of the *mec*A gene, which encodes the penicillin-binding protein 2a, inferring methicillin resistance [87]. In this assay, two primer sets were used in a single reaction for a multiplex PCR. One primer set detected the presence of the *mec*A gene, while the other was a universal 16S rRNA primer set used as an internal control to identify potential false-negative samples [87].

Typing can also be done by arbitrarily primed PCR where a nonspecific primer is used to prime the PCR. The primers used in this procedure are usually small (6–12 bp), and the annealing temperature is carried out at low stringency. The PCR analysis of *Enterobacter cloacae* and *Leptospira interrogans* provided results similar to those of more standard genotypic analyses [88,89]. When several arbitrary primers in separate reactions are used to characterize isolates, the analysis is called *random amplified polymorphic DNA analysis* (RAPD). Group A *streptococcus (Streptococcus pyogenes)* isolates have been typed using RAPD analysis [90]. The relationship of *L. monocytogenes* isolates implicated in food-borne outbreaks of listeriosis have been investigated using 19 primers in a RAPD analysis [91]. Rather than using random primers, other investigators have used restriction fragment length polymorphism (RFLP) for genotyping. In this procedure, the pathogen's DNA or a PCR product is digested with endonuclease restriction enzymes, which results in a DNA fragment profile when separated by gel electrophoresis. *Helicobacter pylori* isolates were analyzed by RFLP using a specific *ure*C gene PCR product with three separate endonuclease digestions, *Hha*I, *Mbo*I, or *Mse*I [92]. This typing method provided a reproducible scheme for the study of *H. pylori* infections and proved that infection with more than one isolate is not rare [92].

6.2.3. Approaches to Parasite Infection Diagnosis

Many of the approaches used to detect diseases, described in the previous sections, have also been used to detect parasites. Genus- and type-specific 16S rRNA probes have been developed for the northern hemisphere zoonotic disease tularemia *(Franscisella tularensis)* [93]. Previously, strain identification of this disease was performed by biochemical and virulence tests that were time consuming and expensive. Repetitive DNA probes have been used to diagnose bovine babesiosis *(Babesia bigenina),* lymphatic filariasis *(Brugia malayi, Brugia pahangi),* and malaria *(Plasmodium falciparum)* [94–96]. Several groups have used a 21 base pair repeat region from *P. falciparum* to develop a hybridization probe. A comparison of a probe to this region and an organelle DNA repeat probe in the diagnosis of malaria was performed. The results indicated that the 21 base pair repeat probe was more sensitive than the organelle probe. The organelle probe reacted with a broad range, suggesting that a combination of the two would be useful [97].

A repetitive DNA probe to detect *Trypanosoma evansi,* which causes a wasting disease in the domestic animal Surra, was compared with a minicircle DNA or kinetoplast DNA probe. Kinetoplast DNA can have up to 5,000 copies per cell, which should make for a good diagnostic probe. The kinetoplast DNA probe for trypanosomes, however, was found to be variable with strains; thus the repetitive DNA probe provided more consistent hybridizations [98]. With leishmania, a PCR test was developed based on kinetoplast DNA that proved useful in detecting and distinguishing among strains in human biopsy material [99].

6.3. EVALUATION OF THE TEST

Once a nucleic acid test is developed, the test must be evaluated for sensitivity and specificity in the presence of the disease. This is usually accomplished by comparison with current diagnostic tests or "gold" standards; however, the "gold standards" are often not the best comparisons. The reason that a nucleic acid test was developed in the first place was because the "gold" standard is fraught with problems. One review of the evaluation of a diagnostic test with imperfect standards described an example where the "gold" standard underestimated the sensitivity and specificity of a new test [100]. Seven studies were conducted on a new latex immunoagglutination test for *Clostridium difficile*–associated diarrhea. Six of the seven studies used stool cytotoxicity as their comparison. One study used a clinical definition of *C. difficile*–associated diarrhea, indicating that the other studies underestimated the performance of the new test [100].

The ultimate purpose of a diagnostic test is to determine the presence of a disease or infection. Therefore, the comparison of the new test with the presence of the disease, as clinically defined, provides the best comparison. This is difficult because clinical definitions in many cases are inconclusive, resulting in the comparison of new tests existing tests. There are a few conditions that should be met in order to minimize the biases caused by imperfect standards and personal opinions. The evaluator must be aware that the prevalence of a disease favors the clinical diagnosis of the disease. When clinical diagnosis is used, the comparison should be made using different definitions of the disease. Likewise, if available, the new test should be compared with more than one current test. The evaluator of the new test must then consider the problems associated with both the old and the new tests when interpreting the comparison. The final consideration is the usefulness of the test to clinicians, diagnosticians, and other scientists.

This chapter discusses several approaches to pathogen detection using nucleic acid technology. There are several commercial nucleic acid–based tests available for diagnosis and research. The high sensitivity, versatility, and specificity make these desirable tests. Nucleic acid tests only require the presence of a target nucleic acid that is the product of a pathogen. The presence of the nucleic acid does not directly indicate that the pathogen is still present in a viable form [9]. Caution is then needed in interpreting the results of these tests, and, when possible, the results should be confirmed by culture tests. The tremendous effort directed toward this area will no doubt ensure that some nucleic acid tests will stand up to the thorough evaluation of end users.

ACKNOWLEDGMENTS

The author thanks Drs. B. Beaty, S. Brodie, J. Mecham, W. Tabachnick, and H. Van Campen for critical review of an earlier draft of this chapter. The author especially thanks Dr. C. Dangler for critical review of several earlier drafts of this chapter.

REFERENCES

1. Albandar, J.M., Olsen, I.: Nucleic acid probes as potential tools in oral microbial epidemiology. Community Dent. Oral Epidemiol. 18:88 (1990).

2. Paul, P.S.: Applications of nucleic acid probes in veterinary infectious diseases. Vet. Microbial. 24:409 (1990).

3. Purchase, H.G.: Future applications of biotechnology in poultry. Avian Dis. 30:47 (1985).

4. Smith, T.F.: Rapid diagnosis of viral infections. Adv. Exp. Med. Biol. 263:115 (1990).

5. Joens, L.A., Marquez, R.: The diagnosis of swine dysentery using a labeled nucleic acid probe. Proc. 10th Congress International Pig Veterinary Society, Rio de Janeiro, 1988.

6. Karch, H., Meyer, T.: Evaluation of oligonucleotide probes for identification of shiga-like-toxin-producing *Escherichia coli*. J. Clin. Microbiol. 27:1180 (1989).

7. Maddox, C.W., Wilson, R.A.: High technology diagnostics: Detection of enterotoxigenic *Escherchia coli* using DNA probes. J. Vet. Med. Assoc. 188:57 (1986).

8. Moseley, S.L., Eicheverria, P., Serivatana, J., Tirapat, C., Chaicumpa, W., Sakuidalpeara, T., Falkow, S.: Identification of enterotoxigenic *Escherichia coli* by colony hybridization using three enterotoxin gene probes. J. infect. Dis. 145:863 (1982).

9. MacLachlan, N.J., Nunamaker, R.A., Katz, S.B., Sawyer, M.M., Akita, G.Y. Osburn, B.I., Tabachnick, W.J.: Detection of bluetongue virus in the blood of inoculated calves: Comparison of virus isolation, PCR assay, and *in vitro* feeding of *Culicoides variipennis*. Arch. Virol. 136:1 (1994).

10. Unger, E.R., Budgeon, L.R., Myerson, D., Brigati, D.J.: Viral diagnosis by *in situ* hybridization. Description of a rapid simplified colorimetric method. Am. J. Surg. Pathol. 10:1 (1986).

11. Dangler, C.A., Dunn, S.J., Squire, K.R., Stott, J.L., Osburn, B.I.: Rapid identification of bluetongue virus by nucleic acid hybridization in solution. J. Virol. Methods 20:353 (1988).

12. Wilson, W.C.: Development and optimization of a hybridization assay for epizootic hemorrhagic disease viruses. J. Virol. Methods 30:173 (1990).

13. Dunn, A.R., Hassell, J.A.: A novel method to map transcripts: Evidence for homology between an adenovirus mRNA and discrete multiple regions of the viral genome. Cell 12:23 (1977).

14. Ranki, M., Palva, A., Virtanen, M., Lacksonen, M., Soderlund, H.: Sandwich hybridization as a convenient method for the detection of nucleic acids in crude samples. Gene 21:77 (1983).

15. Schoepp, R.J., Bray, J.F., Olson, K.E., El Hussein, A., Holbrook, F.R., Blair, C.D., Roy, P., Beaty, B.J.: Detection of bluetongue virus serotype 17 in *Culicoides variipennis* by nucleic acid blot and sandwich hybridization techniques. J. Clin. Microbiol. 28:1952 (1990).

16. Dahlen, P., Syvanen, A.C., Hurskainen, P., Kwiatowski, M., Sund, L., Ylikoski, J., Soderlund, H., Lovgren, T.: Sensitive detection of genes by sandwich hybridization and time-resolved fluorometry. Mol. Cell. Probes 1:159 (1987).

17. Chandler, L.J., Blair, C.D., Beaty, B.J.: Detection of Dengue-2 viral RNA by reversible target capture hybridization. J. Clin. Microbiol. 31:2641 (1993).

18. Katz, J.B., Alstad, A.D., Gustafson, G.A., Moser, K.M.: Sensitive identification of blue-

tongue virus serogroup by a colorimetric dual oligonucleotide sorbent assay of amplified viral nucleic acid. J. Clin. Microbiol. 31:3028 (1993).

19. Wilson, W.C.: Development of nested-PCR tests based on sequence analysis of epizootic hemorrhagic disease viruses non-structural protein 1 (NS1). Virus Res. 31:357 (1994).

20. Haase, A.T., Retzel, E.F., Staskus, K.A.: Amplification and detection of lentiviral DNA inside cells. Proc. Natl. Acad. Sci. U.S A. 87:4971 (1990).

21. Embretson, J., Zupancic, M., Beneke, J., Till, M., Wolinsky, S., Ribas, J.L., Burke, A., Hasse, A.: Analysis of human immunodeficiency virus-infected tissues by amplification and *in situ* hybridization reveals latent and permissive infections at single-cell resolution. Proc. Natl. Acad. Sci. U.S.A. 90:357 (1993).

22. Bagasra, O., Hauptman, S.P., Lischner, H.W., Sachs, M., Pomerantz, R.J.: Detection of human immunodeficiency virus type 1 provirus in mononuclear cells by *in situ* polymerase chain reaction. N. Engl. J. Med. 326:1385 (1992).

23. Bagasra, O., Pomerantz, R.J.: Human immunodeficiency virus type I provirus is demonstrated in peripheral blood monocytes *in vivo:* A study utilizing an *in situ* polymerase chain reaction. AIDS Res. Hum. Retrovirus 9:69 (1993).

24. Nuovo, G.J., Gallery, F., MacConnel, P., Becker, J., Bloch, W.: An improved technique for the *in situ* detection of DNA after polymerase chain reaction amplification. Am. J. Pathol. 139:1239 (1991).

25. Wilson, W.C., Fukusho, A., Roy, P.: Diagnostic complementary DNA probes for genome segments 2 and 3 of epizootic hemorrhagic disease virus serotype 1. Am. J. Vet. Res. 51:855 (1990).

26. Einsele, H., Vallbracht, A., Jahn, G., Kandolf, R., Muller, C.A.: Hybridization techniques provide improved sensitivity for Hcmv detection and allow quantitation of the virus in clinical samples. J. Virol. Methods 26:91 (1989).

27. Ambinder, R.F., Charache, P., Staal, S., Wright, P., Forman, M., Hayward, D.S., Hayward, G.S.: The vector homology problem in diagnostic nucleic hybridization of clinical specimens. J. Clin. Microbiol. 24:16 (1986).

28. Diegutis, P.S., Keirman, E., Burnett, L., Nightingale, B.N., Cossart, Y.E.: Detection of bovine enteric coronavirus in clinical specimens by hybridization with cDNA probes. J. Clin. Microbiol. 23:779 (1986).

29. Verbeek, A., Dea, S., Tijssen, P.: Detection of bovine enteric coronavirus in clinical specimens by hybridization with cDNA probes. Mol. Cell Probes. 4:107 (1990).

30. Montone, K.T., Budgeon, L.R., Brigati, D.J.: Detection of Epstein Barr virus genomes by *in situ* DNA hybridization with a terminally biotin labeled synthetic oligonucleotide probe from the EBV Not I and Pst I tandem repeat regions. Mod. Pathol. 3:89 (1990).

31. van den Brule, A.J.C., Claas, E.C.J., du Maine, M., Melchers, W.J.G., Helmerhorst, T., Quint, W.G.V., Lindeman, J., Meijer, J.L.M., Walboomers, J.M.M.: Use of anticontamination primers in the polymerase chain reaction for the detection of human papilloma virus genotypes in cervical scrapes and biopsies. J. Med. Virol. 29:20 (1989).

32. Telenti, A., Aksamit, A.J. Jr., Proper, J., Smith, T.F.: Detection of JC virus DNA by polymerase chain reaction in patients with progressive multifocal leukoencephalopathy. J. Infect. Dis. 162:858 (1990).

33. Wilson, W.C.: Detection of epizootic hemorrhagic disease virus in *Culicoides varipennis* (Diptera: Ceratopogonidae). J. Med. Entomol. 28:742 (1991).

34. Jestin, A., Foulon, T., Pertuiset, B., Blanchard, P., Labourdet, M.: Rapid detection of pseudorabies virus genomic sequences in biological samples from infected pigs using polymerase chain reaction DNA amplification. Vet. Microbiol. 23:317 (1990).

35. Demmler, G.J., Buffone, G.J., Schimbor, C.M., May, R.A.: Detection of cytomegaloma in urine from newborns by using polymerase chain reaction DNA amplification. J. Infect. Dis. 158:1177 (1988).

36. Wilde, J., Elden, J., Yolken, R.: Removal of inhibitory substances from human fecal specimens for detection of group A rotaviruses by reverse transcriptase and polymerase chain reactions. J. Clin. Microbiol. 28:1300 (1990).

37. van Kerckhoven, I., Franien K., Peeters, M., Beenhouwer, H.D., Piot, P., van der Groen, G.: Quantification of human immunodeficiency virus in plasma by RNA PCR, viral culture, and p24 antigen detection. J. Clin. Microbiol. 32:1669 (1994).

38. Kämmerer, U., Kunkel, B., Korn, K.: Nested PCR for specific detection and rapid identification of human picornaviruses. J. Clin. Microbiol. 32:285 (1994).

39. Moe, C.L., Gentsch, J., Ando, T., Grohmann, G., Monroe, S.S., Jiang, X., Wang, J., Estes, M.K., Seto, Y., Humphrey, C., Stine, S., Glass, R.I.: Application of PCR to detect Norwalk virus in fecal specimens from outbreaks of gastroenteritis. J. Clin. Microbiol. 32:642 (1994).

40. Sakamoto, K., Punyahotra, R., Mizukoshi, N., Ueda, S., Imagawa, H., Sugiura, T., Kamada, M., Fukusho, A.: Rapid detection of African horsesickness virus by the reverse transcriptase polymerase chain reaction (RT-PCR) using the amplimer for segment 3 (VP3 gene). Arch. Virol. 136:87 (1994).

41. Stone-Marschat, M., Carville, A., Skowronek, A., Laegreid, W.W.: Detection of African horse sickness virus by reverse transcription-PCR. J. Clin. Med. 32:697 (1994).

42. Akita, G.Y., Chinsangaram, J., Osburn, B.I., Ianconescu, M., Kaufman, R.: Detection of bluetongue virus serogroup by polymerase chain reaction. J. Vet. Diagn. Invest. 4:400 (1992).

43. McColl, K.A., Gould, A.R.: Detection and characterisation of bluetongue virus using the polymerase chain reaction. Virus Res. 21:19 (1991).

44. Wade-Evans, A.M., Mertens, P.P., Bostock, C.J.: Development of the polymerase chain reaction for the detection of bluetongue virus in tissue samples. J. Virol. Methods 30:15 (1990).

45. Wilson, W.C., Chase, C.C.L.: Nested and multiplex polymerase chain reactions for the identification of bluetongue virus infection in the biting midge, *Culicoides-variipennis*. J. Virol. Methods 45:39 (1993).

46. Aradaib, I.E., Akita, G.Y., Osuburn, B.I.: Detection of epizootic hemorrhagic disease virus serotypes 1 and 2 in cell culture and clinical samples using polymerase chain reaction. J. Vet. Diagn. Invest. 6:143 (1994).

47. Aradaib, I.E., Wilson, W. C., Cheney, I.W., Osburn, B.I.: Application of PCR for specific-identification of epizootic hemorrhagic disease virus serotype 2. J. Vet. Diagn. Invest. 7:388–392 (1995).

48. Wilson, W.C.: Development of nested-PCR tests based on sequence analysis of epizootic hemorrhagic disease viruses non-structural protein 1 (NS1). Virus Res. 31:357 (1994).

49. Gouvea, V., Santos, N., Timenetsky, M.C.: VP4 typing of bovine and porcine group A rotaviruses by PCR. J. Clin. Microbiol. 32:1333 (1994a).

50. Gouvea, V., Santos, N., Timenetsky, M.C.: Identification of bovine and porcine rotavirus G types by PCR. J. Clin. Microbiol. 32:1338 (1994b).

51. Xu, L., Harbour, D., McCrae, M.A.: The application of polymerase chain reaction to the detection of rotaviruses in faeces. J. Virol. Methods 27:29 (1990).

52. Sneller, L.N., Coelen, R.J., Mackenzie, J.S.: A one-tube, one manipulation RT-PCR reaction for detection of Ross River virus. J. Virol. Methods 40:255 (1992).

53. Lee, H.L., Ting, L.J., Shien, J.H., Shieh, H.K.: Single-tube noninterrupted reverse transcription-PCR for detection of infectious bursal disease virus. J. Clin. Microbiol. 32:1268 (1994).

54. Liu, H.-J., Giambrone, J.J., Dormitorio, T.: Detection of genetic variations in serotype I isolates of infectious bursal disease virus using polymerase chain reaction and restriction endonuclease analysis. J. Virol. Methods 48:281 (1994).

55. Chen, S., Tabrizi, S.N., Fairley, C.K., Borg, A.J., Garland, S.M.: Simultaneous detection and typing strategy for human papillomaviruses base on PCR and restriction endonuclease mapping. BioTechniques 17:138 (1994).

56. Vorndam, V., Kuno, G., Rosado, N.: A PCR-restriction enzyme technique for determining dengue virus subgroups within serotypes. J. Virol. Methods 48:237 (1994).

57. van den Velde, C., Verstraete, M., Van Beers, D.: Fast multiplex polymerase chain reaction on boiled clinical samples for rapid viral diagnosis. J. Virol. Methods 30:215 (1990).

58. Lawrence, G.L., Gilkerson, J., Love, D.N., Sabine, M., Whalley, J.M.: Rapid, single-step differentiation of equid herpesviruses 1 and 4 from clinical material using the polymerase chain reaction and virus-specific primers. J. Virol. Methods 47:59 (1994).

59. Larzul, D., Chevrier, D., Thiers, V., Guesdon, J.L.: An automatic modified polymerase chain reaction procedure for hepatitis virus DNA detection. J. Virol. Methods 27:49 (1990).

60. Porter-Jordan, K., Rosenberg, E.I., Keiser, J.F., Gross, J.D., Ross, A.M., Nasim, S., Garrett, C.T.: Source of contamination in polymerase chain reaction assay. J. Med. Virol. 30:85 (1990).

61. Paabo, S., Higuchi, R.G., Wilson, A.C.: Ancient DNA and the polymerase chain reaction. J. Biol. Chem. 264:9709 (1989).

62. Kwok, S., Higuchi, R.: Avoiding false positives with PCR. Nature 339:237 (1989).

63. Rosa, P.A., Schwan, T.G.: A specific and sensitive assay for the Lyme disease spirochete *Borrelia burgdorferi* using the polymerase chain reaction. J. Infect. Dis. 160:1018 (1989).

64. Thaker, S.R., Dutta, S.K., Adhya, S.L., Mattingly-Napier, B.L.: Molecular cloning of *Ehrlichia risticii* and development of a gene probe for the diagnosis of Potomac horse fever. J. Clint Microbiol. 28:1963 (1990).

65. Bobo, L., Coutlee, F., Yolken, R.H., Quinn, T., Viscidi, R.P.: Diagnosis of *Chlamydia trachomatis* cervial infection by detection of amplified DNA with an enzyme immunoassay. J. Clin. Microbiol. 28:1968 (1990).

66. Mahbubani, M.H., Bej, A.K., Miller, R., Haff, L., DiCesare, J., Atlas, R.M.: Detection of *Legionella* with polymerase chain reaction and gene probe methods. Mol. Cell. Probes 4:175 (1990).

67. Kelly, D.J., Marana, D.P., Stover, C.K., Oaks, E.V., Carl, M.: Detection of *Rickettsia tsutsugamushi* by gene amplification using polymerase chain reaction techniques. Ann. N.Y. Acad. Sci. 590:564 (1990).

68. Engleberg, N.C.: Nucleic acid probe tests for clinical diagnosis—Where do we stand? ASM News 57:183 (1991).

69. Iqbal, Z., Chaichanasiriwithaya, W., Rikihisa, Y.: Comparison of PCR with other tests for early diagnosis of canine Ehrlichiosis. J. Clin. Microbiol. 32:1658 (1994).

70. Leys, E.J., Griffen, A.L., Strong, S.J., Fuerst, P.A.: Detection and strain identification of *Actinobacillus actinomeycetemcomitans* by nested PCR. J. Clin. Microbiol. 32:1288 (1994).

L 71. Makino, S.-l., Okada, Y. Maruyama, T., Ishikawa, K., Takahashi, T., Nakamura, M., Ezaki, T., Morita, H.: Direct and rapid detection of *Erysipelothrix rhusiopathiae* DNA in animals by PCR. J. Clin. Microbiol. 32:1526 (1994).

72. Weiss, J., Mecca, J., da Silva, E., Gassner, D.: Comparison of PCR and other diagnostic techniques for detection of *Helicobacter pylori* infection in dyspeptic patients. J. Clin. Microbiol. 32:1663 (1994).

73. Greisen, K., Loeffelholz, M., Purohit, A., Leong, D.: PCR primers and probes for the 16S rRNA gene of most species of pathogenic bacteria, including bacteria found in cerebrospinal fluid. J. Clin. Microbiol. 32:335 (1994).

74. Frazier, M. E., Mallavia, L.P., Samuel, J.E., Baca, O.G.: DNA probes for the identification of *Coxiella burnetti* strains. Ann. N.Y. Acad. Sci. 590:445 (1990).

75. Sanchez-Pescador, R., Stempien, M.S., Urdea, M.S.: Rapid chemiluninescent nucleic acid assays for detection of TEM-1 beta-lactamase-mediated penicillin resistance in *Neisseria gonorrhoeae* and other bacteria. J. Clin. Microbiol. 26:1934 (1988).

76. Frankel, G., Riley, L., Giron, J.A., Valmassoi, J., Friedmann, A., Strockbine, N., Falkow, S., Schoolnik, G.K.: Detection of *Shigella* in feces using DNA amplification. J. Infect. Dis. 161:1252 (1990).

77. Kapperud, G., Dommarsnes, K., Skurnik, M., Hornes, E.A.: Synthetic oligonucleotide probe and a cloned polynucleotide probe based on the Yopa gene for detection and enumeration of virulent *Yersinia-Enterocolitica*. Appl. Environ. Microbiol. 56:17 (1990).

78. Bej, A.K., Mahbubani, M.H., Miller, R., Dicesare, J.L., Haff, L., Atlas, R.M.: Capture probes for detection of bacterial pathogens and indicators in water. Molec. Cell. Probes 4:353 (1990).

79. Clark-Curtiss, J.E., Docherty, M.A.: A species-specific repetitive sequence in *Mycobacterium leprae* DNA. J. Infect. Dis. 159:7 (1989).

80. Williams, D.L., Gillis, T.P. Booth, R.J., Looker, D., Watson, J.D.: The use of a specific DNA probe and polymerase chain reaction for the detection of *Mycobacterium leprae*. J. Infect. Dis. 162(1):193 (1990).

81. Hartskeerl, R.A., de Wit, M.Y.L., Klatser, P.R.: Polymerase chain reaction for the detection of *Mycobacterium leprae*. J. Gen. Microbiol. 135:2357 (1989).

82. Eisenach, K.D., Crawford, J.T., Bates, J.H.: Repetitive DNA seqeunces as probes for *Mycobacterium tuberculosis*. J. Clin. Microbiol. 26:2240 (1988).

83. Eisenach, K.D., Cave, M.D., Bates, J.H., Crawford, J.T.: Polymerase chain reaction amplification of a repetitive DNA sequence specific for *Mycobacterium tuberculosis*. J. Infect. Dis. 161:977 (1990).

84. Fries, J.W.U., Patel, R.J., Piessens, W.F., Wirth, D.F.: Genus-specific and species-specific DNA probes to identify Mycobacteria using the polymerase chain reaction. Mol. Cell. Probes 4:87 (1990).

85. Hermans, P.W.M., Schuitema, A.R.J., Van-Soolingen, D., Verstynen, C.P.H.J., Bik,

E.M., Thole, J.E.R., Kolk, A.H.J., van-Embden, J.D.A.: Specific detection of *Mycobacterium tuberculosis* complex strains by polymerase chain reaction. J. Clin. Microbiol. 28:1204 (1990).

86. Nagai, S., Someno, S., Yagihashi, T.: Differentiation of toxigenic from nontoxigenic isolates of *Pasteurella multocida* by PCR. J. Clin. Microbiol. 32:1004 (1994).

87. Geha, D.J., Uhl, J.R., Gustaferro, C.A., Persing, D.H.: Multiplex PCR for identification of methicillin-resistant staphylococci in the clinical laboratory. J. Clin. Microbiol. 32:1768 (1994).

88. Grattard, F., Pozzetto, B., Berthelot, P., Rayet, I., Ros, A., Lauras, B., Gaudin, O.G.: Arbitrarily primed PCR, ribotyping, and plasmid pattern analysis applied to investigation of a nosocomial outbreak due to *Enterobacter cloacae* in a neonatal intensive care unit. J. Clin. Microbiol. 32:596 (1994).

89. Perolat, P., Merien, F., Eillis, W.A., Baranton, G.: Characterization of *Leptospira* isolates from serovar hardjo by ribotyping, arbitrarily primed PCR, and mapped restriction site polymorphisms. J. Clin. Microbiol. 32:1949 (1994).

90. Seppälä, H., Qiushui, H., Österblad, M., Huovinen, P.: Typing of group A streptococci by random amplified polymorphic DNA analysis. J.Clin. Microbiol. 32:1945 (1994).

91. Czajka, J., Batt, C.A.: Verification of causal relationships between *Listeria monocytogenes* isolates implicated in food-borne outbreaks of Listeriosis by randomly amplified polymorphic DNA patterns. J. Cin. Microbiol. 32:1280 (1994).

92. Fujimoto, S., Marshall, B., Blaser, M.J.: PCR-based restriction fragment length polymorphism typing of *Helicobacter pylori*. J. Clin. Microbiol. 32:331 (1994).

93. Forsman, M., Sandstrom, G., Jaurin, B.: Identification of *Francisella* species and discrimination of type A and type B strains of *F. tularensis* by 16S rRNA analysis. Appl. Environ. Microbiol. 56:949 (1990).

94. Buening, G.M., Barbet, A., Myler, P., Mahan, S., Nene, V., McGuire, T.C.: Characterization of a repetitive DNA probe for *Babesia bigemina*. Vet. Parasitol. 36:11 (1990).

95. Mucenski, C.M., Guerry, P., Buesing, M., Szarfman, A., Trosper, J., Walliker, D., Watt, G., Sangalang, R., Ranoa, C.P., Tuazon, M., Majam, O.R., Quakyl, I., Scheibel, L.W., Cross, J.H., Perine, P.L.: Evaluation of a synthetic oligonucleotide probe for diagnosis of *Plasmodium falciparum* infections. Am. J. Trop. Med. Hyg. 35:912 (1986).

96. Poole, C.B., Williams, S.A.: A rapid DNA assay for the species specific detection and quantification of *Brugia* in blood samples. Mol. Biochem. Parasitol. 40:129 (1990).

97. Holmberg, M., Vaidya, A.B., Shenton, F.C., Snow, R.W., Greenwood, B.M., Wigzell, H. Petterson, U.: A comparison of two DNA probes, one specific for *Plasmodium falciparum* and one with wider reactivity, in the diagnosis of malaria. Trans. R. Soc. Trop. Med. Hyg. 84:202 (1990).

98. Viseshakul, N., Panyim, S.: Specific DNA probe for the sensitive detection of *Trypanosoma evansi*. Southeast Asian J. Trop. Med. Public Health 21:21 (1990).

99. Rodgers, M.R., Popper, S.J., Wirth, D.F.: Amplification of kinetoplast DNA as a tool in the detection and diagnosis of *Leishmania*. Exp. Parasitol. 71:267 (1990).

100. Valenstein, P.N.: Evaluating diagnostic tests with imperfect standards. J. Clin. Pathol. 93:252 (1990).

CHAPTER 7

DIAGNOSTIC NUCLEIC ACID PROBE TECHNOLOGY IN VETERINARY MEDICINE

JARASVECH CHINSANGARAM and BENNIE I. OSBURN
Department of Veterinary Pathology, Microbiology, and Immunology,
School of Veterinary Medicine,
University of California,
Davis, CA 95616

7.1. INTRODUCTION

Approaches to animal disease diagnostics are based on a variety of factors from different disciplines. Clinical signs associated with certain diseases provide important indicators of the particular types of diseases affecting animal species. Although clinical signs are important indicators, they are often somewhat limited in providing a definitive diagnosis. A second level of diagnostic capability is based on pathologic changes in the tissues of affected animals. Pathologic abnormalities can include changes in hematological or biochemical values and specific morphological lesions that may be associated with certain specific diseases. In many instances, a number of different disease agents may precipitate similar morphological changes in various body organs. An obvious and more definitive diagnostic parameter is the actual isolation or characterization of the infectious or chemical agents that are responsible for the disease.

Conventional technology for identifying infectious agents utilizes methods of culturing microorganisms, followed by classifying the isolates by either biochemical or serological procedures [1]. Viruses may be isolated by animal inoculation, by culturing in laboratory animal species or embryonated chicken eggs, or by direct inoculation and recovery of agents in cell or tissue cultures. The isolation step serves

Nucleic Acid Analysis: Principles and Bioapplications, pages 131–155
© 1996 Wiley-Liss, Inc.

as an initial indication of a transferable pathogen and frequently enhances further identification by amplifying the pathogen. Confirmatory diagnosis is often based on immunological reagents, commonly the application of fluorescent antibody or neutralization assays that have been developed to the specific agents in question. These methods have been utilized in diagnosing disease for the past 20–30 years. One notable drawback of this sequence of primary isolation followed by secondary characterization is that it may require considerable periods of time before a definitive diagnosis can be made.

Molecular biology has provided useful technological approaches for analyses of infectious disease agents. The techniques employed in these analyses detect specific genetic materials of microorganisms in infected tissues or in extracted preparations and can also detect alterations in genes of animals. Such gene alterations often contribute to defective gene product expression in animals. In addition, the techniques of molecular biology also provide researchers alternative and more effective means to study microorganisms and animal genetics at the molecular level, the knowledge of which will render the development of better disease diagnosis, treatment, prevention, and possibly eradication.

7.2. NUCLEIC ACID TECHNOLOGY FOR VETERINARY MEDICINE

7.2.1. Restriction Mapping and Restriction Fragment Length Polymorphism

DNA contains specific sequences that are recognized by restriction enzymes. Most type II restriction endonucleases recognize a particular sequence in DNA molecules that have a twofold axis of rotational symmetry (sometimes mistakenly called *palindromic sequences*) and then cleave these DNA molecules at or near the recognition site [2]. The resulting DNA fragments of different sizes are separated by gel electrophoresis. The profile (map) of the migration pattern of these DNA fragments is based on separation by molecular weights. Similarly, some endoribonucleases cleave RNA molecules at specific base sites. For instances, ribonuclease A cleaves RNA at the 3′ end of pyrimidine molecules and ribonuclease T1 at the 3′ end of guanine nucleotides [3].

Different microorganisms, organisms, and individuals of the same species possess DNA molecules of various sequence compositions. The differences in DNA sequences can be determined by cleavage with several restriction endonucleases. The DNA can be digested and the resulting DNA fragments of different sizes separated on an agarose gel and visualized in the gel by ethidium bromide staining. Such a technique produces restriction fragment length polymorphisms (RFLPs), which permits evaluation of genetic relationships of DNA sequences. It also provides information regarding genetic defects since the differences in the gene sequences of patients suffering from the genetic disease can be compared with those of normal individuals. The gene of a patient may contain a sequence recognized by one particular enzyme that does not exist in a normal individual's gene or vice versa. Because a gene contains a large number of bases, the analysis of RFLPs of a specific sequence in that gene can be very difficult. A DNA probe can be used in hybridization assays to locate the fragments derived from the comparable gene segment. In microbiological studies, this technique is used to differentiate and identify strains of pathogens capable of causing disease in animals.

7.2.2. Nucleic Acid Profiles on Polyacrylamide Gels

Both RNA and DNA can be separated on polyacrylamide gels (PAGE). Nucleotide chains are separated based on their molecular weights by PAGE and then visualized in gel by silver nitrate staining. In a two-dimensional electrophoretic system, however, nucleotide chains are separated by both molecular weight and charge, resulting in the appearance of a unique pattern called a "nucleotide fingerprint" [4]. The migration pattern of the genetic material on a gel (electrophoretic pattern) is helpful in classifying some microorganisms. In some cases, similarities in size and charge may result in comigration of heterologous gene sequences. RNA genome segments of viruses in the family Reoviridae (bluetongue virus, reovirus, and rotavirus) are usually compared and studied by this technique because their natural segmental genomes facilitate the use of PAGE for their genetic analysis. By using this technique, the differences between groups and strains of these viruses can be identified. Moreover, these viruses sometimes exchange their genome segments within their groups (reassortment), a situation that can be detected by PAGE.

7.2.3. Cloning Nucleic Acid

Cloning genes is a means to derive specific gene segments independent of many other genes of the parental organism. The cloned gene has several applications in molecular biology, such as diagnostic probes, templates for sequence analysis, and as the coding unit for gene expression. The first step in cloning is the isolation of the desired gene segment. Restriction endonuclease digestion, mechanical shearing, or chemical degradation of the genome yields several DNA fragments that can be separated by PAGE or agarose gel electrophoresis. In RNA cloning, a reverse transcription reaction is performed to produce a cDNA (complementary DNA) from the target RNA. The DNA or the cDNA of interest is then inserted into a plasmid or other vectors to serve as a vehicle for transforming bacteria. The amplification of the inserted gene is achieved during the growth period of bacteria. Following the cloning manipulations, the specificity of the inserted gene must be identified by nucleic acid hybridization. The cloned gene can be a partial or a full-length copy of the template gene. A partial clone, if not too short, is usually enough if a hybridization probe is to be produced or a partial expression of a gene is required. However, the full-length clone is often necessary for the expression when a proper folding of the expressed protein is needed.

7.2.4. Sequence Analysis

The two methods of sequencing normally used include the Maxam and Gilbert method and the dideoxy or chain-termination method (Sanger method). Both methods provide populations of radiolabeled oligonucleotides of various lengths that will be resolved by electrophoresis under conditions that can discriminate DNA molecules that differ in length by as little as one nucleotide. In the Maxam and Gilbert method, the DNA is radioactively labeled, and chemical reagents are used to alter specific bases of the DNA. These altered bases are removed and the DNA strand at that point will be cleaved with piperidine, creating several DNA fragments of different lengths, each of which has the specific base (A, T, C, or G) at the 3' end depending on the type of chemical reagents used. The sets of labeled fragments are generated and analyzed on a sequencing gel.

The dideoxy method is the most commonly used method for sequence determination. The DNA is cloned into a vector (M13mp, pGEM, etc.) to produce single-stranded DNA targets. The complementary strands are produced in the presence of the Klenow fragment of DNA polymerase I, oligonucleotide primers, radiolabeled free nucleotides (dNTP), and specific 2'3'-dideoxynucleoside triphosphates (ddNTP). Four reactions for A, T, C, and G are performed using one kind of ddNTP (ddATP, ddTTP, ddCTP, and ddGTP) in one tube. The addition of ddNTP into a growing complementary strand will stop the elongation process at that point. After the determined incubation period, the DNA in each mixture is denatured and electrophoresed on a sequencing gel side by side. The bands are detected by autoradiography, and the sequence of the particular DNA is determined. Alternatively, the bands can be blotted to a membrane and detected using a nonradioactive assay. Sequencing can also be performed from double-stranded DNA targets by including an additional denaturation step before the sequencing reaction, although the single-stranded sequencing technique usually yields a better result.

Analogues of dNTPs are often used in the sequencing reaction to resolve the compression problem caused by the presence of secondary structures in DNA molecules [3]. Nucleotide analogues such as dITP (2'-deoxyinosine-5'-triphosphate) and 7-deaza-dGTP (7-deaza-2'-deoxyguanosine-5'-triphosphate) are normally used with the enzyme Sequenase (an enzyme modified from bacteriophage T7 DNA polymerase). *Taq* DNA polymerase is another enzyme that is very useful for sequencing because it works at high temperatures which interfere with the formation of secondary structure [2,3].

7.2.5. Nucleic Acid Hybridization

Hybridization technology has been widely used to detect the presence of complementary nucleic acid sequences. The target nucleic acid is either immobilized to a solid support or suspended in the solution. The labeled probe, which has a complementary sequence to the target, will anneal to the target under the proper conditions, and the resulting double-stranded probe–target complex will be detected by detection systems.

Probes are acquired from the cloning procedure or can be synthesized if the known sequences are available. Probes can be labeled with radioactive or nonradioactive labels [2,3,5]. Radioactive probes are normally labeled with ^{32}P by methods such as nick translation, primer extension, or end labeling. In addition, RNA probes can be prepared from DNA templates by using labeled ribonucleoside triphosphate as the substrate and the enzyme RNA polymerase. Biotin is often used in nonradioactive labeling systems. The biotinylated nucleotides are incorporated into polynucleotide probes, and these biotin-labeled probes can be detected in a color-generating system. An alternative method (chemiluminescence) directly cross-links the enzyme horseradish peroxidase to the probes. After the hybridization reaction, reagents are added to create light, which will be detected by autoradiography similar to the detection of radioactive probes. The nonradioactive labeling system is very convenient, although the sensitivity of the test is usually lower than that of radioactive labeling system.

Southern Blot Hybridization. This technique was developed by Southern [6] to detect DNA immobilized on filter paper. DNA fragments are separated on an agarose gel, denatured into single strands by alkali treatment, and then transferred to the filter membrane by capillary action. The DNA molecules are immobilized on the membrane by baking at 80°C in a vacuum environment. The membrane is then subjected to hybridization.

The membrane is placed in a plastic bag containing prehybridization buffer and incubated to eliminate nonspecific signal. The labeled probes in hybridization buffer are then added to replace the first buffer. After incubation, the membrane is washed (posthybridization wash), and the signal is detected by autoradiography. The optimum temperature, salt concentration, and hybridization time must be adjusted empirically.

Northern Blot Hybridization. Northern blot hybridization assays are used as diagnostic tests for viral RNAs and messenger RNAs. Target RNA separated on an agarose gel can be transferred to a membrane by capillary action from an agarose gel to a membrane, fixed, and then hybridized to the probes.

In the detection of some viral RNAs, the RNA molecules are separated on a PAGE and the genome profile analyzed. Thereafter, the transfer of RNA molecules is achieved by a technique called *electroblotting*. A high electrical field generated from a power supply will cause the movement of the RNA molecules that have a negative charge from a PAGE to a membrane placed in front of the positive pole. The RNA molecules are then fixed to the membrane for the hybridization as described above.

Dot Blot Hybridization. The use of dot blot hybridization for diagnosis is common. Although the principle of this assay is the same as those of the Southern and Northern blot hybridization assays, this test does not separate out the genome segments of the sample tested prior to hybridization. The nucleic acid is extracted from infected tissues or cells, concentrated, and denatured. An aliquot of the nucleic acid is dotted onto a membrane and the membrane hybridized with the labeled probe. The advantage of this test is the rapidity with which it can be performed.

Many nucleic acid samples can also be placed onto a membrane at the same time by a technique referred to as slot blot [7,8]. The samples are denatured and loaded in the slots of a vacuum filtration manifold on top of the membrane and blotted onto the membrane under a vacuum. The membrane is fixed and hybridized to the labeled probe.

In Situ Hybridization. This test involves the detection of the genetic material within cells on glass slides under hybridization conditions [9,10,12–15]. The cells can be fixed cell cultures, fixed blood cells, frozen tissues, or formalin-fixed tissues. The nonradioactive labeled probes of the color-generating system (biotin-avidin-enzyme system) can be detected by viewing under the light microscope. The advantages of this system are that the morphology of the cells and tissues remains intact and the site of target nucleic acid (i.e., site of viral replication or a gene on a chromosome) can be identified. In addition, the histopathology of the cells can be correlated with the presence of the genetic material of the infectious agent detected by the hybridization.

Solution Hybridization. In solution hybridization both the target and the labeled probe move freely in the liquid phase, resulting in a greater possibility that complementary nucleotides will meet and bind each other [2]. Therefore, it requires less time to complete the hybridization. Separation of the single-stranded nucleic acid from bound double-stranded nucleic acid is very critical before the detection can be performed. Digestion of the free single-stranded nucleic acid with an appropriate nuclease facilitates the reaction. Hydroxyapatite (HA) can also be used to separate these two entities, since the HA binds to the single-stranded nucleic acid and the double-stranded nucleic acid at different salt concentrations.

New techniques and equipment have been adapted to perform the test with less time and higher sensitivity [16]. For example, DNA or RNA can be transferred from an agarose gel to the membrane by a technique called *vacuum blotting*. This technique requires less than 1 hour to perform rather than the conventional several hours or overnight. Nylon membranes are often used in place of the nitrocellulose membranes, which are very fragile. UV cross-linkers are now used in many laboratories to fix the nucleic acid to the membranes. Only a few seconds are required to cross-link the nucleic acid permanently to the membranes, whereas it takes 2 hours for immobilization using a vacuum baker. Hybridization in hybridization bottles rotating in a special oven helps to resolve the problem of radioactive contamination. Furthermore, the nonradioactive hybridization protocol is very simple and can be performed on a benchtop without radioactive precautions in a laboratory that is not designated for radioactive compounds.

7.2.6. Oligonucleotide Synthesis

Oligonucleotides can be chemically synthesized from individual nucleotides via a DNA synthesizer [5]. The automated DNA synthesis is done in the direction of 3′ → 5′. During each base addition, the binding sites of the nucleotides are first protected and later deprotected when the binding of two nucleotides is needed so that the desired binding can be regulated. The synthesized oligonucleotide is then isolated from chemicals and failure sequences by precipitation, PAGE separation, HPLC, or the commercially available oligonucleotide purification columns. Oligonucleotides of less than 150 bases in length can be readily synthesized in this manner. A probe for any gene or for the genetic material of any infectious agent can be synthesized if a short amino acid sequence of its protein is known. This strategy is often used to find a particular gene that codes for the protein of interest so that it can be cloned and studied. During the synthesis, labeled nucleotide can be used to produce a labeled probe for diagnostic purposes. Primers for sequencing and for PCR and antisense oligonucleotides can be readily synthesized if the sequence is known or available in computerized gene data bank programs. Furthermore, mutations can be introduced into these molecules to study the effects of particular mutations.

7.2.7. Polymerase Chain Reaction

Polymerase chain reaction (PCR) is a very powerful tool to amplify enzymatically a target nucleic acid sequence. It involves cyclic amplification of the target DNA in the presence of a pair of synthetic oligonucleotide primers complementary to sequences flanking the target DNA and *Taq (Thermus aquaticus)* DNA polymerase [3,5]. These primers anneal to opposite strands of the target DNA and are oriented so that the DNA synthesis by the *Taq* polymerase proceeds across the region between the primers. The RNA target requires an additional cDNA synthesis step using reverse transcriptase enzyme before the cyclic amplification can be performed. Each cycle of PCR includes (1) heat denaturation of the target DNA to produce the single-stranded DNA, (2) annealing of complementary oligonucleotide primers to the target single-stranded DNA, and (3) elongation from the primers on the target DNA mediated by *Taq* DNA polymerase. After 30 cycles of PCR, highly specific amplification of 270 millionfold is obtained [5]. The introduction of the thermostable *Taq* polymerase by Sakai et al. [17] has resolved the previous problem of PCR that required the addition of the fresh Klenow fragment of *Escherichia coli* polymerase I after the denaturation step of each cycle.

The application of PCR for diagnosis of suspected nucleic acid has been very successful. It provides an alternative diagnostic method with higher sensitivity, rapidity, and specificity. In theory, only one piece of the target DNA is required for the amplification. For diagnostic purposes, the PCR products are visualized on an agarose gel stained with ethidium bromide or on PAGE stained with silver nitrate. The size of PCR products can be estimated from a molecular weight standard or compared with the size of positive control of the target. The specificity of PCR products is confirmed by restriction mapping or by hybridization to a complementary nucleic acid probe for the target in either a radioactive or a nonradioactive hybridization reaction. An alternative confirmation method is to amplify an internal portion of PCR products using another set of primers [18]. By using this technique (called *nested PCR*), only the actual target containing sequences complementary to both sets of primers will provide expected PCR products. Normally, both PCR hybridization and nested PCR will increase the sensitivity of the detection in addition to specificity verification. In some instances, single primers of arbitrary sequences (mostly of short sequences) can be used in the detection of organisms [19–22]. This technique, termed *random amplification of polymorphic DNA* (RAPD), is especially useful when the sequence of target DNA is not available. In a PCR reaction at low stringency, these single arbitrary primers randomly bind to and amplify the target DNA, resulting in several PCR products that can be separated on an agarose gel. The unique profile of DNA (PCR products) obtained from an organism by RAPD can be established. This profile is then used for later identification of that particular organism.

In situ PCR is a modified PCR technique that enables detection of targets on microscope slides [23]. The nucleic acid present in the samples is denatured on slides, and the PCR reaction is carried out using a thermal cycler specifically designed for this purpose. A quantitative PCR technique has recently been developed in which a known amount of internal standard is included into the same reaction [24–26]. The internal standard contains sequences complementary to detection primers but differs in size from the actual target. Primers simultaneously amplify both the internal standard and the actual target, resulting in two PCR products that are distinguished by their size. The original amount of the actual target is then obtained by comparison to the amount of the internal standard originally added to the reaction.

PCR technology can also be applied to other areas such as cloning, sequencing, *in vitro* mutagenesis, and genetic studies [25,27]. It provides the opportunity to amplify and to clone the gene that was previously impossible to do because of the low number of the rare sequence. Its high sensitivity, however, can be a drawback, since serious precautions have to be taken to avoid contamination with even trace amounts of nucleic acid that will otherwise result in amplification of the contaminating nucleic acid and lead to an improper diagnosis. Several precautions and techniques have been suggested to prevent or eliminate this contamination problem [3,28–31].

7.3. PRODUCTS AVAILABLE FOR USE IN VETERINARY MEDICINE

7.3.1. Nucleic Acid Products Used for Diagnostic Purposes

Restriction Maps. In the past, differences between genes were usually detected at the phenotypic level based on the mendelian theory. Blood group antigens, for example, were the mendelian markers in different human populations. Later, more informative studies of protein polymorphisms were introduced. Electrophoresis detects the differences between migratory patterns of proteins based on the amino acid sequence. Substitution of an amino acid reflects the change at the nucleic acid level. However, a change in nucleic acid sequence is not always reflected at the amino acid level, and the change at amino acid level is not always detected by electrophoretic pattern. In addition, the protein polymorphisms will not provide any information if the difference of the nucleic acid is in the noncoding region of the gene. This region, including intervening sequence, flanking sequence, and regulatory sequence, is also of importance.

Although the sequence of a gene is very useful in genetic analysis, sequencing is very tedious and cumbersome. The development of the restriction mapping technique bypasses the need for gene sequencing in genetic studies. Derangements in the genes such as gene deletions/insertions, gene inversions, gene rearrangements, or point mutations, can be identified from RFLPs. The PCR amplification has recently been incorporated into the mapping technique to amplify selectively the segments containing the DNA polymorphisms [25,27]. The PCR products can be digested to produce DNA fragments called *amplified sequence polymorphisms* (ASPs) or PCR fragment length polymorphisms (PCRFLPs).

The RFLP technique was employed to study the presence of endogenous avian leukosis virus loci in commercial broiler lines [32]. The result of this analysis indicated that the endogenous virus RFLP patterns for these lines were highly complex and contained many loci unreported in White Leghorn layers. The correlation of a high frequency RFLP at the 3' end of the pigeon pro alpha 2 collagen gene detected by the restriction endonuclease *Eco*RI and the genetic susceptibility of pigeons to spontaneous atherogenesis was examined [33]. However, the result suggested no linkage between them. This technique was also used to detect a genetic defect (deficiency of uridine monophosphate synthase [DUMPS]) in bovine embryos [34]. The restriction mapping technique, however, has some limitations. For instance, incomplete digestion by restriction enzymes due to the presence of enzyme inhibitors or other contaminants is sometimes problematic. Second, although several enzymes are used, a point mutation that does not produce, change, or destroy the recognition site will not be detected. Third, the resulting fragment from enzyme digestion created by the change of nucleic acid is sometimes too small or too similar in size to other fragments to be detected. In addition, methylation of cytidine bases will inhibit the digestion at that point even though the recognition site exists.

PAGE Profiles. PAGE analysis has proved to be very useful for double-stranded RNA virus diagnosis [35–38]. Viruses such as reoviruses, bluetongue viruses (BTV), and epizootic hemorrhagic disease viruses (EHDV) possess 10 double-stranded RNA genome segments, and rotaviruses have 11 double-stranded RNA genome segments. The migration pattern (electrophoretic pattern) of these genome segments is used to differentiate and classify these viruses into groups or electropherotypes [37,39]. The samples to be examined can be cell culture or embryonated chicken egg infected with viruses (BTV and EHDV) or feces from infected animals (reoviruses and rotaviruses). The sample is mixed with sodium dodecyl sulfate (SDS) and proteinase K to dissociate the viral proteins. The protein is then removed from the samples by phenol extraction and the nucleic acid is run on a PAGE. The concentrated PAGE (CPAGE) has been used to detect and analyze rotaviruses when there is a low amount of viruses in the samples [35]. The extracted RNA is concentrated by ethanol precipitation before PAGE analysis.

In rotavirus studies, PAGE has been employed to characterize the viruses of non-group A since these viruses do not possess a triplet character (i.e., the migrations of genome segments 7, 8, and 9 are so similar that they are grouped together on the gel) of group A viruses [37,40]. In addition, the standard electropherotypes have been proposed as a means to differentiate these nongroup A rotaviruses into electropherogroups B, C, D, E, F, and G since the antibodies used to characterize them have not been readily available and their morphology cannot be distinguished by electron microscopy (EM). PAGE is often coupled to the hybridization reaction for diagnostic purposes. In BTV Northern blot hybridization technique, 10 genome segments of BTV are separated on PAGE, electroblotted onto a nylon membrane, and hybridized with a specific probe [7,8].

Nucleic Acid Hybridization. Specific identification of target genetic sequences by hybridization requires that the genetic material be made accessible to the probe. There are several advantages to the use of this technique for diagnosis. For example, no cultivation of infectious agents is normally necessary. This is very useful when the target organism cannot be grown, grows poorly, or requires a long complicated procedure to be cultured. Although antibody detection can be done for many diseases, the antibody titers do not always indicate active infection of animals unless paired sera are tested. The rapidity, accuracy, and sensitivity of nucleic acid tests are usually better than those of conventional diagnostic methods as long as optimization is achieved. Furthermore, a specific probe can be designed and produced so that it can identify all serotypes or only some serotypes of significance of an infectious agent. Tables 7.1 and 7.2 display some examples of nucleic acid probes available for veterinary diagnosis. For genetic studies, the probe designed from the protein of interest is used to identify either normal or abnormal genes that will then be cloned

---

TABLE 7.1. Examples of Nucleic Acid Hybridization Probes for Diagnostic Purposes

Viruses	Bacteria
Aquareovirus [67]	*Anaplasma* [83,84]
Avian leukosis virus [14]	*Campylobacter* [41,85]
Bluetongue [7,8,38,68]	*Leptospira* [86]
Bovine coronavirus [69]	*Mycobacterium* [87]
Bovine herpesvirus 4 [70]	*Mycoplasma* [88,89,90]
Bovine leukosis virus [71]	
Bovine and porcine rotaviruses [72–75]	Parasites
Bovine virus diarrhea [76]	*Babesia* [91]
Canine distemper virus [15]	*Eimeria* [92]
Chicken anemia virus [77]	*Eperythrozoa* [93]
Equine encephalosis [78]	*Theileria* [94]
Foot-and-mouth disease [42]	*Toxoplasma* [95]
Infectious bovine rhinotracheitis [9]	*Trypanosoma* [96,97]
Infectious bursal disease [79]	
Ovine lentivirus [80]	
Porcine parvovirus [81]	
Rabies [12,82]	

TABLE 7.2. Diagnostic Tests With Application to Food Safety

Campylobacter hybridization probe and oligonucleotide primers for polymerase chain reaction [41,98–100]

Toxoplasma hybridization probe and oligonucleotide primers for polymerase chain reaction [95,101]

Rotavirus hybridization probe and oligonucleotide primers for polymerase chain reaction [11,16,72–75,102,103]

Listeria hybridization probe and oligonucleotide primers for polymerase chain reaction [104,105]

Coliform bacteria oligonucleotide primers for polymerase chain reaction [106]

---

for a further study. When a hybridization probe is used, a condition of hybridization referred to as stringency is very critical. The low stringency favoring mismatch base-pairing (for example, A with G and T with C) is allowed at low temperature, low formamide concentration, and high salt concentration, whereas the high *stringency* condition is obtained in the opposite ways. The degree of stringency will determine the degree of mismatching allowed in that hybridization reaction.

Ng et al. [41] studied genes contributing to tetracycline resistance in *Campylobacter coli* (*C. coli*) and *C. jejuni* isolated from swine and cattle. By using a DNA probe of 1.8 kb length, these genes were detected as 50 kb plasmids except in one C. coli strain in which it appeared to be chromosomally mediated. A dot blot hybridization technique was used to detect foot-and-mouth disease virus (FMDV) in esophageal-pharyngeal fluids from experimentally infected cows [42]. The probe designed from the sequence, including the viral polymerase sequence, detected FMDV types A, O, and C. The probe proved very useful for FMDV carrier-state detection since it could detect the virus from the samples at 180 and 560 days postinfection while the infective virus could not be recovered from these samples. These studies also indicated the presence of a high ratio of noninfective viral mutants in FMDV carrier cattle.

Rabies virus genomic RNA and mRNA were detected in paraffin-embedded mouse and human brain tissues by in situ hybridization [12]. ^{3}H-labeled single-stranded RNA probe demonstrated the difference in the amount of mRNA in infected mouse and human brain tissues. In mouse brain, there was a greater amount of signal for the mRNAs than for genomic RNA, while it was opposite in human brain. The finding suggested either a relative block at the level of transcription or greater loss of mRNAs than of genomic RNA during the agonal period in the human. An alternative explanation was that loss of RNA in tissues might have occurred, preferentially affecting mRNAs, in autolysed infected neurons prior to fixation. *In situ* hybridization was used to detect African horse sickness virus in cell culture, proviral DNA of avian leukosis virus in tissue samples, and canine distemper virus in brain tissue [10,14,15]. *In situ* hybridization was tested for the ability to detect BTV nucleic acids in blood mononuclear cells [43]. The signal was detected in BTV-infected cultured mononuclear cells; however, the same protocol failed to identify BTV in freshly isolated blood mononuclear cells from an experimentally infected heifer.

PCR. Both DNA and RNA can be efficiently amplified when PCR protocol is optimized and PCR inhibitors eliminated from the samples. However, PCR of DNA target normally provides better results because an additional cDNA synthesis step is not required. Samples can be collected from blood, cells, fluids, or even paraffin-embedded tissues [25]. Several parameters such as the concentration of *Taq* DNA polymerase, reverse transcriptase (for RNA target), dNTPs, magnesium, and primers are very critical for PCR optimization [25,27]. Time and temperature of the cycle and size and design of the primers are of equal importance. Special methods are sometimes included in the protocol to purify the target partially before amplification since some samples contain inhibitors of the PCR reaction. After nucleic acid extraction, DNA or RNA binding material can be used to absorb the target out of the solution and resuspend it in a desired buffer solution. Ion binding beads can also be used when the problem is due to the presence of ions interfering with the reaction.

　　PCR can be used to detect microorganisms that are difficult and time consuming to identify by conventional diagnosis. Examples of PCR used for diagnostic purposes are shown in Table 7.3. DNA of alcelaphine herpesviruses (AHV-1, malignant catarrhal fever and AHV-2) was amplified from bovine blood buffy coat specimens [18]. A two-stage (nested) PCR consisting of two pairs of primers was employed. In the second stage, the second set of primers internal to the first set was used to amplify the target obtained from the first-stage PCR. The specificity of PCR products was confirmed by restriction endonuclease digestion. An RT/PCR for BTV detection was developed that could detect RNA of all U.S. serotypes (2, 10, 11, 13, and 17) and a spectrum of Israeli serotypes (2, 4, 6, 10, and 16) from cell culture [44].

TABLE 7.3. Examples of Polymerase Chain Reactions Used for Diagnostic Purposes

Viruses	Bacteria
African horse sickness [107,108]	*Campylobacter* [98–100]
Avian leukosis virus [14]	Enterotoxigenic *Escherichia*
Bluetongue [44,109,110]	*coli* [127]
Bovine leukemia virus [111,112]	*Leptospira* [128]
Bovine rotavirus [11,16,102,103]	*Mycoplasma* [129,130]
Bovine virus diarrhea [45,113,114]	*Renibacterium salmoninarum*
Borna disease virus [115]	[131]
Epizootic hemorrhagic disease virus [116,117]	*Salmonella* [132]
Equine arteritis virus [118]	
Equine influenza virus [119]	Parasites
Infectious bursal disease [46]	*Eperythrozoa* [133]
Marek's disease virus [120]	*Theileria* [134]
Malignant catarrhal fever [18,121]	*Toxoplasma* [101]
Ovine lentivirus [80]	*Trypanosoma* [97,135]
Porcine parvovirus [122]	
Porcine reproductive and respiratory syndrome	
virus [123]	
Pseudorabies [124]	
Rabies [125,126]	

The sensitivity of the test was at 5×10^2–10^3 infectious units (10 viral particles per one infectious unit) when the PCR product was visualized on an agarose gel. When dot blot hybridization was incorporated into the protocol the sensitivity of five infectious units was achieved. PCR protocols were developed to provide rapid and highly sensitive assays for the detection of group A and group B bovine rotaviruses in fecal samples [11,16]. In addition, PCR detection for group B bovine rotavirus overcame the problems in group B rotavirus diagnosis due to the inability to grow this virus in cell culture and the limited availability of antibodies to this virus. PCR was also used to detect bovine viral diarrhea virus (BVDV) and was found to provide clearer identification of persistently infected animals than DNA hybridization under similar conditions [45]. PCR detection of infectious bursal disease was performed with formalin-fixed paraffin-embedded tissue sections [46]. The result from this study demonstrated the power of the PCR technology in diagnosis even when a fresh sample was not available.

7.3.2. Implications of Technologic Application

Restriction Maps and Sequence Analysis. Restriction endonuclease mapping and sequence analysis permits studies on genetic relationships of pathogenic organisms for classification and epidemiology purposes. By comparing the maps and sequences of different isolates possessing different characters such as virulence and tissue tropism, the molecular pathogenesis of pathogenic organisms can be analyzed. Specific DNA fragments derived from the endonuclease cleavage can be isolated, studied, and cloned for diagnostic purposes or for gene expression. For instance, DNA fragments that have type-specific or group-specific sequences can be cloned as diagnostic probes while the fragments containing coding regions for immunogenic proteins can be cloned and expressed for vaccine production or for the production of antigens and/or eventual antibodies that can be used for diagnostic purposes or for research.

The genetic exchange of *Trypanosome brucei* was studied [47]. Two trypanosome clones from *T. brucei rhodesiense* and *T. brucei brucei* were cotransmitted through tsetse flies and 10 resulting clones were isolated. The nuclear and kinetoplast DNA (kDNA) polymorphisms of these clones were analyzed. Five different recombinant genotypes were found, indicating that genetic exchange occurred between these two clones in tsetse flies. The restriction endonuclease DNA cleavage patterns of eight isolates of malignant catarrhal fever–associated herpesviruses from a blue, a white-tailed and a white-bearded wildebeest, a cape hartebeest, a greater kudu, a sika deer, an ibex, and a domestic cow were examined [48]. By using the restriction endonucleases *Hin*dIII and *Eco*RI, these viruses were assigned to two distinct groups. The first group was composed of isolates from the blue wildebeest, the sika deer, and the ibex, whereas the remaining five isolates were in the second group. In another study, DNA of an Australian isolate of equine rhinopneumonitis virus (EHV-4) was analyzed using a restriction endonuclease mapping technique and compared with the previously published data of an English isolate [49]. The restriction maps of these two isolates were found to be similar. This technique was also used to study the correlation between RFLPs of host prion protein gene and incubation period and susceptibility to the disease [50,51]. This relationship was found in rodents and sheep, although that for bovine spongiform encephalopathy (BSE) has not been determined. Hyperkalemic periodic paralysis (HYPP) is a genetic disease in Quarter horses that is associated with hyperkalemia and can be induced by ingestion of potassium. By sequence analysis, the linkage between this disease and the sodium channel gene was demonstrated [52]. A similar linkage is also found in HYPP in humans.

Mapping and sequencing of the entire human genome is in progress. The Human Genome Project will provide understanding of the causes and processes of genetic disorders. It will accelerate the studies of development, aging, and cancer as well as the development of a precise diagnosis and treatment of genetic problems at the DNA level. At the meeting of the Committee on Biotechnology of the U.S. Animal Health Association (1991), a genome mapping project for cattle, sheep, and swine was planned for genetic improvement of livestock. The project will concentrate on identification of 150–250 markers spaced approximately equally along each genome. It will provide information useful for the production of transgenic animals with desired traits and the identification of disease-resistant animals. The project will include sequencing of selected areas as necessary for better definition of specific traits and the identification of genes controlling "rate-limiting" processes in animal production such as ovulation and marbling. Recent studies using microsatellite on short terminal repeals (STRs) are being applied successfully to parentage analysis in horses.

PAGE Profiles. PAGE analyses have long been used to study the epidemiology of BTV and to study the reassortment of their genetic material both *in vitro* and *in vivo* [39,53,54]. The genome profiles of reovirus, rotavirus, and BTV were utilized to study the pathogeneses of these viruses. Because each genome segment encodes for one protein, the difference in the migration of a genome segment between two isolates suggests the difference of that particular encoded viral protein product. From these studies, it was found that the M2 gene of reovirus was associated with growth and spread of the virus and the S1 gene was related to neurovirulence, genome segment 4 was associated with the virulence of rotavirus, and genome segment 5 was associated with neurovirulence of BTV-11 [53,55,56]. Heidner et al. [4] have utilized oligonucleotide fingerprint analysis to investigate the stability of the BTV genome during prolonged infection in cattle. Three calves were inoculated with plaque-purified BTV-10 and viruses were isolated at 35–56 days postinoculation. These isolates were passaged once in cell culture and labeled with ^{32}P, and the labeled RNA was then extracted. The RNA genome segments were separated on a PAGE, and each segment was purified and cleaved by T1 ribonuclease. Each segment was subjected to the two-dimensional PAGE, first through a urea-containing 10% PAGE at pH 3.5 and then through a 20% PAGE at pH 8.3, and finally autoradiographed. The fingerprint analysis demonstrated that the predominant population of circulating virus was not different from the original virus.

Nucleic Acid Hybridization. Hybridization techniques can be used to study the genetic relationships of nucleic acid sequences. At high stringency, one sequence will only anneal to a perfectly matched target. At lower stringency, on the other hand, a specific probe from one species can hybridize to the target of another species if sufficient sequence similarity exists. This strategy provides an opportunity to select, clone, and study a particular gene of various animals and microorganisms when a sequence of the gene of a related animal species or a related microorganism is available.

A new therapeutic approach to diseases is the application of antisense DNA and RNA [57–60]. Antisense RNA is transcribed from plasmids containing the insertion of a gene in reverse orientation, and antisense DNA is usually a synthesized oligonucleotide. Their sequences are complementary to all or a part of the mRNA. The hybrid formed between antisense DNA or RNA and target mRNA will inhibit the translation process taking place on that mRNA. This technique is expected to be used to inhibit the expression of viral genes or oncogenes during the translation stage. In addition, it can be used to study the role of a specific protein by examining the effect of a blockage of the production of that protein. Ribozyme or reagents can be linked to antisense DNA or RNA to cause damage or modifications to the target. Inhibition of in vitro bovine leukemia virus expression was demonstrated using ribozyme method [57]. An intercalating agent such as acridine is sometimes linked to these antimessenger molecules, providing higher affinity for their target molecules. Verspierin et al. [59] coupled an acridine to the synthesized 9-mer oligonucleotide complementary to part of the terminal sequence present at the 5′ end of all mRNAs of *T. brucei*. This acridine-coupled antisense DNA had a lethal effect that was not seen with unmodified oligonucleotide or with an acridine-linked oligomer not complementary to mRNA.

PCR. PCR has been widely investigated for diagnostic purposes because of its extreme sensitivity. However, there are several applications of PCR in other areas such as cloning and sequencing [5,25,27]. In PCR cloning, the target is amplified using primers flanking the region of interest, and the amplified PCR products are then ligated to plasmid vectors for bacterial transformation. To improve the efficiency of the cloning, primers containing restriction sites at the 5′ ends are used. PCR products are cleaved with restriction enzymes to create cohesive ends that can be more efficiently ligated to the plasmids. PCR products can be directly sequenced without having to clone into bacteria. Sequencing can be performed with either single- or double-stranded templates. When single-stranded templates are needed, different amounts of each of two primers will be used in the reaction to produce many more folds of one strand than the other. The PCR sequencing reaction is similar to the conventional dideoxy sequencing method except that *Taq* polymerase is used in place of the Klenow fragment. Finally, direct sequencing can also be performed from RNA targets by using the enzyme thermostable reverse transcriptase.

For diagnosis, other methods that are derived from PCR have been developed. One method is referred to as the *ligase chain reaction* (LCR) [61,62]. This technique employs two probes whose annealing sites are adjacent to each other on the target sequence. The ligase enzyme will ligate these two probes only when they are exactly aligned. Therefore, only the right target will allow the ligation of probes. Ligated products are then amplified in a thermal cycle similar to that of PCR using a thermostable ligase enzyme [63,64]. Ligated products are separated from LCR probes by gel electrophoresis. Alternatively, LCR probes can be modified to facilitate the detection of ligated products. In this method, one LCR probe is modified to enable its immobilization to a solid support while the other probe is modified with a detectable label. After the reaction is complete, only ligated products will be captured and detected using an appropriate detection system. This technique was used in the detection of bovine leukocyte adhesion deficiency [61]. Another method is to amplify RNA probes after hybridization to the targets [25,65,66]. RNA probe is ligated to the specific RNA sequence recognized by enzymes such as phage T7 RNA polymerase or Q-β replicase. After hybridization to the targets and elimination of unbound probes, the enzymes will start producing millions of copies of RNA molecules containing the probes that can be easily detected on an agarose gel stained with ethidium bromide.

7.4. COMMERCIALIZED PRODUCTS

The number of nucleic acid probes that have made it from the research laboratory to a commercialized product for use as diagnostic reagents are few in number. In the United States a pilot study utilizing PCR technology for diagnosing cases of *Mycobacterium pseudotuberculosis* has been carried out. At this time, this product is not commercially available. A few reagents and tests utilizing PCR technology are being commercialized in Europe. Oligonucleotide primers to bluetongue viruses, epizootic hemorrhagic disease viruses, rotaviruses, bovine virus diarrhea virus and to the bovine leukemia virus have been developed for use with PCR technology. Currently there is considerable interest on the part of domestic animal breed organizations to use nucleic acid analyses for parentage and gene-deficiency disease diagnosis.

The reasons for the slow acceptance of this technology are many. Although nucleic acid hybridization and PCR technologies are highly specific and often produce results more rapidly than conventional cultural technology, the personnel that perform and interpret the assays must have in depth scientific background and training to make the correct diagnosis. The personnel in most diagnostic units are not sufficiently trained, nor do they have the background to implement or use these tests. Many laboratories lack the equipment, and the high cost of each test continues to discourage acceptance and use of the technology. National and foreign regulatory agencies have only recently approved tests utilizing either hybridization or PCR technology for diagnostic purposes. Until these regulatory bodies accept the technology, sales of products will depend on in-country demand.

In the United States, all reagents for diagnostic purposes were produced and distributed by the U.S. Department of Agriculture until approximately 6 years ago. As a result, there are very few commercial enterprises in the United States that produce and distribute diagnostic kits. Only one company has ventured into nucleic acid technology for diagnostic purposes for domestic animals. As the refinement of technologies and the development of nonradioactive labels for identifying nucleic acids continue to improve, there should be a major shift toward utilization of these technologies for identification of animal diseases and food-borne pathogens. There needs to be encouragement on the part of the regulatory agencies and diagnostic laboratories to various commercial companies to develop these products.

7.5. SUMMARY

Major technological advances in nucleic acid research have impacted significantly on improved diagnosis for animal genetics, animal diseases, and food-borne pathogens. The technology, including PAGE, nucleic acid hybridization, Northern and Southern blot hybridizations, gene cloning, nucleic acid synthesis, sequencing, and PCR, have led to revelations in animal health research laboratories. Numerous reports are now in the literature that demonstrate that reagents are available or can easily be made available for diagnostic purposes. The slow acceptance of this technology can be attributed to a lack of educational background, appropriate equipment, experience, and regulatory acceptance of the technology. Furthermore, there are very few commercial companies that are committed to producing and distributing this technology. On the other hand, a number of European countries have accepted the technology and are rapidly commercializing products produced in research laboratories.

REFERENCES

1. Dangler, C.A., Osburn, B.I., Veterinary products. Biotechnology 7b:207–225 (1989).
2. Old, R.W., Primrose, S.B., Principles of Gene Manipulation. Oxford, Blackwell Scientific, 1989.
3. Maniatis, T., Fritsch, E.F., Sambrook, J., Molecular Cloning. Cold Spring Harbor, NY, Cold Spring Harbor Laboratory, 1989.
4. Heidner, H.W., MacLachlan, N.J., Fuller, F.J., Richards, R.G., Whetter, L.E.: Bluetongue virus genome remains stable throughout prolonged infection of cattle. J. Gen. Virol. 69:2629–2636 (1988).
5. Ausubel, F.M., Brent, R., Kingston, R.E., Moore, D.D., Seidman, J.G., Smith, J.A., Struhl, K. (eds.): Current Protocols in Molecular Biology. New York, Wiley Interscience, 1989.
6. Southern, E.M.: Detection of specific sequences among DNA fragments separated by gel electrophoresis. J. Mol. Biol. 98:503–517 (1975).
7. Chinsangaram, J., Hammami, S., Osburn, B.I.: Detection of bluetongue virus using a cDNA probe derived from genome segment 4 of bluetongue virus serotype 2. J. Vet. Diagn. Invest. 4:8–12 (1992).

8. de Mattos, C.C., de Mattos, C.A., Osburn, B.I., Dangler, C.A., Chuang, R.Y., Doi, R.H.: Recombinant DNA probe for serotype-specific identification of bluetongue virus 17. AJVR 50:536–541 (1989).

9. Ayers, V.K., Collins, J.K., Blair, C.D., Beaty, B.J.: Use of *in situ* hybridization with a biotinylated probe for the detection of bovine herpesvirus-1 in aborted fetal tissue. J. Vet. Diagn. Invest. 1:231–236 (1989).

10. Brown, C.C., Meyer, R.F., Grubman, M.J.: Identification of African horse sickness virus in cell culture using a digoxigenin-labeled RNA probe. J. Vet. Diagn. Invest. 6:153–155 (1994).

11. Chinsangaram, J., Akita, G.Y., Osburn, B.I.: Detection of bovine group B bovine rotaviruses in feces by polymerase chain reaction. J. Vet. Diagn. Invest. 6:302–307 (1994).

12. Jackson, A.C., Wunner, W.: Detection of rabies virus genomic RNA and mRNA in mouse brains by using *in situ* hybridization. J. Virol. 65:2839–2844 (1991).

13. Jackson, A.C.: Detection of rabies virus mRNA in mouse brain by using *in situ* hybridization with digoxigenin-labelled RNA probes. Mol. Cell. Probes 6:131–136 (1992).

14. Van Woelsel, P.A., Van Blaaderen, A., Moorman, R.J., De Boer, G.F.: Detection of proviral DNA and viral RNA in various tissue early after avian leukosis virus infection. Leukemia 6:135s–137s (1992).

15. Zurbriggen, A., Muller, C., Vandevelde, M.: *In situ* hybridization of virulent canine distemper virus in brain tissue, using digoxigenin-labeled probes. AJVR 54:1457–1461 (1993).

16. Chinsangaram, J., Akita, G.Y., Osburn, B.I.: PCR detection of group A bovine rotaviruses in feces. J. Vet. Diagn. Invest. 5:516–521 (1993).

17. Saiki, R.K., Gelford, D.H., Stoffel, S., Scharf, S.J., Higuchi, R., Horn, G.T., Mullis, K.B., Erlich, H.A.: Primer-directed enzymatic amplification of DNA with a thermostable DNA polymerase. Science 239:487–491 (1988).

18. Katz, J., Seal, B., Ridpath, J.: Molecular diagnosis of alcelaphine herpesvirus (malignant catarrhal fever) infections by nested amplification of viral DNA in bovine blood buffy coat specimens. J. Vet. Diagn. Invest. 3:193–198 (1991).

19. Corney, B.G., Colley, J., Djordjevic, S.P., Whittington, R., Graham, G.C.: Rapid identification of some *Leptospira* isolates from cattle by random amplified polymorphic DNA fingerprinting. J. Clin. Microbiol. 31:2927–2932 (1993).

20. Lawrence, L.M., Harvey, J., Gilmour, A.: Development of a random amplification of polymorphic DNA typing method for *Listeria monocytogenes*. Appl. Environ. Microbiol. 59:3117–3119 (1993).

21. Procunier, J.D., Fernando, M.A., Barta, J.R.: Species and strains differentiation of *Eimeria* spp. of the domestic fowl using DNA polymorphisms amplified by arbitrary primers. Parasitol. Res. 79:98–102 (1993).

22. Williams, J.G.K., Kubelik, A.R., Livak, K.J., Rafalski, J.A., Tingey, S.V.: DNA polymorphisms amplified by arbitrary primers are useful as genetic markers. Nucleic. Acids. Res. 18:6531–6535 (1990).

23. Gosden, J., Hanratty, D.: PCR *in situ*: A rapid alternative to in situ hybridization for mapping short, low copy number sequences without isotopes. Biotechniques 15:78–80 (1993).

24. Clementi, M., Menzo, S., Bagnarelli, P., Manzin, A., Valenza, A., Varaldo, P.E.:

Quantitative PCR and RT-PCR in virology. PCR Methods Applications 2:191–196 (1993).

25. Innis, M.A., Gelfand, D.H., Sninsky, J.J., White, T.J.: PCR Protocols. San Diego: Academic Press, 1990..

26. Vanden Heuvel, J.P., Tyson, F.L., Bell, D.A.: Construction of recombinant RNA templates for use as internal standards in quantitative RT-PCR. Biotechniques 14:395–398 (1993).

27. Erlich, H.A.: PCR Technology. New York: W.H. Freeman, 1992.

28. Longo, M.C., Berninger, M.S., Hartley, J.L.: Use of uracil DNA glycosylase to control carry-over contamination in polymerase chain reactions. Gene 93:125–128 (1990).

29. McPherson, M.J., Quirke, P., Taylor, G.R.: PCR: A Practical Approach. New York: Oxford University Press, (1993).

30. Rys, P.N., Persing, D.H.: Preventing false positives: quantitative evaluation of three protocols for inactivation of polymerase chain reaction amplification products. J. Clin. Microbiol. 31:2356–2360 (1993).

31. Walder, R.Y., Hayes, J.R., Walder, J.A.: Use of PCR primers containing a 3'-terminal ribose residue to prevent cross-contamination of amplified sequences. Nucleic Acids Res. 21:4339–4343 (1993).

32. Boulliuo, A., Le Pennec, J.P., Hubert, G., Donal, R., Smiley, M.: Restriction fragment length polymorphism analysis of endogenous avian leukosis viral loci: Determination of frequencies in commercial broiler lines. Poultry Sci. 70:1287–1296 (1991).

33. Boyd, C.D., Song, J., Kniep, A.C., Park, H.-S., Fastnacht, C., Smith, E.C., Smith, S.C.: A restriction fragment length polymorphism in the pigeon pro α2(1) collagen gene: Lack of an allelic association with an atherogenic phenotype in pigeons genetically susceptible to the development of spontaneous atherosclerosis. Conn. Tissue Res. 26:187–197 (1991).

34. Schwenger, B., Tammen, I., Aurich, C.: Detection of the homozygous recessive genotype for deficiency of uridine monophosphate synthase by DNA typing among bovine embryos produced *in vitro.* J. Reprod. Fertil. 100:511–514 (1994).

35. Hammami, S., Castro, A.E., Osburn, B.I.: Comparison of polyacrylamide gel electrophoresis, an enzyme-linked-immunosorbent assay, and an agglutination test for the direct identification of bovine rotavirus from feces and coelectrophoresis of viral RNA's. J. Vet. Diagn. Invest. 2:184–190 (1990).

36. Lozano, L.-F., Hammami, S., Castro, A.E., Osburn, B.I.: Interspecies polymorphism of double-stranded RNA extracted from reoviruses of turkeys and chickens. J. Vet. Diagn. Invest. 4:74–77 (1992).

37. Saif, L.J., Theil, K.W.: Viral Diarrheas of Man and Animals. Boca Raton: CRC Press, 1990.

38. Squire, K.R.E., Chuang, R.Y., Chuang, L.F., Doi, R.H., Osburn, B.I.: Detecting bluetongue virus RNA in cell culture by dot hybridization with a cloned genetic probe. J. Virol. Methods 10:59–68 (1985).

39. Squire, K.R.E., Osburn, B.I., Chuang, R.Y., Doi, R.H.: A survey of electropherotype relationships of bluetongue virus isolates from the western United States. J. Gen. Virol. 64:2103–2115 (1983).

40. Bridger, J.C.: Novel rotaviruses in animals and man. In Novel Diarrhoea Viruses. Ciba Foundation Symposium 128. Chichester, England: John Wiley & Sons Ltd., 1987.

41. Ng, L.-K., Stiles, M.E., Taylor, D.E.: DNA probes for identification of tetracycline resistance genes in *Campylobacter* species isolated from swine and cattle. Antimicrob. Agents Chemother. 31:1669–1674 (1987).

42. Rossi, M.S., Sadir, A.M., Schudel, A.A., Palma, E.L.: Detection of foot-and-mouth disease virus with DNA probes in bovine esophageal-pharingeal fluids. Arch. Virol. 99:67–74 (1988).

43. Dangler, C.A., de la Concha-Bermejillo, A., Stott, J.L., Osburn, B.I.: Limitations of in situ hybridization for the detection of bluetongue virus in blood mononuclear cells. J. Vet. Diagn. Invest. 2:303–307 (1990).

44. Akita, G.Y., Chinsangaram, J., Osburn B.I., Ianconescu, M., Kaufman, R.: Detection of bluetongue virus serogroup by polymerase chain reaction. J. Vet. Diagn. Invest. 4:400–405 (1992).

45. Brock, K.V.: Detection of persistent bovine viral diarrhea virus infections by DNA hybridization and polymerase chain reaction assay. Arch. Virol. (Suppl 3):199–208 (1991).

46. Wu, C.C., Lin, T.L.: Detection of infectious bursal disease virus in digested formalin-fixed paraffin-embedded tissue sections by polymerase chain reaction. J. Vet. Diagn. Invest. 4:452–455 (1992).

47. Gibson, W.C.: Analysis of genetic cross between *Trypanosoma brucei rhodesiense* and *T. b. brucei,* Parasitology 99:391–402 (1989).

48. Shih, L.M., Zee, Y.C., Castro, A.E.: Comparison of genome of malignant catarrhal fever-associated herpesviruses by restriction endonuclease analysis. Arch. Virol. 109:145–151 (1989).

49. Nagesha, A.S., McNeil, J.R., Ficorilli, N., Studdert, M.J.: Cloning and restriction endonuclease mapping of the genome of an equine herpesvirus 4 (equine rhinopneumonitis virus), strain 405/76. Arch. Virol. 124:379–387 (1992).

50. Hunter, N., Foster, J.D., Benson, G., Hope, J.: Restriction fragment length polymorphisms of the scrapie-associated fibril protein (PrP) gene and their association with susceptibility to natural scrapie in British sheep. J. Gen. Virol. 72:1287–1292 (1991).

51. Ryan, A.M., Womack, J.E.: Somatic cell mapping of the bovine prion protein gene and restriction fragment length polymorphism studies in cattle and sheep. Anim. Genet. 24:23–26 (1993).

52. Rudolph, J.A., Spier, S.J., Byrns, G., Hoffman, E.P.: Linkage of hyperkalaemic periodic paralysis in quarter horses to the horse adult skeletal muscle sodium channel gene. Anim. Genet. 23:241–250 (1992).

53. Carr, M.A., de Mattos, C.C., de Mattos, C.A., Osburn, B.I.: Association of bluetongue virus gene segment 5 with neuroinvasiveness. J. Virol. 68:1255–1257 (1994).

54. Stott, J.L., Oberst, R.D., Channel, M.B., Osburn, B.I.: Genome segment reassortment between two serotypes of bluetongue virus in a natural host. J. Virol. 61:2670–2674 (1987).

55. Offit, P.A., Blavat, G., Greenberg, H.B., Clark, H.F.: Molecular basis of rotavirus virulence: Role of gene segment 4. J. Virol. 57:46–49 (1986).

56. Sharpe, A.H., Fields, B.N.: Pathogenesis of viral infections. N. Engl. J. Med. 312:486–497 (1985).

57. Cantor, G.H., McElwain, T.F., Birkebak, T.A., Palmer, G.H.: Ribozyme cleaves rex/tax

mRNA and inhibits bovine leukemia virus expression. Proc. Natl. Acad. Sci. U.S.A. 90:10932–10936 (1993).

58. Toulmé, J.J., Hélène, C.: Antimessenger oligodeoxyribonucleotides: An alternative to antisense RNA for artificial regulation of gene expression—a review. Gene 72:51–58 (1988).

59. Verspieren, P., Cornelissen, A.W.C.A., Thuong, N.T., Hélène, C., Toulmé, J.J.: An acridine-linked oligodeoxynucleotide targeted to the common 5′ end of trypanosome mRNAs kills cultured parasites. Gene 61:307–315 (1987).

60. Weintraub, H.M.: Antisense RNA and DNA. Sci. Am. 262:40–46 (1990).

61. Batt, C.A., Wagner, P., Wiedmann, M., Luo, J., Gilbert, R.: Detection of bovine leukocyte adhesion deficiency by nonisotopic ligase chain reaction. Anim. Genet. 25:95–98 (1993).

62. Winn-Deen, E.S., Batt, C.A., Wiedmann, M.: Non-radioactive detection of *Mycobacterium tuberculosis* LCR products in a microtitre plate format. Mol. Cell. Probes 7:179–186 (1993).

63. Barany, F., Gelfand, D.H.: Cloning, overexpression and nucleotide sequence of a thermostable DNA ligase-encoding gene. Gene 109:1–11 (1991).

64. Barany, F.: Genetic disease detection and DNA amplification using cloned thermostable ligase. Proc. Natl. Acad. Sci. U.S.A. 88:189–193 (1991).

65. Cahill, P., Foster, K., Mahan, D.E.: Polymerase chain reaction and Q beta replicase amplification. Clin. Chem. 37:1482–1485 (1991).

66. Kramer, F.R., Lizardi, P.M.: Replicatable RNA reporters. Nature 339:401–402 (1989).

67. Subramanian, K., Lupiani, B., Hetrick, F.M., Samal, S.K.: Detection of aquareovirus RNA in fish tissues by nucleic acid hybridization with a cloned cDNA probe. J. Clin. Microbiol. 31:1612–1614 (1993).

68. Roy, P., Ritter, G.D., Jr., Akashi, H., Collisson, E., Inaba, Y.: A genetic probe for identifying bluetongue virus infections *in vivo* and *in vitro*. J. Gen. Virol. 66:1613–1619 (1985).

69. Verbeek, A., Tijssen, P.: Biotinylated and radioactive cDNA probes in the detection by hybridization of bovine enteric coronavirus. Mol. Cell. Probes 2:209–223 (1988).

70. Galik, P.K., van Santen, V.L., Stringfellow, D.A., Bird, R.C., Wright, J.C., Smith, P.C.: Development of a DNA probe for identification of bovine herpesvirus 4. AJVR 54:653–659 (1993).

71. Stott, M.L., Thurmond, M.C., Dunn, S.J., Osburn, B.I., Stott, J.L.: Integrated bovine leukosis proviral DNA in T helper and T cytotoxic/suppressor lymphocytes. J. Gen. Virol. 72:307–315 (1991).

72. Hussein, H.A., Parwani, A.V., Rosen, B.I., Lucchelli, A., Saif, L.J.: Detection of rotavirus serotypes G1, G2, G3, and G11 in feces of diarrheic calves by using polymerase chain reaction–derived cDNA probes. J. Clin. Microbiol. 31:2491–2496 (1993).

73. Parwani, A.V., Rosen, B.I., Flores, J., McCrae, M.A., Gorziglia, M., Saif, L.J.: Detection and differentiation of bovine group A rotavirus serotypes using polymerase chain reaction–generated probe to the VP7 gene. J. Vet. Diagn. Invest. 4:148–158 (1992).

74. Parwani, A.V., Rosen, B.I., McCrae, M.A., Saif, L.J.: Development of cDNA probes for typing group A bovine rotaviruses on the basis of VP4 specificity. J. Clin. Microbiol. 30:2717–2721 (1992).

75. Rosen, B.I., Parwani, A.V., Gorziglia, M., Larralde, G., Saif, L.J.: Characterization of full-length and polymerase chain reaction-derived partial-length Gottfried and OSU gene 4 probes for serotypic differentiation of porcine rotaviruses. J. Clin. Microbiol. 30:2644–2652 (1992).

76. Brock, K.V., Brian, D.A., Rouse, B.T., Potgieter, L.N.: Molecular cloning of complementary DNA from a pneumopathic strain of bovine viral diarrhea virus and its diagnostic application. Can. J. Vet. Res. 52:451–457 (1988).

77. Allan, G.M., Smyth, J.A., Todd, D., McNulty, M.S.: *In situ* hybridization for the detection of chicken anemia virus in formalin-fixed, paraffin-embedded sections. Avian Dis. 37:177–182 (1993).

78. Viljoen, G.J., Huismans, H.: The characterization of equine encephalosis virus and the development of genomic probes. J. Gen. Virol. 70:2007–2015 (1989).

79. Hathcock, T.L., Giambrone, J.J.: Tissue-print hybridization using a non-radioactive probe for the detection of infectious bursal disease virus. Avian Dis. 36:202–205 (1992).

80. Johnson, L.K., Meyer, A.L., Zink, M.C.: Detection of ovine lentivirus in seronegative sheep by *in situ* hybridization, PCR, and cocultivation with susceptible cells. Clin. Immunol. Immunopathol. 65:254–260 (1992).

81. Krell, P.J.: Salas, T., Johnson, R.P.: Mapping of porcine parvovirus DNA and development of a diagnostic DNA probe. Vet. Microbiol. 17:29–43 (1988).

82. Ermine, A., Tordo, N., Tsiang, H.: Rapid diagnosis of rabies infection by means of a dot hybridization assay. Mol. Cell. Probes 2:75–82 (1988).

83. Ambrosio, R.E., Visser, E.S., Koekhoven, Y., Kocan, K.M.: Hybridization of DNA probes to *A. marginale* isolates from different sources and detection in *Dermacentor andersoni* ticks. Onderstepoort J. Vet. Res. 55:227–229 (1988).

84. Goff, W., Barbet, A., Stiller, D., Palmer, G., Knowles, D., Kocan, K.M., Gorham, J., McGuire, T.: Detection of *Anaplasma marginale*–infected tick vectors by using a cloned DNA probe. Proc. Natl. Acad. Sci. U.S.A. 85:919–923 (1988).

85. Larson, D.J., Wesley, I.V., Hoffman, L.J.: Use of oligodeoxynucleotide probes to verify *Campylobacter jejuni* as a cause of bovine abortion. J. Vet. Diagn. Invest. 4:348–351 (1992).

86. VanEys, G.J., Zaal, J., Schoone, G.J., Terpstra, W.J.: DNA hybridization with hardjobovis-specific recombinant probes as a method for type discrimination of *Leptospora interrogans serovar hardjo*. J. Gen. Microbiol. 134:567–574 (1988).

87. Hurley, S.S., Splitter, G.A., Welch, R.A.: Development of a diagnostic test for Johne's disease using a DNA hybridization probe. J. Clin. Microbiol. 27:1582–1587 (1989).

88. McCully, M.A., Brock, K.V.: Development of a DNA hybridization probe for the detection of *Mycoplasma bovis.* J. Vet. Diagn. Invest. 4:464–467 (1992).

89. Stipkovits, L., Belak, S., McGwire, B.S., Ballagi-Pordany, A., Santha, M.: Rapid identification of *Mycoplasma gallisepticum* using a simple method of nucleic acid hybridization. Mol. Cell. Probes 2:339–344 (1988).

90. Zhao, S., Yamamoto, R.: Species-specific recombinant DNA probes for *Mycoplasma meleagridis*. Vet. Microbiol. 35:179–185 (1993).

91. Jasmer, D.P., Reduker, D.W., Goff, W.L., Stiller, D., McGuire, T.C.: DNA probes distinguish geographical isolates and identify a novel DNA molecule of *Babesia bovis*. J. Parasitol. 76:834–841 (1990).

92. Profaus-Juchelka, H., Liberator, P., Turner, M.: Identification and characterization of cDNA clones encoding antigens of *Eimeria tenella*. Mol. Biochem. Parasitol. 30:233–241 (1988).

93. Oberst, R.D., Hall, S.M., Schoneweis, D.A.: Detection of *Eperythrozoon suis* DNA from swine blood by whole organism DNA hybridizations. Vet. Microbiol. 24:127–134 (1990).

94. Chen, P.P., Conrad, P.A., ole-MoiYoi, O.K., Brown, W.C., Dolan, T.T.: DNA probes detect *Theileria parva* in the salivary glands of *Rhipicephalus appendiculatus* ticks. Parasitol. Res. 77:590–594 (1991).

95. Savva, D.: Isolation of a potential DNA probe for *Toxoplasma gondii*. Microbios 58:165–172 (1989).

96. Gibson, W.C., Dukes, P., Gashumba, J.K.: Species-specific DNA probes for the identification of African trypanosomes in tsetse flies. Parasitology 97:63–73 (1988).

97. Majiwa, P.A., Thatthi, R., Moloo, S.K., Nyeko, J.H., Otieno, L.H., Maloo, S.: Detection of trypanosome infections in the saliva of tsetse flies and buffy-coat samples from antigenaemic but aparasitaemic cattle. Parasitology 108:313–322 (1994).

98. Giesendorf, B.A., Quint, W.G., Henkens, M.H., Stegeman, H., Huf, F.A., Niesters, H.G.: Rapid and sensitive detection of *Campylobacter* spp. in chicken products by using the polymerase chain reaction. Appl. Environ. Microbiol. 58:3804–3808 (1992).

99. Oyofo, B.A., Thornton, S.A., Burr, D.H., Trust, T.J., Pavlovskis, O.R., Guerry, P.: Specific detection of *Campylobacter jejuni* and *Campylobacter coli* by using polymerase chain reaction. J. Clin. Microbiol. 30:2613–2619 (1992).

100. Wegmuller, B., Luthy, J., Candrian, U.: Direct polymerase chain reaction detection of *Campylobacter jejuni* and *Campylobacter coli* in raw milk and dairy products. Appl. Environ. Microbiol. 59:2161–2165 (1993).

101. MacPherson, J.M., Gajadhar, A.A.: Sensitive and specific polymerase chain reaction detection of *Toxoplasma gondii* for veterinary and medical diagnosis. Can. J. Vet. Res. 57:45–48 (1993).

102. Isegawa, Y., Nakagomi, O., Nakagomi, T., Ishida, S., Uesugi, S., Ueda, S.: Determination of bovine rotavirus G and P serotypes by polymerase chain reaction. Mol. Cell. Probes 7:277–284 (1993).

103. Xu, L., Harbour, D., McCrae, M.A.: The application of polymerase chain reaction to detection of rotaviruses in feces. J. Virol. Methods 27:29–38 (1990).

104. Rossen, L., Holmstrom, K., Olsen, J.E., Rasmussen, O.F.: A rapid polymerase chain reaction (PCR)–based assay for the identification of *Listeria monocytogenes* in food samples. Int. J. Food Microbiol. 14:145–151 (1991).

105. Curiale, M.S., Sons, T., Fanning, L., Lepper, W., McIver, D., Garramone, S., Mazola, M.: Deoxyribonucleic acid hybridization method for the detection of *Listeria* in dairy products, seafoods, and meats: Collaborative study. J. Aoac Int. 77:602–617 (1994).

106. Bej, A.K., Steffan, R.J., DiCesare, J., Haff, L., Atlas, R.M.: Detection of coliform bacteria in water by polymerase chain reaction and gene probes. Appl. Environ. Microbiol. 56:307–314 (1990).

107. Sakamoto, Punyahotra, R., Mizukoshi, N., Ueda, S., Imagawa, H., Sugiura, T., Kamada, M., Fukusho, A.: Rapid detection of African horse sickness virus by the reverse transcriptase polymerase chain reaction (RT-PCR) using the amplimer for segment 3 (VP3 gene). Arch. Virol. 136:87–97 (1994).

108. Stone-Marschat, M., Carville, A., Skowronek, A., Laegreid, W.W.: Detection of African horse sickness virus by reverse transcription-PCR. J. Clin. Microbiol. 32:697–700 (1994).

109. Dangler, C.A., de Mattos, C.A., de Mattos, C.C., Osburn, B.I.: Identifying bluetongue virus ribonucleic acid sequences by the polymerase chain reaction. J. Virol. Methods 28:281–292 (1990).

110. Wade-Evans, A.M., Mertens, P.P.C., Bostock, C.J.: Development of the polymerase chain reaction for detection of bluetongue virus in tissue samples. J. Virol. Methods 30:15–24 (1990).

111. Agresti, A., Wilma, P., Rocchi, M., Meneveri, R., Marozzi, A., Cavalleri, D., Peri, E., Poli, G., Ginelli, E. Use of polymerase chain reaction to diagnose bovine leukemia virus infection in calves at birth. AJVR 54:373–378 (1993).

112. Kelly, E.J., Jackson, M.K., Marsolais, G., Morrey, J.D., Callan, R.J.: Early detection of bovine leukemia virus in cattle by use of the polymerase chain reaction. AJVR 54:205–209 (1993).

113. Hertig, C., Pauli, U., Zanoni, R., Peterhans, E.: Detection of bovine viral diarrhea (BVD) virus using the polymerase chain reaction. Vet. Microbiol. 26:65–76 (1991).

114. Schmitt, B.J., Lopez, O.J., Ridpath, J.F., Galeota-Wheeler, J., Osorio, F.A.: Evaluation of PCR for diagnosis of bovine viral diarrhea virus in tissue homogenates. J. Vet. Diagn. Invest. 6:44–47 (1994).

115. Zimmermann, W., Durrwald, R., Ludwig, H.: Detection of borna disease virus RNA in naturally infected animals by a nested polymerase chain reaction. J. Virol. Methods. 46:133–143 (1994).

116. Aradaib, I.E., Akita, G.Y., Osburn, B.I.: Detection of epizootic hemorrhagic disease virus serotypes 1 and 2 in cell culture and clinical samples using polymerase chain reaction. J. Vet. Diagn. Invest. 6:143–147 (1994).

117. Wilson, W.C.: Development of a nested-PCR test based on sequence analysis of epizootic hemorrhagic disease viruses non-structural protein 1 (NS1). Virus Res. 31:357–365 (1994).

118. St-Laurent, G., Morin, G., Archambault, D.: Detection of equine arteritis virus following amplification of structural and nonstructural viral genes by reverse transcription-PCR. J. Clin. Microbiol. 32:658–665 (1994).

119. Donofrio, J.C., Coonrod, J.D., Chambers, T.M.: Diagnosis of equine influenza by the polymerase chain reaction. J. Vet. Diagn. Invest. 6:39–43 (1994).

120. Silva, R.F.: Differentiation of pathogenic and non-pathogenic serotype 1 Marek's disease viruses (MDVs) by the polymerase chain reaction amplification of the tandem direct repeats within the MDV genome. Avian Dis. 36:521–528 (1992).

121. Hsu, D., Shih, L.M., Castro, A.E., Zee, Y.C.: A diagnostic method to detect alcelaphine herpesvirus-1 of malignant catarrhal fever using the polymerase chain reaction. Arch. Virol. 114:259–263 (1990).

122. Molitor, T.W., Oraveerakul, K., Zhang, Q.Q., Choi, C.S., Ludemann, L.R.: Polymerase chain reaction (PCR) amplification for the detection of porcine parvovirus. J. Virol. Methods 32:201–211 (1991).

123. Suarez, P., Zardoya, R., Prieto, C., Solana, A., Tabares, E., Bautista, J.M., Castro, J.M.: Direct detection of the porcine reproductive and respiratory syndrome (PRRS) virus by

reverse transcription-polymerase chain reaction (RT-PCR). Arch. Virol. 135:89–99 (1994).

124. Dangler, C.A., Deaver, R.E., Kolodziej, C.M., Rupprecht, J.D.: Genotypic screening of pseudorabies virus strains for the thymidine kinase deletions by use of the polymerase chain reaction. AJVR 53:904–908 (1992)..

125. Ermine, A., Larzul, D., Ceccaldi, P.E., Guesdon, J.L., Tsiang, H.: Polymerase chain reaction amplification of rabies virus nucleic acids from total mouse brain RNA. Mol. Cell. Probes 4:189–191 (1990).

126. Sacramento, D., Bourhy, H., Tordo, N.: PCR technique as an alternative method for diagnosis and molecular epidemiology of rabies virus. Mol. Cell. Probes 5:229–240 (1991).

127. Olive, D.M.: Detection of enterotoxigenic *Escherichia coli* after polymerase chain reaction amplification with a thermostable DNA polymerase. J. Clin. Microbiol. 27:261–265 (1989).

128. VanEys, G.J., Gravekamp, C., Gerritsen, M.J., Quint, W., Cornelissen, M.T., Schegget, J.T., Terpstra, W.J.: Detection of *Leptospira* in urine by polymerase chain reaction. J. Clin. Microbiol. 27:2258–2262 (1989).

129. Kempf, I., Blanchard, A., Gesbert, F., Guittet, M., Bennejean, G.: Comparison of antigenic and pathogenic properties of *Mycoplasma iowae* strains and development of a PCR-based detection assay. Res. Vet. Sci. 56:179–185 (1994).

130. Zhao, S., Yamamoto, R.: Amplification of *Mycoplasma iowae* using polymerase chain reaction. Avian Dis. 37:212–217 (1993).

131. Leon, G., Maulen, N., Figueroa, J., Villanueva, J., Rodriguez, C., Vera, M.I., Krauskopf, M.: A PCR-based assay for the identification of the fish pathogen *Renibacterium salmoninarum*. FEMS Microbiol. Lett. 115:131–136 (1994).

132. Widjojoatmodjo, M.N., Fluit, A.C., Torensma, R., Keller, B.H., Verhoef, J.: Evaluation of the magnetic immune PCR assay for rapid detection of *Salmonella*. Eur. J. Clin. Microbiol. Infect. Dis. 10:935–938 (1991).

133. Gwaltney, S.M., Hays, M.P., Oberst, R.D.: Detection of *Eperythrozoon suis* using the polymerase chain reaction. J. Vet. Diagn. Invest. 5:40–46 (1993).

134. Tanaka, M., Onoe, S., Matsuba, T., Katayama, S., Yamanaka, M., Yonemichi, H., Hiramatsu, K., Baek, B.K., Sugimoto, C., Onuma, M.: Detection of *Theileria sergenti* infection in cattle by polymerase chain reaction amplification of parasite-specific DNA. J. Clin. Microbiol. 31:2565–2569 (1993).

135. Masiga, D.K., Smyth, A.J., Hayes, P., Bromidge, T.J., Gibson, W.C.: Sensitive detection of trypanosomes in tsetse flies by DNA amplification. Int. J. Parasitol. 22:909–918 (1992).

CHAPTER 8

DIAGNOSTIC NUCLEIC ACID PROBES IN HUMAN MEDICINE: APPLICATIONS, PRESENT AND FUTURE

JAMES VERSALOVIC AND JAMES R. LUPSKI
Department of Molecular and Human Genetics (J. V., J. R. L.)
and Department of Pediatrics (J. R. L.), Baylor College of Medicine,
Houston, TX 77030

8.1. INTRODUCTION

Identification of specific biological molecules associated with pathological conditions facilitates disease diagnosis in human medicine. Specific biological products may be detected by various methods depending on the chemical nature of the substrate, or of the substrate alteration, requiring detection. Questions regarding the identification of biological molecules implicated in human disease can be stated as follows: What is the chemical composition of the biological molecule? How can one specifically detect such a molecule? Which disease-specific variation must be recognized? Is the nature of the change that one is trying to detect a matter of presence or absence of a substrate, a quantitative difference (e.g., chromosomal DNA amplification), or a qualitative change (e.g., point mutation)?

Elements of a complex mixture of biological molecules can be separated and distinguished by size or by charge differences in an electrical field. Biological marker molecules, such as proteins or other organic byproducts, which are present at detectable levels in specific human disease states, can be measured. Absolute levels of specific proteins or organic byproducts may be assayed after separation by various techniques, including electrophoresis, chromatography, and mass spec-

Nucleic Acid Analysis: Principles and Bioapplications, pages 157–202
© 1996 Wiley-Liss, Inc.

trometry [1]. Immunoglobulin protein molecules may be present in elevated levels in human serum as a sign of underlying malignancies or infections, thereby suggesting specific human disease states. These serum protein abnormalities may be detected by conventional protein electrophoresis [1]. Elevated levels of smaller organic compounds such as porphyrins [2] and uric acid [3] may be observed after resolution by high pressure liquid chromatography and aid in the diagnosis of porphyrias or gout, respectively.

In addition to measuring absolute or relative levels of biomolecules, qualitative changes that result in the formation of abnormal biological molecules may be useful for disease diagnosis. The first example of such, and that which signals the beginnings of molecular medicine, is the detection of sickle cell hemoglobin (HbS) by migration differences in an electrical field [4]. The migration difference is due to a mutation leading to an amino acid substitution (Glu →Val) [5] that presumably changes the surface charge of the protein. HbS migrates toward the anode more slowly than wild-type hemoglobin during electrophoresis, [1,4], allowing effective separation and visualization of the abnormal protein. However, hemoglobin is a special case because it represents the blood protein in greatest concentration. Many biological molecules involved in human diseases are present in minute quantities in complex mixtures or in tissues that are not readily accessible.

Methods of identification must recognize biomolecules in complex mixtures (e.g., blood) and provide detectable signals with a favorable signal-to-noise ratio. Enzyme–substrate interactions satisfy the specificity criterion. Chromogenic [6,7] or fluorescent substrates can be used in assays to identify specific enzymes in body fluids. High levels of specific enzymes such as creatine kinase and lactate dehydrogenase have been associated with acute myocardial infarction or skeletal muscle necrosis [8]. These levels can be monitored with specific enzyme tests [8]. However, enzyme-based detection systems lack versatility since enzyme–substrate interactions must be previously defined. Many pathologic processes require detection of components with undefined or undetectable enzymatic activities.

Immunological methods are based on the recognition of structural determinants, antibody binding to epitope, and satisfy the specificity criterion. Immunoassays offer improved versatility relative to biochemical methods because serum is readily available. Purified proteins can be used directly to produce specific antibodies *in vivo*. The antigen–antibody precipitin reaction has formed the basis for the detection of specific protein antigens and antibodies by immunoprecipitation [9,10]. Immunoprecipitation and immunoelectrophoresis have been utilized in serological studies of infectious diseases, autoimmunity, and serum protein detection [9]. For almost 20 years, monoclonal antibody technology [11] has increased the potential for specific antibody-mediated recognition and its utilization in diagnosis. Antibodies can be raised against specific epitopes *in vivo* and provide previously unmatched specificity. Immunochemistry and enzymology technologies have been combined by using antibody-mediated recognition for the identification of specific enzymes implicated in human disease [6].

During the past 20 years recombinant DNA technology developed from basic discoveries in nucleic acid biochemistry and bacterial genetics. Discoveries of restriction endonucleases, ligases, cloning vectors such as phages, and plasmids allowed scientists to manipulate nucleic acid molecules (DNA or RNA). Hydrogen bond–mediated base-pairing between nucleic acid probes and complementary target DNA sequences enabled the detection of specific DNA sequences. An alternative to antibody-mediated antigen detection was now available for identification of human disease-specific biological molecules. Specific DNA probes could be cloned or synthesized *de novo*. Nucleic acid probes may be utilized to detect an infectious pathogen's DNA sequence or distinguish a mutant human allele from the wild-type allele at a single genetic locus [12].

A principal advantage of DNA probe technology is its universality. DNA molecules that are isolated from different sources possess relatively uniform biochemical properties. DNA–DNA hybridization techniques can be standardized regardless of the source of the probe or target DNA [13]. In contrast, enzymes may require different buffer conditions even though the biochemical activities are identical. For example, bacterial and human glycerol kinase require different assay conditions [14–16].

Nucleic acid hybridization technology refers to specific annealing of variable length DNA or RNA probes with complementary target sequences to produce a specific signal. Specificity of interaction lies at the level of primary DNA sequence, the relative order of bases, adenine (A), guanine (G), cytosine (C), and thymine (T), in a linear sequence instead of spatial three-dimensional macromolecular conformation. This property of nucleic acid hybridization technology may enable greater potential specificity for nucleic acid probes relative to monoclonal antibodies. Specificity of identification is dependent on the possession of a probe length of sufficient size to permit recognition of a specific sequence in a genomic background (genome size is ~10^6 bp in eubacteria versus ~10^9 bp in humans). Once cloned DNA is available, limitless supplies of diagnostic nucleic acid probes (DNA probes) can be made by *in vivo* cloning in bacteria or DNA amplification *in vitro*. Alternatively, oligonucleotide probes can be rapidly produced with current semiautomated nucleic acid synthesis technology.

This chapter focuses on the application of DNA probe technology to particular areas of human disease diagnosis, including infectious diseases and oncology. The reader is referred to other chapters with respect to neonatal screening and DNA diagnosis of inherited human diseases and forensics. A broad methodological approach to infectious disease diagnosis, both human and veterinary, is described elsewhere in this volume. Finally, detailed discussions of DNA probe methods and technical considerations are described in the first section of this book (Chapters 1–4).

8.2. METHODS

8.2.1. DNA Hybridization Methods in Medicine

Hybridization between radioisotopically labeled DNA probes and clinical samples containing a target DNA sequence represented the first application of nucleic acid technology to human medicine [12]. Widespread application of DNA probe technology to human disease awaited refinement of techniques and characterization of potential target DNA sequences.

DNA–DNA hybridization was first used extensively in medical microbiology to investigate taxonomic structures of pathogenic bacteria [17]. Genomic DNA samples from different bacterial species were mixed and annealed with each other at various rates. Renaturation kinetics were evaluated to distinguish levels of relatedness between pathogenic bacteria. No DNA cloning or detailed characterization of the probe or target sequences was required for these studies. However, during the past 20 years many new bacterial and viral DNA sequences have been characterized and represent a new pool of potential target sequences for hybridization.

The development of restriction fragment length polymorphisms (RFLPs) permitted the isolation of polymorphic markers linked to specific human diseases [18]. Since the primary DNA sequence can vary at multiple nucleotide positions between individuals and chromosomal homologues, human DNA fragments of different sizes could be obtained after restriction digestion and hybridization with labeled DNA probes. This RFLP principle was employed in the first example of specific human disease detection by nucleic acid hybridization [19,20]. Briefly, the point mutation in the sixth codon of β-globin gene from homozygous (SS) sickle cell anemia patients abolishes a restriction site in the DNA. Digestion of genomic DNA with a specific restriction enzyme, whose recognition sequence is destroyed by the sickle cell mutation, prior to Southern hybridization [21] yielded differently sized DNA fragments in the samples from affected patients [19]. Once genetic linkage was established by molecular approaches using positional cloning methods [22], specific human disease genes could be cloned and DNA sequence alterations associated with disease could be identified. Cloned gene probes or allele-specific oligonucleotide (ASO) probes [23] were subsequently used for direct detection of DNA sequence changes associated with human disease phenotypes [24].

The relative stability of DNA and the universality of DNA methods are probably responsible for the current prevalence of DNA–DNA hybridization methods. RNA–DNA duplexes are more stable than DNA–DNA duplexes, and the advent of improved RNAse inhibitors may increase the utility of replicatable RNA probes (see discussion of the Qβ replicase system below). Pitfalls of conventional DNA hybridization methods include relative lack of sensitivity and utilization of radioactively labeled probes. Sensitivity limitations may be alleviated by either replacement or augmentation of hybridization techniques with newer DNA amplification methods (see next section). Importantly, the relative lack of sensitivity of DNA hybridization may be advantageous in specific diagnostic dilemmas when DNA amplification techniques may be too sensitive. Amplification techniques (e.g., the polymerase chain reaction [PCR]) may be prone to false-positive results secondary to contaminant detection, especially with nonsterile tissue and body fluids. Incorporation of improved chemiluminescent, fluorescent, or enzymatic (colorimetric) labels in nucleic acid probes, instead of radioisotopes, will be necessary before these techniques are routinely used in clinical laboratories.

8.2.2. Nucleic Acid Amplification Methods

DNA amplification methods surpass the sensitivity of hybridization methods since the substrate DNA is amplified exponentially. That is, amplification does not proceed in a linear manner but instead the amount of product doubles with each cycle and a chain reaction occurs. Amplification of a specific DNA segment may exceed 10^6 copies if more than 20 cycles (2^{20}) are performed. DNA and RNA targets can be amplified efficiently by various techniques. Human genomic analysis often requires amplification of messenger RNA targets because these molecules contain relevant coding sequences without noncoding introns. Messenger RNA detection may be accomplished by a combination of reverse transcription and PCR (RT-PCR) [25]. Nucleic acid amplification methods include hybridization to a specific target and amplification for sensitive detection. Specific amplification and detection are made possible through the use of synthetic oligonucleotide primers (usually 20–30mers for the PCR) or replicatable probes of sufficient length to prevent recognition of nonspecific sequences in clinical samples.

The advent of PCR technology enabled exponential amplification of specific DNA segments *in vitro* [26–29]. PCR refers to the use of specific primers (probes) that hybridize to specific DNA sequences and permits the amplification of DNA fragments lying between two primer sequences. Previously, defined DNA segments could only be amplified in bacterial cells by the biological cloning of DNA in plasmid vectors. Cell-mediated cloning of DNA required extensive manipulation of DNA. Heterologous DNA sequences (e.g., human) may not be tolerated by the primary bacterial cloning vehicle, *Escherichia coli*. RNA targets can also be amplified by an adaptation known as reverse transcription PCR (RT-PCR) [25]. Messenger RNA targets are first copied to cDNA by reverse transcriptase, RNA–DNA duplexes are degraded by RNaseH, and double-stranded cDNA is synthesized by reverse transcriptase. These DNA duplexes are subsequently amplified by thermostable *Taq* DNA polymerase as in conventional PCR.

PCR technology was greatly simplified following the application of a thermostable DNA polymerase from the thermophilic eubacterium *Thermus aquaticus* [25–30]. PCR procedures became semiautomated because the steps required for PCR, including denaturation of the template DNA, annealing (hybridization) of the primers to the template, and extension of DNA sequences from the 3′ ends of the primers, were performed in a single thermal cycler without replenishment of DNA polymerase. This improvement made PCR an attractive tool for the clinical laboratory where convenience and ease of use represent primary concerns [31].

DNA amplification methods present challenges for template DNA purification of clinical samples because amplification techniques, in contrast to hybridization techniques, require *in vitro* enzymatic activity. DNA isolation and extraction procedures need to be circumvented for rapid clinical diagnostic applications. Direct methods of DNA isolation from clinical samples need to be developed, but success rates vary with different clinical sources. Urine specimens from newborn infants revealed 18%–25% false-negative results with PCR-based detection of cytomegalovirus (CMV) in one study [32]. A large-scale study of more than 2,000 blood specimens correlated CMV viremia by culture with 100% detection by PCR [33]. In contrast, approximately 50% of viruria-positive, viremia-negative samples were negative by PCR [33]. More recently, 100% of patients were CMV PCR-positive when peripheral blood leukocytes were used versus a 55% positivity fraction when plasma was the substrate [34]. PCR-based detection of *Borrelia burgdorferi* from 10 patients, with active central nervous system infection as measured by intrathecal antibody responses, produced nine positive results from urine samples but only four positive results with cerebrospinal fluid (CSF) samples [35]. Sampling methods may adversely affect amplification results. PCR amplifications failed to yield any DNA products with heparinized plasma even after DNA extraction and precipitation [36]. Identical PCR reactions with serum or sodium citrate–treated plasma yielded positive PCR results in most patients [36]. Sample treatment with the chaotropic agent guanidium thiocyanate reduced false-negative rates for mycobacterial detection [37]. Rapid sputum treatment with heat detergent enabled PCR detection of *Mycobacterium tuberculosis* in 94% of fluorochrome smear-positive specimens [38].

Alternative DNA amplification methods also derive specificity from the hybridization of DNA probes with specific target DNA sequences. The ligase chain reaction (LCR) technology utilizes the specificity of the DNA target template and probe to establish a specific ligation [39, 40]. One oligonucleotide matches a specific mutated [variant] base at its 3' end, and a second oligonucleotide probe matches the adjacent sequence on the same strand of target DNA. Ligation of these sequences will only occur if the specific mutated (variant) base is present. A second pair of oligonucleotide probes matching the opposite strand is included and enables exponential amplification of the ligated target sequence. Multiple cycles of denaturation, renaturation, and ligation with a thermostable DNA ligase enabled detection within 30 minutes [40]. This method is faster than PCR because no polymerase extension is required. Its usefulness has been demonstrated in direct mutation detection of the β-globin gene of sickle cell anemia patients [40]. This technique offers increased speed and comparable sensitivity relative to PCR [40].

RNA replicase, or transcription-based systems, amplify RNA templates and offer further alternatives to DNA amplification. The single-stranded coliphage Qβ, encodes an RNA replicase that recognizes specific RNA replication sequences with defined secondary structures (e.g., MDV-1 template RNA) [41]. Recombinant RNA molecules may be constructed that contain the essential RNA replication sequences and adjacent probe sequences that do not disturb the RNA secondary structure required for replication. These RNA probes hybridize with specific RNA or DNA targets, and the probes are amplified exponentially following the addition of Qβ replicase [42]. In contrast to PCR, the Qβ probe molecule itself is replicated exponentially as single-stranded RNA (signal amplification). Alternative RNA amplification systems include transcription-based amplification methods that utilize specific promoters and corresponding RNA polymerases to amplify nucleic acid targets. One technique, 3SR amplification [43], employs three enzymes, reverse transcriptase, RNaseH, and T7 RNA polymerase. Both oligonucleotide primers contain T7 promoter sequences that permit alternating cycles of cDNA formation by reverse transcriptase and amplification by T7 polymerase at a single temperature (isothermal reaction). The 3SR method has been used successfully to detect HIV RNA in human plasma samples [44].

8.2.3. Dual Hybridization/Amplification Methods

Conventional nucleic acid hybridization and amplification techniques may be combined in a single diagnostic application. Two approaches that combine conventional DNA–DNA hybridization and DNA amplification techniques are described below.

Rate-limiting steps for PCR-based diagnostics include manipulation of individual tubes and analysis of PCR reactions. Development of the Perkin-Elmer thermal cycler 9600 permits the use of 96 well microtiter plates. PCR-based DNA amplification with biotinylated primers, and subsequent annealing to capture probes covalently bound to well surfaces in the microtiter plates, enables specific target DNA duplex detection. Sequential DNA amplification and hybridization significantly reduce the complexity of the template DNA containing the target. Reduced target complexity decreases the chances of nonspecific hybridization in the final step. Subsequent addition of avidin-linked horseradish peroxidase permits rapid colorimetric detection of individual wells. This combination of DNA amplification and hybridization has been used for the detection of infectious pathogens [45–47].

The multiple potential alleles present at specific genetic loci provide additional challenges for nucleic acid probe technology and automation [48]. If 20 different mutations in a single oncogene need to be screened simultaneously in cancer patients, separate reactions may be necessary that utilize different mutant probes in each sample. However, the "reverse" dot blot [49] technique enables many different ASO probes to be immobilized on a single membrane filter. Human DNA samples may be amplified first and then applied to this ASO filter. A single hybridization reaction following DNA amplification permits screening of many different mutations simultaneously at reduced cost. With robotics automation, multiplex PCR analysis of the cystic fibrosis gene, CFTR, and ASO hybridization was used to screen 24 mutations simultaneously in 10 exons per assay [50].

8.3. INFECTIOUS DISEASES

8.3.1. General Overview

Diagnosis of infectious diseases depends on either detection of the pathogen or assay of the human immune response to the pathogen. Limitations with current techniques include (1) delayed results secondary to cultivation requirements, (2) an inability to distinguish closely related but different microbes, (3) an inability to detect previously unknown pathogens, (4) cross-reactivity and other false-positive results, and (5) sensitivity (false negatives).

Direct detection of the pathogen usually requires cultivation of the microorganism. The current exception to the cultivation requirement is microscopic smear analysis. Culture-based methods represent the basis for biochemical (enzyme) and antibiotic resistance profiles. Microbial culture in selective media historically signified the most sensitive detection method available since only a single organism must be present in the sample to yield growth. An important consideration is the length of time required for an organism to grow sufficiently in available media. Organisms with short doubling times in commercially available cultivation systems make prompt detection possible. A disadvantage of culture-based methods concerns organisms that either grow slowly or fail to grow with conventional media formulations. Many bacterial species require specialized conditions such as low oxygen tension or completely anaerobic environments. Latent viruses represent nonreplicative viruses in human cells; culture-based methods fail to diagnose these diseases unless the virus can be induced to replicate. Laboratory cultivation also introduces hazards for technical personnel.

Another current diagnostic limitation is the inability to distinguish closely related, but different, microorganisms. One specific organism from a group of related (though different) species may cause human disease. The ability to distinguish related microorganisms has primary importance in epidemiological analysis. Cell morphology was one of two primary criteria used to group the gastric pathogen *Helicobacter pylori* [51]. This visual method of identification is potentially misleading since cell morphology of distinct bacterial species may be similar. Even though they all possess a peculiar spiral shape, *H. pylori* appears to represent an extremely heterogeneous group of organisms [52,53]. *Borrelia burgdorferi*, a spirochete, had also been defined by morphology and limited serology. However, *B. burgdorferi* now has been resolved into three separate genospecies that are responsible for different clinical manifestations [54].

Present detection methods lack the ability to characterize previously unknown pathogens promptly. Molecular probes proved their utility in the rapid detection and characterization of previously undescribed pathogens. Microorganisms with a peculiar morphology were detected by smear analysis of patients with a common symptomatology [55,56]. DNA probes complementary to conserved regions of eubacterial 16S rDNA were used to amplify DNA segments by PCR directly from clinical samples. Sequencing of these 16S rDNA PCR products revealed a eubacterium related to the genus *Rickettsia* [56,57]. This new organism, closely related to the species *Rochalimaea quintana* [57], could be studied more rapidly by using the same cultivation conditions and antibiotics as were previously utilized with closely related organisms. Another recent example is the isolation of the gram-positive actinomycete bacterium associated with Whipple disease by direct PCR amplification of a 16S rDNA target sequence [58]. Whipple disease represented an established diagnosis for 85 years and had been associated for many years with an uncultivatable bacillus. Oligonucleotide primer–mediated DNA amplification and sequencing finally enabled this bacterium to be unambiguously identified as a new species *Trophyrema whippelii* [58]. *T. whippelii*–specific oligonucleotide primers could subsequently identify the cause of Whipple's disease by molecular analysis of peripheral blood without biopsy [59]. This example stresses the importance of rational microbial classification schemes based on the highly conserved 16S and 5S rRNA genes [60–62].

Antigen cross-reactivity refers to the recognition of multiple antigens with related structural epitopes by a single antibody. Common epitopes in different proteins may limit the specificity of antibody-based detection methods. For example, false-positive HIV test results have been obtained from potential blood donors who were vaccinated for influenzae virus [63]. A false-positive HIV ELISA result and indeterminant Western blot were also associated with recent rabies vaccination [64]. Cross-hybridization, the recognition of multiple DNA molecules by a single DNA probe, is due to the sharing of common primary DNA sequence motifs and may present a challenge for DNA probe technology. Recently a 20mer oligonucleotide probe failed to distinguish two clinically important mycobacterial species, *Mycobacterium tuberculosis* and *Mycobacterium intracellulare* [65]. DNA sequencing of this conserved intragenic target revealed only two mismatches in the 20 base pair probe–target DNA duplex. These two differences were localized at one end of the probe–target duplex and failed to destabilize the mismatched duplex sufficiently. However, new target DNA sequences can be obtained relatively rapidly. Modification or replacement of DNA probes is generally easier than replacement of specific antibodies.

Passive transfer of maternal antibodies to neonates across the placenta represents a significant limitation of antibody-based detection methods for newborn diagnosis. Currently used serologic tests, such as the rapid plasma reagin (RPR) test, for congenital syphilis are insensitive and nonspecific [66]. Elevated false-positive rates of serologic tests are largely due to the passive transfer of polyclonal antibodies from mother to neonate. Direct detection of the pathogen by a DNA probe assay would eliminate this problem. PCR-based assays that directly detect *Treponema pallidum* in amniotic fluid [67], sera [67], whole blood [68], and cerebrospinal fluid [69,70] have been developed. PCR was significantly more sensitive than IgM immunoblotting for detection of *T. pallidum* in CSF samples from symptomatic neonates, whereas IgM immunoblotting was more effective at detecting infection in asymptomatic infants [71]. Passive maternal antibody transfer also limits the utility of HIV test results during the first year of life (see below).

Sensitivity of detection remains a paramount issue with clinical samples. Failure to detect a pathogen may prevent effective treatment. Historically, direct pathogen culture represents the most sensitive method available for detection. However, many potential microbial pathogens are currently unable to be cultured and hence are unknown. One obvious alternative to culture methods is the Gram stain and cell morphology inspection of bacteria in a microscopic smear. However, the poor sensitivities of smear techniques are a significant disadvantage. Approximately 10^5 organisms per milliliter are required for visibility on a Gram stain. An exception is blood smear analysis for *Plasmodium* parasites in the diagnosis of malaria. In this case smear analysis remained the sensitivity standard even after the advent of molecular probes [72,73]. Therefore, relative sensitivity differences depend on the mode of analysis and on the specific disease in question. With respect to malaria, the responsible microorganisms replicate efficiently and undergo significant amplification in abundant red blood cells prior to overt symptomatology, thus facilitating smear analysis. DNA amplification techniques permit sensitive and specific detection of microbial pathogens without the absolute need for cultivation.

8.3.2. Bacteriology

Most current methods of bacterial identification require cultivation of the organism of interest. Examples of techniques for characterizing bacteria include antibiotic resistance profiling, biochemical testing, and growth in selective media. Organisms with rapid growth rates may yield positive results in 12–24 hours. However, many bacterial pathogens are fastidious and fail to grow in standard media formulations.

Fastidious bacterial pathogens, and organisms with extended doubling times, represent natural targets for molecular diagnostic techniques. Organisms of the *Mycobacterium tuberculosis* complex are fastidious with respect to their nutritional requirements and require extended periods for cultivation. Due to significant increases in the numbers of immunocompromised patients, mycobacterial infections have re-emerged with increasing frequency and severity. Outbreaks of *M. tuberculosis* result in serious consequences for HIV-infected populations [74–76]. Multidrug-resistant *M. tuberculosis* increased in incidence during the past several years and poses serious public health risks [77, 78]. Multidrug-resistant mycobacteria are also affecting the nonimmunocompromised population, and this may reflect the fact that immunocompromised individuals can act as a reservoir for such organisms. Traditional cultivation methods for mycobacteria require several weeks [79]. Radiometric methods (BACTEC) for small vial cultures decreased *M. tuberculosis* detection times from 19.4 days to 8 days [80]. Disease dissemination may occur during these cultivation periods. Specific DNA sequences have been used as olgionucleotide primer-binding sites to amplify DNA fragments from samples containing *M. tuberculosis* [37, 81–84]. Rapid sonication-based lysis of mycobacterial cells [85] and DNA amplification methods could conceivably allow same-day detection. The sensitivity of PCR amplification of *M. tuberculosis* DNA sequences from sputum smears surpassed that of Ziehl-Neelsen–stained smears in one study [86]. While mycobacterial amplification methods remain in development, commercial DNA probes (Gen-Probe) based on 16S rDNA sequences are currently used by clinical laboratories in hybridization methods. These probes distinguish two important species, *M. avium* and *M. intracellulare*, which were difficult to distinguish by previous microbiological methods [87,88]. DNA probes were combined with the BACTEC culture system for rapid, specific diagnosis of *M. tuberculosis* and *M. avium* complex organisms [89,90]. A method analogous to reverse dot blots (see discussion of methods), microdilution plate hybridization, was used to identify 22 different mycobacterial species from clinical samples [91]. Microtiter plates were coated with different, immobilized species-species DNA probes in each well so multiple species were screened on individual plates. Alternatively, direct PCR amplification followed by restriction digestion allowed mycobacterial species identification without hybridization [92].

Obligate intracellular bacterial pathogens require a eukaryotic cell host for reproduction and survival. Intracellular pathogens require elaborate mammalian cell culture systems for propagation in the laboratory. Molecular diagnostic techniques that obviate the need for such elaborate cell culture systems provide enormous benefits of speed and convenience. Examples of intracellular bacterial pathogens include the genera *Chlamydia, Ehrlichia*, and *Rickettsia. Chlamydia trachomatis* represents the most common sexually transmitted bacterial pathogen in the United States, and culture techniques require at least 48 hours for detection [93]. Rapid diagnostic methods could likely decrease transmission rates by enabling antibiotic treatment before repeated sexual encounters of affected individuals. DNA sequences complementary to genes for cell surface proteins and DNA sequences from common plasmids have been used as oligonucleotide primer-binding sites in PCR-based detection of the genitourinary pathogen *C. trachomatis* [94–96]. Colorimetric detection methods have been developed for *Chlamydia* detection. Biotinylated primers containing the *lac* operator are used in PCR amplifications, and nested PCR products are immobilized on streptavidin-coated magnetic beads. *Chlamydia*-specific PCR products are detected in solution after addition of a β-galactosidase/*lac* repressor fusion protein [97]. An alternative solid phase–based, colorimetric detection system was developed. Biotinylated PCR products anneal to immobilized species-specific oligonucleotide probes in microtiter wells. Streptavidin conjugated with horseradish peroxidase was added prior to colorimetric detection in solution. This test utilizes the common cryptic plasmid of *C. trachomatis* as the DNA target and was used to diagnose *C. trachomatis* urethritis in men [98] and endocervicitis in women [99]. Interestingly, the sensitivity of PCR as compared with culture was greater with male urethral swabs [98] than with endocervical swabs [99], and results were obtained much more rapidly (same day) with PCR. In another study, PCR was shown to be more sensitive for *C. trachomatis* detection in male and female genital specimens than the rapid enzyme immunoassay (Chlamydiazyme; Abbott Laboratories, Abbott Park, IL) [100].

Acute bacterial meningitis may result in severe morbidity and death. Previously healthy children can die from meningitis resulting from an infection with *Neisseria meningitidis* in 48–72 hours. Rapid diagnostic techniques that circumvent the need for cultivation and directly assay the presence of bacteria in CSF samples could help reduce potential fatalities. Oligonucleotide primers matching *N. meningitidis*–specific sequences may allow rapid diagnosis and treatment of infected patients [101]. DNA probes that match a universal bacterial DNA sequence (e.g., conserved 16S rRNA sequences) [102] would permit the detection of any bacterial species in otherwise sterile CSF. Hospital treatment may be avoided for children lacking bacterial meningitis (presumably viral meningitis). PCR-based amplification of specific DNA sequences from *B. burgdorferi* (the causative agent of Lyme disease) detected several examples of asymptomatic central nervous system (CNS) infection [103]. Antibiotic treatment could be initiated in these patients prior to onset of meningitis.

8.3.3. Virology

Multiple opportunities are available for the application of DNA probes for the detection of human viral diseases. Like the obligate intracellular bacteria, human viruses require human host cells and necessitate mammalian cell culture systems for propagation *in vitro*. Human viruses may integrate into human genomic DNA. Viruses may be latent and require nucleic acid–based probes for detection within clinical samples.

A dramatic story in infectious diseases was the emergence of the human immunodeficiency virus (HIV). Following the discovery of HIV as the causative agent of AIDS in 1983 [104], much effort has been expended on detection of this lethal virus in random samples from the human population and tissue and blood banks [105]. Currently, the primary HIV screening test is the assay of HIV-specific antibodies in human serum [106]. Typically a "window period" of 6–8 weeks transpires before antibody detection [105]. Occult HIV infections have been detected by PCR with primers matching HIV *gag* sequences in seronegative patients who were also antigen negative [107]. HIV transmission from a seronegative organ donor to several recipients was reported [108]. Cross-reactivity problems with antibody-based assays caused significantly high false-positive rates, especially in populations where the HIV prevalence was low [109]. Rapid solid phase–based microtiter plate DNA probe assays have been combined with PCR amplification to rival the speed and screening potential of current enzyme-linked immunosorbent assays (ELISAs used for HIV antibody detection) [45,110,111] (see discussion of bacteriology). As with *Chlamydia*, biotinylated HIV-specific PCR amplification products were hybridized with immobilized DNA probes in microtiter wells. Streptavidin–enzyme conjugates enabled colorimetric detection in solution.

Decrease of immune function in AIDS patients impedes the monitoring of viral load during antiviral treatment by antibody detection. To monitor the levels of transcriptionally active HIV particles, competitive RT-PCR was introduced to monitor HIV load [112–113; for review, see ref. 114]. Unlike previous competitive PCR techniques, competition of PCR product yield occurred in both the reverse transcription and DNA amplification steps. Template DNAs were similar in size (115 vs. 97 bp) and contained identical primer-binding sites. One template is present in the PCR reactions at a defined concentration and permits quantition of HIV RNA in clinical samples [113]. Quantitative crT/PCR assays may be useful in predicting vertical HIV transmission from mother to infant, since high viral loads have been correlated with neonatal infection [115]. Zidovudine (AZT) therapy may be important in pregnant patients with significant HIV loads, since a recent study demonstrated reduction of vertical transmission with treatment during pregnancy [116].

Infants born to HIV-infected mothers possess maternal HIV antibodies that mask their own HIV status and limit the utility of antibody detection in neonates. Alternatively, p24 antigen detection lacks sensitivity compared with viral culture and PCR-based detection [117,118]. The sensitivity of HIV PCR assays exceeded that of HIV cultures and serum IgA in infants less than 6 months of age [119]. False-positive results in the first 2 months of life limit the utility of antigen detection [117]. Oligonucleotide primers matching conserved regions of HIV *gag* and *env* genes have been useful for HIV diagnosis during the first year of life [120–125]. In one study nine infants of HIV seropositive mothers were diagnosed by PCR within the first 38 days [125]. Eight of nine HIV-infected neonates were PCR positive within 48 hours of birth, while 30 of 32 HIV-infected infants were PCR positive by 62 days of age [126]. Infants of HIV seropositive mothers can benefit from the application of DNA probes for disease detection and prompt institution of treatment.

Design of PCR-based tests requires consideration of genetic variation in viral isolates from different geographical regions. HIV replicates its RNA genome by reverse transcription, which lacks the fidelity of DNA polymerase–mediated replication. Oligonucleotide primers matching two sequences, used first in France, failed to amplify several HIV isolates from Africa [127]. Lack of HIV DNA amplification occurred because DNA polymorphisms arose within the primer-binding sites.

Human herpesviruses represent significant causes of sexually transmitted, congenital, and opportunistic viral infections [128,129]. The primary method for herpes simplex virus detection is viral culture because serologies are often uninformative [128,129]. However, many cases of genital herpes are missed by viral culture, leading to a high false-negative rate [130]. CMV, a herpesvirus family member, is more difficult to detect by viral culture because longer generation times delay diagnosis. Typically viral cultures require 2 weeks from time of sampling to time of reporting to the clinician [129].

Molecular diagnostic methods permit the timely assessment of CMV viral load in a single day. PCR has been useful for CMV detection in primary and reactivated disease [131; for review, see ref. 132]. PCR-based amplification of CMV DNA in post-transplantation patients permitted diagnosis prior to antigen detection [133], viral culture [134], or cytology [134]. In liver transplant recipients, CMV DNA was detected by PCR amplification on average of 13 days prior to the onset of CMV disease symptoms [135]. For CMV pneumonitis, PCR amplification of CMV DNA in bronchoalveolar lavage (lung) specimens does not provide more information for the clinician than viral culture. Both methods have specificity and sensitivity values approaching 100% [134]. PCR-based methods provide similar amounts of information much faster than viral culture. Instead of 2 weeks for viral culture, single-day assessment is possible by PCR [134]. Is PCR too sensitive to provide clinically useful information? Studies with CMV detection demonstrated that PCR-based methods do provide positive results in cases where disease is absent [133]. However, viral culture yields a similar number of positive results in the absence of disease [136]. As described previously, additional complications are the relative difficulties of PCR amplification with patient samples from different body sources. The sensitivity and specificity of PCR varied significantly depending on the source, with blood specimens being troublesome compared with urine and bronchoalveolar lavage fluid [137].

DNA probe–based methods that allow the assessment of viral load during antiviral therapy provide important advantages. Immune responses to CMV are delayed, and serological studies are useful only retrospectively. Clinical specimens contain significant levels of antiviral compounds during therapy. Viral cultures during therapy result in false-negative data because the antiviral drugs in the clinical specimen presumably inhibit active viral replication in culture [138,139]. DNA probes used in PCR applications are not susceptible to this inhibition and represent useful approaches for monitoring viral load during therapy [138,139]. Hepatitis C virus (HCV) infection is also an important issue for DNA probe detection since cDNA PCR-positive patients are often seronegative with second generation ELISAs [140]. By RT-PCR, HCV RNA levels have been correlated with clinical response and relapse during interferon-α therapy [141], and greater serum HCV RNA levels correspond to liver disease severity [142].

Human papillomaviruses (HPV) can cause benign skin and genital warts [143]. These viruses have been linked to malignant diseases of the genital tract, namely, squamous cell carcinoma of the cervix [144,145] and penile carcinoma [146,147]. HPV types 6 and 11 are found predominantly in benign epithelial lesions [143]. In contrast, HPV types 16 and 18 are associated with malignant carcinomas [143,145, 148, 149]. PCR amplification with primers matching conserved open reading frames was followed by hybridization with variable type-specific DNA probes [149]. Alternatively, conserved sequence primers can be used to amplify HPV DNA fragments that are digested by restriction enzymes to produce type-specific DNA fingerprint patterns [150]. Restriction site polymorphisms vary between HPV types, and differently sized digestion products are created. DNA probe technology provides HPV type resolution unmatched by other methods. Cell culture cannot be used to cultivate or classify intact HPV particles [143]. Monoclonal antibodies are only available for a minority of documented HPV types [143] and fail to detect any virus in the majority of biopsies. As the grade of the lesion becomes more severe, intact viruses are not produced and antibodies are unable to detect HPV antigens [143]. The irony is that, as the severity of disease increases, the likelihood of viral antigen detection decreases. DNA hybridization techniques do not rely on the presence of intact viral particles but instead on the presence of viral genomes. Detection of specific HPV types by DNA probe *in situ* hybridization to "punch" skin biopsy specimens can provide important prognostic information for patients [151]. In a recent study, PCR of preneoplastic cervical intraepithelial lesions was more sensitive than either immunohistochemistry or *in situ* hybridization, detecting HPV DNA in 71% and 86% of low-grade and high-grade lesions, respectively [152].

8.3.4. Mycology and Parasitology

Accurate and rapid detection of opportunistic pathogens has become more important with the increased numbers of immunocompromised patients. The advent of HIV prevalence, increasing numbers of invasive medical procedures, and rising numbers of elderly patients and premature infants have resulted in a larger immunocompromised patient population. Opportunistic pathogens have become significant causes of morbidity and mortality in such patients. Fungal and protozoal pathogens signify important opportunistic pathogens.

The most common opportunistic fungal pathogens are the single-celled yeasts of the genus *Candida*. *C. albicans* represents the most commonly isolated yeast from clinical specimens [153]. Immunocompromised individuals are especially susceptible to systemic *Candida* infections. In contrast to filamentous fungi (e.g., *Aspergillus*), yeasts are susceptible to antifungal agents, and effective treatment follows prompt identification. Blood cultures from patients with proven disseminated candidiasis are often negative [154]. Antibody detection represents a limited strategy in the immunocompromised patient population. *C. albicans*–specific DNA probes have been used for colony hybridization [155], and species-specific *Candida* oligonucleotide probes have been used to distinguish four different species by hybridization [156]. Species-specific PCR-based detection, based on mitochondrial [157] and ribosomal DNA sequences [158], has been developed for *C. albicans*.

Malaria represents one of the most prevalent infectious diseases in the world [159]. Various species of the protozoal genus *Plasmodium* cause malaria. Diagnosis of malaria is currently performed by morphological analysis of blood smears [159]. Since *Plasmodium* infects the most numerous cellular component of blood, erythrocytes, smear analysis remains the sensitivity standard [72]. However, in the early stages of disease it is difficult to distinguish *Plasmodium falciparum* and *Plasmodium vivax* by morphology [159]. Serology is not a useful alternative because antibody titers rise several weeks after acute illness [159]. Since sensitive detection methods are less important in malarial diagnosis, DNA hybridization probes should be quite useful for speciation with blood samples obtained from acutely ill patients. Species-specific DNA probes for *P. falciparum* [73, 160] and *P. vivax* [161] demonstrated efficacy in separate trials.

8.3.5. DNA Probes and Molecular Epidemiology

Epidemiological investigations of infectious disease outbreaks require the distinction of closely related strains within a single bacterial or viral species [162, 163]. Historically, methods based on phenotypic detection such as serotyping (serology), phage susceptibility typing, and biotyping (biochemical testing) have been used to discriminate clinical bacterial isolates [164].

Previous methods were based on phenotypic characters. A limitation of phenotypic trait-based testing results from an absolute dependence on gene expression. Gene expression may vary with storage and cultivation conditions. Even with standardized conditions, only 57.7% of *Enterobacter cloacae* clinical isolates could be assigned to a specific serotype [165]. Another study revealed that seven of eight ciprofloxacin-resistant *Staphylococcus aureus* isolates were nontypeable by phage susceptibility [166]. In contrast, these *S. aureus* isolates were typeable by rRNA gene probes. Less robust molecular methods such as plasmid profiling also present significant limitations. In a study of *Helicobacter pylori*, only 51% of isolates contained plasmids [53]. Plasmids represent nonessential, autonomously replicating, extrachromosomal elements. Plasmids are often freely exchanged between strains in nature by conjugation [167]. Therefore chromosomal DNA targets represent a preferential target for DNA probe–based detection [163].

Initially, randomly cloned DNA probes were used for typing bacterial species. Randomly cloned, radioisotopically labeled DNA probes reproducibly distinguished different isolates of *Salmonella* [168]. Genomic DNA was digested with restriction enzymes and hybridized with random probes after Southern transfer [21]. Following autoradiography specific DNA band patterns or "fingerprints" distinguished different strains [168]. A similar approach utilized a pool of chromosomal DNA probes to distinguish isolates of *M. tuberculosis* [169].

The utility of a DNA probe depends on its primary sequence, which may vary significantly within a single gene. Specific single-copy T6 gene probes were hybridized against digested genomic DNA of streptococcal species [170]. Probes matching different regions of the T6 gene recognized different strains previously characterized by serotyping [170]. The T6 probe matching the amino terminus recognized 10 of 25 serotypes, whereas another probe matching the carboxy terminus only recognized 3 of 25 serotypes. The number of different strains recognized by a specific probe correlates directly with the extent of conservation of the target DNA sequence, or complementarity to probe DNA sequence, in different isolates. Alternatively, PCR-based DNA amplification of a single-copy gene product was followed by restriction digestion and agarose gel electrophoresis for *C. trachomatis* typing [171]. Differences revealed by hybridization and amplification methods may reflect the extent of restriction site polymorphisms in a chromosomal DNA region.

Repeated DNA sequences provide useful DNA probe targets for epidemiological analysis [163]. By definition, repeated DNA sequences are conserved and present in multiple copies in the genome. Microorganisms contain several repetitive DNA sequences in their genomes [172]. In contrast to many randomly cloned probes, these sequences are sufficiently conserved in various organisms. Their multiple occurrences in each genome permit simultaneous scanning for DNA polymorphisms at different genomic loci. Ribotyping refers to the application of ribosomal RNA (rRNA) gene probes for DNA fingerprinting [173,174]. rRNA genes are clustered in operons that are repeated seven times in *Escherichia coli* [175]. Natural abundance of rRNA in *E. coli* cells permits easy access to nucleic acid probes. End-labeled rRNA probes annealed to digested *Hemophilus influenzae* biogroup *aegyptius* DNA and produced DNA fingerprints following autoradiography [176]. These DNA fingerprints confirmed the clonal nature of the causative agent of Brazilian purpuric fever [176,177]. Ribotyping when combined with pulsed field gel electrophoresis (PFGE) also helped confirm that cystic fibrosis patients were persistently infected with single strains of *Pseudomonas cepacia* [178].

DNA probes matching a specific low-copy repetitive insertion sequence (IS) element, IS*986* (1,355 bp in length), distinguished strains of *M. tuberculosis* with greater reproducibility and resolving power than was previously possible [179,180]. These techniques employed the use of chemiluminescent DNA probes in hybridization methods with digested chromosomal DNA [179,181,182]. IS*986*-based mycobacterial typing confirmed the source identity of a recent *M. tuberculosis* outbreak in a California residence for patients with AIDS [76]. Another repetitive DNA sequence, Ca3, has been utilized in a similar approach to type the fungal pathogen *C. albicans* [183,184].

Distance variations between repeated DNA elements in bacteria have been exploited to produce specific DNA fingerprints [185; for reviews, see refs. 186,187]. PCR amplification between repeated DNA sequences yields DNA fragments of various sizes depending on the intervening distances. PCR products can be size-fractionated by agarose gel electrophoresis to produce specific DNA fingerprints [185]. Transfer RNA [tRNA] genes are conserved and separated by various distances. Oligonucleotide primers matching tDNA genes amplified DNA fragments between tDNA genes [188]. tDNA-PCR produced single bands which vary in size depending on the distances separating conserved tDNA sequences. tDNA-PCR distinguishes streptococcal [189] and staphylococcal [190] species but failed to distinguish strains within a species [189,190].

The high copy number, extragenic repetitive DNA sequences REP and ERIC (IRU) are conserved in many bacterial species [185]. These sequences are dispersed in the genomes of *E. coli* and *Salmonella typhimurium* [191–194]. REP-like and ERIC-like sequences are located in different positions and separated by various distances dependent on the particular bacterial strain and species [191–194]. PCR amplification between oligonucleotide primers matching these repeated DNA elements (rep-PCR) yielded strain-specific DNA fingerprints that distinguished pathogenic clones of a single species [185]. Fingerprinting bacterial strains by rep-PCR has been useful in multiple applications [for reviews, see refs. 186, 187], including strain identification [195], typing soil bacteria [196], epidemiologic analyses of bacterial pathogens [197–199], and analysis of the clonal nature of penicillin-resistant *Streptococcus pneumoniae* [200]. A similar approach has been applied to fingerprinting cytomegalovirus. CMV contains a genomic region, the L–S junction, with interspersed DNA repeats. Oligonucleotide primers matching these conserved repeats amplified differently sized DNA fragments to produce strain-specific DNA fingerprints of CMV isolates [201]. Epidemiologically linked CMV isolates yielded similar DNA fingerprints [201]. Rapid rep-PCR typing methods have been developed that bypass genomic DNA isolation and permit the use of cell suspensions [202]. The application of fluorescent oligonucleotide primers in fluorophore-enhanced rep-PCR (FERP) [203] enables automation of DNA fingerprinting applications of microbial pathogens with computer-aided data analysis.

Two PCR-based DNA fingerprinting alternatives include direct variable repeat PCR (DVR-PCR) and randomly amplified polymorphic DNA (RAPD) [204] or arbitrary priming PCR (AP-PCR) [205]. DVR-PCR [206] utilizes direct repeat clusters that vary with respect to adjacent inter-repeat sequences to generate PCR patterns that distinguish *M. tuberculosis* strains. RAPD [for review, see ref. 207] or AP-PCR, does not require sequence information as oligonucleotide primers of random sequence are used to establish DNA fingerprints of different microbial pathogens.

8.3.6. DNA Probes and Antibiotic Resistance

Nucleic acid probes may detect antibiotic resistance genes directly. Conventional antibiotic susceptibility tests are conducted in agar or broth media and rely on the ability of bacteria to grow and multiply in the presence of the antibiotic. The principal advantage of these conventional tests is that the assay detects the relevant phenotype directly. Clinicians are mainly interested in whether the pathogen is susceptible or resistant. The precise mechanism of antibiotic resistance is irrelevant to the physician. However, conventional antibiotic susceptibility tests are dependent on gene expression, are labor intensive, and require cultivation of the organism. An important question to consider also is: how will one detect antibiotic resistance in organisms that cannot be currently cultured?

Methicillin-resistant *Staphylococcus aureus* (MRSA) emerged as a significant public health problem in the 1980s [208]. Few therapeutic alternatives exist for patients infected with these organisms. Methicillin resistance in staphylococcal species is due to the presence of a specific gene, *mecA* [209,210], in resistant isolates. This gene encodes a penicillin-binding protein [211] and is present in almost all MRSA isolates studied. *mecA* is not present in susceptible staphylococcal isolates. The *mecA* gene was cloned [211] and sequenced [212], and recently PCR-based detection of this gene correlated strongly with conventional antibiotic susceptibility results [209,210]. Forty-two of 46 *mecA*-positive staphylococcal isolates were methicillin-resistant *S. aureus* or *S. epidermidis* [209]. Another study utilized PCR-based detection of *mecA* in staphylococci, and 31 of 33 resistant isolates were *mecA* positive [210]. Both studies used methodologies whereby bacterial cells were lysed directly and results were obtained in a single day. Other antibiotic resistance determinants on plasmids, transposons, or chromosomes could potentially be identified in a similar manner.

In contrast to gene detection, mutations within specific genes have also been correlated with antimicrobial resistance in bacteria. The majority of missense mutations conferring rifampicin resistance in *M. tuberculosis* were confined to a restricted region of the *rpoB* gene encoding the β-subunit of RNA polymerase [213,214, 215]. A recent study found that greater than 90% of rifampicin-resistant strains have missense mutations within a 69 bp region of the *rpoB* gene. Additionally, single base substitutions within a single-stranded loop of 23S rRNA were associated with clarithromycin resistance in *M. intracellulare* [216].

8.4. ONCOLOGY

8.4.1. Overview

Medical oncology presents many opportunities for the application of DNA probes in clinical medicine. Traditionally, human cancers were diagnosed and evaluated by histopathological examination of biopsy material. Molecular probes could be used to diagnose specific tumors and establish cell lineages of specific malignancies. DNA probes may also permit tumor staging or grading based on the correlation of specific molecular markers with grades of the tumor. DNA amplification methods could be combined with cytology to establish specific diagnoses [217]. As invasive biopsies become more difficult to justify in a cost-conscious clinical setting, direct

DNA amplification of cytology and body fluid samples will be increasingly important. DNA probes may be used directly with biopsy material for *in situ* hybridization techniques to extract both molecular and morphological data from a single specimen [218]. Glass slides prepared for histological analysis could subsequently be used for PCR-based mutation detection [219].

8.4.2. Molecular Cytogenetics

Molecular cytogenetics refers to the application of DNA probes to chromosomal analysis. Applications of specific DNA probes to cytogenetic analysis may overcome limitations of levels of resolution using conventional banding techniques. Conventional cytogenetics refers to the examination of chromosomal karyotypes by the use of special dyes and light microscopy (e.g., Q-banding and G-banding). Banding patterns provided by these dyes are distinctive for each chromosome, but resolution limits could obscure the diagnosis with certain chromosomal anomalies. Conventional cytogenetics may fail to detect submicroscopic deletions that may be associated with recognizable clinical syndromes [220]. Several examples of submicroscopic DNA rearrangements, including deletions [221] and duplications, [222, 223] have been demonstrated to cause human disease. Several microdeletions have been correlated with specific human malignancies [224]. Both balanced and unbalanced chromosomal translocations have been associated with human malignancies [224]. Cryptic translocations are invisible by conventional cytogenetics and may be caused by the exchange of similar, telomeric bands [225]. DNA probes are essential for visualization of cryptic translocations [225]. A limitation of conventional cytogenetics includes the inability to identify certain derivative or marker chromosomes. Human malignant cells often possess additional marker chromosomes with unknown origins [224]. Information on marker chromosome origins may enable precise characterization of cancer cell karyotypes.

Ready availability of blood samples and the ability to culture hemopoietic malignancies have permitted extensive cytogenetic analysis of human leukemias and lymphomas. A limitation of conventional cancer cytogenetics is the requirement for tissue culture cultivation prior to karyotype preparation. Tissue culture methods may be associated with the introduction of artifactual chromosomal aberrations. Mean percentages of abnormal metaphase cells increased with time of tissue culture [226]. Also, many solid tumors remain uncharacterized by cytogenetics because they are refractory to tissue culture. The most prevalent human cancers, including breast, lung, and colon, are solid, epithelial tissue-derived (carcinoma) tumors. Cytogenetic diagnoses of such carcinomas might be possible for the first time. One notable example, prostate cancer, represents the second leading cause of cancer deaths of adult American males and yet lacks substantive cytogenetic data [227].

When added to conventional cytogenetics, DNA probes may extract additional information. Specific deletions, translocations, or marker chromosomes may be described. Tissue culture can be circumvented using DNA probes by *in situ* hybridization of interphase nuclei. Fluorescent DNA probes are used in such methods, and this technique is called fluorescent *in situ* hybridization (FISH) [228,229].

FISH requires the availability of specific DNA probes that recognize DNA targets correlated with disease. Alternatively, primed *in situ* labeling (PRINS) combines probe–chromosome annealing with probe extension by DNA polymerase with labeled nucleotides. PRINS enables sensitive multicolor detection of specific chromosomal sequences during metaphase *in situ* by combining DNA amplification with FISH [230]. Without a complete metaphase spread, as in conventional cytogenetics, information regarding the entire genome is absent from the diagnosis. Rapid increases in human genomic sequence information will make the application of DNA probes to cancer cytogenetics feasible without a reliance on complete karyotypes.

Chromosomal abnormalities may be random events during the multistep genetic process of tumor progression. Other chromosomal events correlate nonrandomly with specific human malignancies. DNA probes may be used to detect specific chromosomes or chromosomal regions in chromosomal "painting" techniques [228, 231–233]. This information enables recognition of specific translocations and marker chromosomes. DNA "painting" probes that match either repeated or single-copy DNA sequences may identify specific chromosomes. Mammalian repetitive DNA sequences, namely, *Alu* and L1, may be used as oligonucleotide primer-binding sites in the PCR to amplify multiple sequences from specific chromosomal regions [234,235]. DNA probes isolated by PCR amplification between *Alu* repetitive DNA sequences and human oncogenes revealed specific leukemia-associated deletions and translocations [236]. These "painting" probes identified specific human chromosomes in hybrid cell lines [234]. Repeated α-centromeric sequences and PCR-amplified chromosome-specific probes unambiguously identified specific human chromosomes and their translocation derivatives [237]. Chromosome-specific α-repeat DNA probes identified chromosomal ploidy abnormalities in interphase samples of solid tumors [238–240]. α-Centromeric repeat probes specific for chromosome 12 identified seven of seven cases by FISH of previously diagnosed chronic lymphocytic leukemia (CLL)–associated trisomy 12 [241]. Five additional cases of CLL-associated trisomy 12 were detected by "α-FISH" that were indeterminate by conventional cytogenetics [241]. Randomly cloned laser-microdissected chromosomal region-specific probes identified specific chromosomal regions and a balanced translocation derivative by FISH [242].

Specific chromosomal translocation events have been previously correlated with human malignancies. Translocation between chromosomes 9 and 22 is tightly correlated with chronic myelogenous leukemia (CML). The 9:22 translocation is present in greater than 90% of patients with CML [243]. This translocation has been characterized at the molecular level and occurs within a defined region in chromosome 22 called the breakpoint cluster region (*bcr*) [244]. The fusion product represents a chimeric gene comprised of *bcr* and the cellular oncogene c-*abl* [245]. Expression of the chimeric mRNA has been demonstrated, and this fusion product causes leukemia in transgenic mice [246]. Conventional cytogenetics identified a marker chromosome, the Philadelphia chromosome [247], which represents the product of the 9:22 translocation [248]. Significant numbers of CML patients and acute lymphoblastic leukemia (ALL) patients lack the Philadelphia chromosome [243]. DNA probes matching the *bcr* region were first used in Southern blots to detect the 9:22 translocation in both Philadelphia-positive and -negative CML patients [249]. *bcr-abl* DNA probes spanning the translocation were applied in *in situ* hybridization of interphase nuclei of CML patients [250]. Several patients who lacked the cytogenetically visible Philadelphia chromosome yielded positive hybridization signals by FISH [250]. FISH with *bcr* and *abl* probes was also useful in the detection of the 9:22 translocation in CML patients with minimal residual disease following bone marrow transplantation [251]. FISH results correlated well with both conventional cytogenetics and PCR amplification of the fusion gene [251].

Sensitive detection and characterization of fusion transcripts at chromosomal translocation breakpoints have facilitated the diagnosis of neoplastic disease. cDNA clones of the *MLL* gene were used to detect *MLL* gene rearrangements at 11q23 translocations in acute lymphoblastic and acute myeloid leukemias [252] by conventional DNA–DNA hybridization methods. *MLL* gene rearrangements identify patients with an especially poor prognosis and who require aggressive chemotherapy or bone marrow transplantation. RT-PCR detection of EWS hybrid transcripts at the t(11:22) translocation confirmed the diagnosis of a Ewing tumor subset comprised of Ewing's sarcomas and peripheral primitive ectodermal tumors [253]. PCR detection of the t(14:18) translocation associated with follicular lymphomas increased the sensitivity of detection to 10^{-6} normal cells, enabling a more thorough evaluation of minimal residual disease in these patients [254].

8.4.3. Oncogenes and Molecular Oncology

Specific point mutations, deletions, and/or DNA amplification events of oncogenes represent molecular events associated with human cancers. In 1976 Varmus and Bishop first established the connection between retroviral oncogenes and their cellular counterparts present in vertebrate genomes [255]. Several years later Weinberg and his colleagues transformed immortalized NIH3T3 cells with genomic DNA from a human bladder carcinoma cell line [256]. The transforming gene contained a point mutation in a cellular analogue of the viral oncogene, *v-ras* [257,258]. A point mutation in a specific gene caused mammalian cells to acquire a transformed phenotype.

The molecular etiology of neoplasia includes two basic categories. A mutated allele of an oncogene may represent a dominant, gain of function mutation leading to neoplastic transformation of a single somatic cell or multiple cells and the development of monoclonal or polyclonal malignancies, respectively. That is, a single allele with a specific mutation may be associated with human neoplasms. The second category consists of recessive mutations wherein a mutation occurs at a specific locus in a somatic cell in an individual who has inherited a germline mutation at this locus in the other chromosomal homologue. In this example human cancers are caused by the loss of the wild-type allele at both copies of a particular genetic locus. Nucleic acid probes can be applied to the detection of specific mutations associated with certain human tumors.

Examples of thoroughly studied dominant oncogenes in human cancers are the cellular *ras* genes, which were initially identified by their similarity to retroviral oncogenes of murine sarcoma viruses [257]. Two of these genes are named H-*ras* and K-*ras*, depending on the specific murine sarcoma virus of origin [259,260]. Mutations in the same three codons, 12, 13, and 61, in both H-*ras* and K-ras have been correlated frequently with human neoplasms [261–264]. Oligonucleotide probes matching specific mutations in both genes have been used to assess the prevalence of specific *ras* mutations in individual human cancers. Identical K-*ras* mutations are present in a variety of human tumors derived from different tissues [263,265,266]. No specific *ras* mutation has been correlated with a single type of tumor, but the prevalence of H-*ras* versus K-*ras* mutations vary in tumors from different tissues [262,264]. Another *ras* homologue, N-*ras*, figures prominently in leukemic disease [for review, see ref. 267]. Mutations in N-*ras* are commonly found in acute myeloid leukemia (AML), acute lymphoblastic leukemia (ALL), and myelodysplastic syndromes [267].

Point mutations in specific oncogenes can be associated with human tumor progression. ASO probes can be used to detect any of these mutations in biopsy or cytology samples from tumor tissues. Since there are only a limited number of point mutations in *ras* associated with human cancers, only a limited number of DNA probes are necessary to detect the prevalent mutations. Reverse dot blots may be used that contain various K-*ras* and H-*ras* ASO probes on a single filter, in addition to other oncogene probes. PCR amplification technology may be used to generate adequate amounts of *ras* DNA from tumor samples, and these amplification products may be used directly for hybridization to the reverse dot blot [268]. A dramatic example of the utility of mutant protooncogene allele screening has been demonstrated with *RET* and MEN-2A associated medullary thyroid carcinoma. Mutation detection within the *RET* gene proved superior to biochemical calcitonin tests for the assessment of thyroid neoplasia in MEN2A patients [269]. Two exons and multiple *RET* mutations were screened simultaneously by restriction digestion of PCR-amplified DNA.

The *neu/c-erbB2* oncogene in human breast cancer represents a distinct mechanism of dominant oncogenesis. Instead of point mutations, increase in gene copy number, or amplification, of the *neu* gene correlated with lower overall survival and shorter time to relapse in patients with breast carcinoma [270]. DNA amplification events were detected by Southern hybridization with *neu*-specific and control single copy gene probes [270]. The *neu* gene was amplified 2–20-fold in 30% of patients [270]. Similar results were obtained in a separate northern African study [271]. Significant levels of *neu* gene amplification were found in breast carcinomas, but not other cancers such as sarcomas [272]. N-*myc* represents another oncogene that is amplified and correlated with advanced stages of neuroblastoma by Southern hybridization [273]. In addition to Southern hybridization, differential PCR can detect such DNA amplification events with *neu*-specific and single-copy control oligonucleotide primers [274]. Quantitative PCR methods yield reproducible and useful information if proper controls are used.

The relevance of tumor suppressor genes to the development of human cancers has been demonstrated during the past decade. Loss of both copies of wild-type alleles leads to the deregulation of cellular growth and proliferation. Mutations and loss of the wild-type allele at the p53 locus on chromosome 17p are associated with different human cancers [275]. Loss of both wild-type p53 alleles in transgenic mice results in viable offspring predisposed to a variety of neoplasms [276]. These results confirm its role as a tumor suppressor gene *in vivo*. Sporadic mutations at the p53 locus are found in a variety of human neoplasms; namely, small cell lung cancer (100%) [277], colon cancer (70%) [277], breast cancer (30%–50%) [277–279], and uterine leiomyosarcomas [280]. Germline p53 mutations have been identified in families with established cancer predispositions [281–283]. Mutations in p53 have been correlated with tumor progression [284–286]. p53 mutations that have been associated with human cancer represent deletions and point mutations dispersed throughout several exons [279]. As with *ras*, ASO DNA probes matching different p53 mutations may be applied to a single filter and used to screen tumors for diagnostic and prognostic information. Another established tumor suppressor gene, the retinoblastoma (Rb) gene, was isolated by examination of interstitial deletions in chromosome 13q. Loss of both alleles by DNA rearrangements predisposes individuals to retinoblastomas and osteosarcomas [287], and can be detected by DNA probes [288].

Recently, loss of heterozygosity at multiple loci has been associated with colorectal and breast cancer progression. Mutations with resultant loss of function of different gene products point to a multistep pathway from wild type to frank malignancy and metastasis. Multiple targets for DNA probes may allow sophisticated prognostic evaluation dependent on particular mutations in specific genes. With respect to colorectal carcinoma, mutations in K-*ras*, p53, DCC, and APC (adenomatous polyposis coli) genes have been associated with tumor progression [for reviews, see refs. 289,290]. More recently, loss of heterozygosity and proposed loss of function at two genetic loci, hMSH2 [for review, see refs. 290,291] and hMLH1 [292], have been associated with hereditary nonpolyposis colorectal cancer (HNPCC). A protein assay based on mutations found within the APC gene facilitated the development of screening assays that detected APC lesions in 87% of familial adenomatous polyposis (FAP) patients [293]. Breast cancer has been associated with loss of heterozygosity at the BRCA1 [294] and BRCA2 loci [295]. As many as 30% of women with a diagnosis of breast cancer before the age of 45 years may carry mutations in these genes [296]. Screening for these mutations in women below the age of 50 years with positive family histories may identify individuals who require frequent breast evaluations and mammograms.

A specific gene, *nm23*, has been linked with a tumor's metastatic potential [297]. Metastasis refers to a cancer cell's ability to invade the bloodstream and migrate to different body locations. Secondary tumors arise by this process and contribute to morbidity and mortality in human cancer. Deletions at the *nm23* locus correlated prospectively with metastasis in patients with colorectal carcinoma [298]. Eight of 11 patients with *nm23* deletions versus 2 of 10 without such deletions developed distant metastases. A somatic homozygous *nm23* deletion was found in a lymph node metastasis of colorectal carcinoma [299]. This finding suggests that the *nm23* gene product may function as a metastasis suppressor gene. DNA probes matching *nm23* sequences may be used to determine prognosis of potentially metastatic tumors.

8.4.4. DNA Probes and Cancer Chemotherapy

Mammalian P-glycoprotein is associated with multidrug resistance to unrelated organic compounds in transformed cell lines [300]. Genes encoding P-glycoproteins, named *mdr* genes [301], have been utilized to evaluate resistance of tumors to cancer chemotherapeutic agents. Initial immunochemical studies indicated that P-glycoprotein expression was an important prognostic indicator for treatment failure in children with soft tissue sarcomas [302]. However, numerous applications with *mdr* gene probes failed to derive a substantial correlation between *mdr* gene amplification or expression and treatment failures [303,304]. Other drug resistance genes correlated with tumor prognosis may be identified in the near future.

8.5. HLA TYPING

The human leukocyte antigen (HLA) system consists of the major histocompatibility complex (MHC) proteins [9,305]. These proteins reside on the cell surface and either are restricted to cells of the immune system (class II) or are present on the surfaces of most nucleated cells (class I). These antigens were initially identified by serology and established as important agents in blood transfusions and tissue transplantations. HLA antigens must be similar or identical to ensure compatibility of tissues exchanged between individuals.

HLA alleles are polymorphic within coding regions, and DNA sequence variations in immunodominant regions enable correlations of serology with DNA typing. HLA alleles evolve rapidly relative to other human genes [306]. A significant degree of polymorphism is desirable in this system since greater numbers of detectable HLA alleles provide opportunities for more specific allogeneic transplantation matches. DNA typing of HLA alleles identifies more gene variants, enabling greater resolution and more specific donor–recipient matches [307]. Historically, HLA phenotyping (especially class II) has been troublesome, so DNA probe applications were attractive. Allele-specific oligonucleotide probes have been useful for HLA DNA typing of blood and tissue samples in clinical laboratories [308–310]. As in molecular oncology, the key challenge is the detection of multiple potential alleles at a single genetic locus. Since large numbers of white blood cells can be obtained in the "buffy coat" of clinical blood samples, DNA amplification may be avoided. However, PCR amplification may be necessary and effective for HLA typing when only serum or plasma samples are available [311]. PCR-based HLA typing methods have rapidly evolved. Different approaches based on PCR amplification of specific exons within both class I and II HLA genes have been used [for review, see ref. 312]. Immunocompromised patients with AIDS or neoplasms who require blood transfusions or bone marrow transplants represent prime candidates for combination HLA gene amplification and ASO reverse dot blot hybridization (PCR-SSO) [313, 314; for reviews, see refs. 312,315]. Also, restriction digestion of

PCR-amplified HLA exons followed by agarose gel electrophoresis (PCR-RFLP) distinguishes HLA alleles [316; for review, see ref. 312].

HLA typing may elucidate susceptibilities to autoimmune disease. Several human autoimmune diseases have been tightly associated with single HLA alleles [305,310]. More than 95% of patients with ankylosing spondylitis (AS) have the HLA-B27 antigen [310]. The HLA-B27 antigen was recognized almost 20 years ago by traditional serological HLA typing methods. However, the predictive value of this antigen for AS is weak because almost 7% of the American white population has this antigen but only 2% of the people with B27 develop the disease [310]. Analysis of four common molecular subtypes [317] failed to establish any B27 subtype associations with AS. However, continued accumulation of DNA sequence data from HLA-B alleles may result in the definition of new molecular subtypes with greater predictive value for AS. Conventional antigen types may fail to provide the resolution necessary for HLA typing correlations with autoimmune disease. HLA alleles may be defined that differ at the DNA sequence level but cannot be distinguished by conventional serology.

A second autoimmune disease, insulin-dependent diabetes mellitus (IDDM; type I), has been associated with a mutation that substitutes an uncharged amino acid for aspartate at position 57 in the HLA-DQ β-chain (DQB1 gene) [318,319] and the presence of arginine at position 52 of the DQ α-chain (DQA1 gene) [319]. Populations with a low incidence of IDDM appear to have a low incidence of non-Asp-57. Therefore, a hypothesis is that the variation in distribution of non-Asp-57 DQB1 and Arg-52 DQA1 alleles explains the variation in incidence of IDDM around the world. Genotypic analysis with DNA probes revealed that human individuals homozygous for both the non-Asp-57 DQB1 and the Arg-52 DQA1 alleles were at highest relative risk for IDDM [320]. Individuals with aspartate at position 57 of DQβ were resistant to disease regardless of the DQA1 allele status [320]. Allele-specific oligonucleotide PCR primers have been used to identify a non-Asp-57 HLA-DQw8 allele in IDDM patients [321]. Since these HLA alleles are polymorphic, disease associations and susceptibility assessments could be evaluated by reverse dot blot hybridizations with PCR-amplified HLA DNA. Furthermore, a mitochondrial DNA mutation was associated with maternal transmission of IDDM, relative insulin resistance, and sensory hearing loss [322]. As with human malignancies, multiple gene targets may be important with specific disease processes and offer sophisticated prognostic information.

8.6. CONCLUSIONS

A paradigmatic shift is occurring in human medicine in the 1990s. The modern explosion of genomic DNA sequence information from both microbial and human sources will provide unlimited opportunities for the application of DNA probes in clinical medicine. This new branch of molecular medicine will emphasize rapid detection of infectious diseases and human disease susceptibility testing by identification of involved molecules. Tremendous growth of the molecular knowledge base will support the shift from "crisis-oriented" medicine to "preventive" medicine. This knowledge base shift was reflected by the dedication to molecular medicine of all of the American Medical Association (AMA) journal issues published during November of 1993. Acute bacterial infections may be diagnosed in a matter of hours as an emergency room procedure and emerging antibiotic resistance patterns in hospital patients may be detected before overt clinical resistance is observed. Sensitive detection of oncogene mutations or HLA types that predispose individuals to cancer or autoimmune disease, respectively, will permit the combination of "preventive" disease monitoring with recommended lifestyle changes. Nucleic acid probe technology will play an instrumental role in the medical revolution known as the advent of molecular medicine.

REFERENCES

1. McBride, J.H.: Amino acids and proteins. In Howanitz, J.H., Howanitz, P.J. (eds.): Laboratory Medicine: Test Selection and Interpretation. New York: Churchill Livingstone, 1991.

2. Ford, R.E., Ou, C.-N., Ellefson, R.D.: Liquid-chromatographic analysis for urinary porphyrins. Clin Chem. 27:397–401 (1981).

3. Becker, M.A., Roessler, B.J.: Hyperuricemia and gout. In Scriver, C.R., Beaudet, A.L., Sly, W.S., Valle, D. (eds.): The Metabolic and Molecular Bases of Inherited Disease (7th ed., Volume I). New York: McGraw-Hill, 1995.

4. Pauling, L., Itano, H.A., Singer, S.J., Wells, I.C.: Sickle cell anemia, a molecular disease. Science 110:543–458 (1949).

5. Ingram, V.M.: Gene mutations in human hemoglobin: The chemical difference between normal and sickle cell hemoglobin. Nature 180:326–328 (1957).

6. Landt, Y., Valdya, H.C., Porter, S.E., Whalen, K., McClellan, A., Amyx, C., Parvin, C.A., Kessler, G., Nahm, M.H., Dietzler, D.N., Ladenson, J.H.: Semi-automated direct colorimetric measurement of creatine kinase isoenzyme MB activity after extraction from serum by use of a CK-MB-specific monoclonal antibody. Clin Chem. 34:575–581 (1988).

7. Gallimore, M.J., Friberger, P.: Chromogenic peptide substrate assays and their clinical applications. Blood Rev. 5:117–127 (1991).

8. Fink, D.J., and Harker, J.A.: Pulmonary and cardiac function. In Howanitz, J.H., Howanitz, P.J. (eds.): Laboratory Medicine: Test Selection and Interpretation. New York: Churchill Livingstone, 1991.

9. Roitt, I.: Essential Immunology, 5th ed. Oxford: Blackwell Scientific, 1983.

10. Phillips, T.M.: Immune complex assays: Diagnostic and clinical application. Crit. Rev. Clin. Lab. Sci. 27:237–264 (1989).

11. Kohler, G., Milstein, C.: Continuous cultures of fused cells secreting antibody of predefined specificity. Nature 256:495–497 (1975).

12. Orkin, S.H.: Molecular genetics and inherited human disease. In Scriver, C.R., Beaudet, A.L., Sly, W.S., Valle, D. (eds.): The Metabolic Basis of Inherited Disease. New York: McGraw-Hill, 1989.

13. Sambrook, J., Fritsch, E.F., Maniatis, T.: Molecular Cloning: A Laboratory Manual (2nd ed.). Cold Spring Harbor, NY: Cold Spring Harbor Laboratory, 1989.

14. Lupski, J.R., Zhang, Y.H., Rieger, M., Minter, M., Hsu, B., Ooi, B.G., Koeuth, T., McCabe, E.R.B.: Mutational analysis of the *Escherichia coli glpFK* region with Tn5 mutagenesis and the polymerase chain reaction. J. Bacteriol. 172:6129–6134 (1990).

15. McCabe, E.R.B., Fennessey, P.V., Guggenheim, M.A., Miles, B.S., Bullen, W.W., Sceats, D.J., Goodman, S.I.: Human glycerol kinase deficiency with hyperglycerolemia and glyceroluria. Biochem. Biophys. Res. Commun. 78:1327–13333 (1977).

16. Pettigrew, D.W., Ma, D.-P., Conrad, C.A., Johnson, J.R.: *Escherichia coli* glycerol kinase. J. Biol. Chem. 263:135–139 (1988).

17. Sanderson, K.E.: Genetic relatedness in the family Enterobacteriaceae. Ann. Rev. Microbiol. 30:327–349 (1976).

18. Botstein, D., White, R.L., Skolnick, M., Davis, R.W.: Construction of a genetic linkage map in man using restriction fragment length polymorphisms. Am. J. Hum. Genet. 32:314–331 (1980).

19. Chang, J.C., Kan, Y.W.: A sensitive test for prenatal diagnosis of sickle cell anemia: Direct analysis of amniocyte DNA with *Mst*II. Trans. Assoc. Am. Physicians 95:71–78 (1982).

20. Kan, Y.W.: Development of DNA analysis for human diseases: Sickle cell anemia and thalassemia as a paradigm. JAMA 267:1532–1536 (1992).

21. Southern, E.M.: Detection of specific sequences among DNA fragments separated by gel electrophoresis. J. Mol. Biol. 98:503–517 (1975).

22. Collins, F.S.: Of needles and haystacks: Finding human disease genes by positional cloning. Clin. Res. 39:615–623 (1991).

23. Conner, B.J., Reyes, A.A., Morin, C., Itakura, K., Teplitz, R.L., Wallace, R.B.: Detection of sickle cell β^s-globin allele by hybridization with synthetic oligonucleotides. Proc. Natl. Acad. Sci. U.S.A. 80:278–282 (1983).

24. Caskey, C.T.: Disease diagnosis by recombinant DNA methods. Science 236: 1223–1229 (1987).

25. Saiki, R.K., Gelfand, D.H., Stoffel, S., Scharf, S.J., Higuchi, R., Horn, G.T., Mullis, K.B., Erlich, H.A.: Primer-directed enzymatic amplification of DNA with a thermostable DNA polymerase. Science 239:487–491 (1988).

26. Saiki, R.K., Scharf, S., Faloona, F., Mullis, K.B., Horn, G.T., Erlich, H.A., Arnheim, N.: Enzymatic amplification of β-globin genomic sequences and restriction site analysis for diagnosis of sickle cell anemia. Science 230:1350–1354 (1985).

27. Mullis, K.B., Faloona, F.A.: Specific synthesis of DNA *in vitro* via a polymerase catalyzed chain reaction. Methods Enzymol. 155:335–350 (1987).

28. Mullis, K.B.: The unusual origin of the polymerase chain reaction. Sci. Am. April:56–65 (1990).

29. Mullis, K.B.: The polymerase chain reaction in an anemic mode: How to avoid cold deoxyribonuclear fusion. PCR Methods Applications 1:1–4 (1991).

30. Chien, A., Edgar, D.B., Trela, J.M.: Deoxyribonucleic acid polymerase from the extreme thermophile *Thermus aquaticus*. J. Bacteriol. 127:1550–1557 (1976).

31. Eisenstein, B.: The polymerase chain reaction—A new method of using molecular genetics for medical diagnosis. N. Engl. J. Med. 322:178–183 (1990).

32. Buffone, G.J., Demmler, G.J., Schimbor, C.M., Greer, J.: Improved amplification of cytomegalovirus DNA from urine after purification of DNA with glass beads. Clin. Chem. 37:1945–1949 (1991).

33. Drouet, E., Michelson, S., Denoyei, G., Colimon, R.: Polymerase chain reaction detection of human cytomegalovirus in over 2000 blood specimens correlated with virus isolation and related to urinary virus excretion. J. Virol. Methods 45:259–276 (1993).

34. Bostrom, L., Brytting, M., Mousavi-Jazi, M., Ringden, O., Llungman, P., Lonnqvist, B., Wahren, B., Sundqvist, V.-A.: PCR detection of CMV DNA in peripheral blood leukocytes and plasma from BMT recipients. Transplant. Proc. 26:1723–1724 (1994).

35. Lebech, A.-M., Hansen, K.: Detection of *Borrelia burgdorferi* DNA in urine samples and cerebrospinal fluid samples from patients with early and late lyme neuroborreliosis by polymerase chain reaction. J. Clin. Microbiol. 30:1646–1653 (1992).

36. Wang, J.-T., Wang, T.-H., Sheu, J.-C., Lin, S.-M., Lin, J.-T., Chen, D.-S.: Effects of anticoagulants and storage of blood samples on efficacy of the polymerase chain reaction assay for hepatitis C virus. J. Clin. Microbiol. 30:750–753 (1992).

37. Brisson Noel, A., Aznar, C., Chureau, C., Nguyen, S., Pierre, C., Bartoli, M., Bonete, R., Pialoux, G., Gicquel, B., Garrigue, G.: Diagnosis of tuberculosis by DNA amplification in clinical practice evaluation. Lancet 338:364–366 (1991).

38. Clarridge, J.E., III, Shawar, R.M., Shinnick, T.M., Plikaytis, B.B.: Large-scale use of polymerase chain reaction for detection of *Mycobacterium tuberculosis* in a routine mycobacteriology laboratory. J. Clin. Microbiol. 31:2049–2056 (1993).

39. Barany, F.: The ligase chain reaction in a PCR world. PCR Methods Applications 1:5–16 (1991).

40. Barany, F.: Genetic disease detection and DNA amplification using cloned thermostable ligase. Proc. Natl. Acad. Sci. U.S.A. 88:189–193 (1991).

41. Haruna, I., Spiegelman, S.: Recognition of size and sequence by an RNA replicase. Proc. Natl. Acad. Sci. U.S.A.: 54:1189–1193 (1965).

42. Lomeli, H., Tyagi, S., Pritchard, C.G., Lizardi, P.M., Kramer, F.R.: Quantitative assays based on the use of replicatable hybridization probes. Clin. Chem. 35:1826–1831 (1989).

43. Fahy, E., Kwoh, D.Y., Gingeras, T.R.: Self-sustained sequence replication (3SR): An isothermal transcription-based amplification system alternative to PCR. PCR Methods Applications 1:25–33 (1991).

44. Bush, C.E., Donovan, R.M., Peterson, W.R., Jennings, M.B., Bolton, V., Sherman, D.G., Vanden Brink, K.M., Beninsig, L.A., Godsey, J.H.: Detection of human immunodeficiency virus type 1 RNA in plasma samples from high-risk pediatric patients by using the self-sustained sequence replication reaction. J. Clin. Microbiol. 30:281–286 (1992).

45. Butcher, A., Salter, L., Kinard, S., Kung, K., McCreedy, B., Spadoro, J.: Evaluation of a single primer pair for the detection of HIV-1 in clinical specimens by PCR amplification: Comparison of a non-radioactive microtiter plate assay with solution hybridization (abstract). Am. Soc. Microbiol. 92:417 (1992).

46. Jungkind, D.L., Silverman, N.S., Bass, C.A., Bondi, J.M.: Evaluation of a new polymerase chain reaction test for detection of *Chlamydia trachomatis* in clinical samples (abstract). Am. Soc. Microbiol. 92:490 (1992).

47. Herman, S.A., Lewinski, C., Lefebvre, J., Marchand, M., Baril, J.G., Dragon, B.: Direct

detection of *Chlamydia trachomatis* in urine, using a rapid, PCR-based diagnostic test (abstract). Am. Soc. Microbiol. 92:488 (1992).

48. Landegren, U., Kaiser, R., Caskey, C.T., Hood, L.: DNA diagnostics—Molecular techniques and automation. Science 242:229–237 (1988).

49. Saiki, R.K., Walsh, P.S., Levenson, C.H., Erlich, H.A.: Genetic analysis of amplified DNA with immobilized sequence-specific oligonucleotide probes. Proc. Natl. Acad. Sci. U.S.A. 86:6230–6234 (1989).

50. DeMarchi, J.M., Beaudet, A.L., Caskey, C.T., Richards, C.S.: Experience of an academic reference laboratory using automation for analysis of cystic fibrosis mutations. Arch. Pathol. Lab. Med. 118:26–32 (1994).

51. Jerris, R.C. *Helicobacter.* In Murray, P.R. (ed.): Manual of Clinical Microbiology (6th edition). Washington, D.C.: American Society for Microbiology, 1995.

52. Prewett, E.J., Bickley, J., Owen, R.J., Pounder, R.E. DNA patterns of *Helicobacter pylori* isolated from gastric antrum, body, and duodenum. Gastroenterology 102:829–833 (1992).

53. Owen, R.J., Bickley, J., Costas, M., Morgan, D.R.: Genomic variation in *Helicobacter pylori*: Application to identification of strains. Scand. J. Gastroenterol. Suppl. 181:43–50 (1991).

54. Marconi, R.T., Garon, C.F.: Development of polymerase chain reaction primer sets for diagnosis of Lyme disease and for species-specific identification of Lyme disease isolates by 16S rRNA signature nucleotide analysis. J. Clin. Microbiol. 30:2830–2834 (1992).

55. Slater, L.N., Welch, D.F., Hensel, D., Coody, D.W.: A newly recognized pathogen as a cause of fever and bacteremia. N. Engl. J. Med. 323:1587–1593 (1990).

56. Relman, D.A., Loutit, J.S., Schmidt, T.M., Falkow, S., Tompkins, L.S.: The agent of bacillary angiomatosis. N. Engl. J. Med. 323:1573–1580 (1990).

57. Relman, D.A., Lepp, P.W., Sadler, K.N., Schmidt, T.M.: Phylogenetic relationships among the agent of bacillary angiomatosis, *Bartonella bacilliformis*, and other alphaproteobacteria. Mol. Microbiol. 6:1801–1807 (1992).

58. Relman, D.A., Schmidt, T.M., MacDermott, R.P., Falkow, S.: Identification of the uncultured bacillus of Whipple's disease. N. Engl. J. Med. 327:293–301 (1992).

59. Lowsky, R., Archer, G.L., Fyles, G., Minden, M., Curtis, J., Messner, H., Atkins, H., Patterson, B., Willey, B.M., McGeer, A.: Brief report: Diagnosis of Whipple's disease by molecular analysis of peripheral blood. N. Engl. J. Med. 331:1343–1346.

60. Woese, C.R. Bacterial evolution. Microbiol. Rev. 51:221–271 (1987).

61. Weisburg, W.G., Barns, S.M., Pelletier, D.A., Lane, D.J.: 16S ribosomal DNA amplification for phylogenetic study. J. Bacteriol. 173:697–703 (1991).

62. Angert, E.R., Clements, K.D., Pace, N.R.: The largest bacterium. Nature 362:239–241 (1993).

63. ASM News: Flu-shot false positive syndrome remains mysterious. ASM News 58:125–126 (1992).

64. Pearlman, E.S., Ballas, S.K.: False-positive human immunodeficiency virus screening test related to rabies vaccination. Arch. Pathol. Lab. Med. 118:805–806 (1994).

65. Fiss, E.H., Chehab, F.F., Brooks, G.F.: DNA amplification and reverse dot blot hybridization for detection and identification of mycobacteria to the species level in the clinical laboratory. J. Clin. Microbiol. 30:1220–1224 (1992).

66. Tramont, E.C.: *Treponema pallidum* (syphilis). In Mandell, G.L., Bennett, J.E., Dolin, R. (eds.): Principles and Practice of Infectious Diseases (4th ed., vol. 2). New York: John Wiley and Sons, Inc., 1990.

67. Grimprel, E., Sanchez, P.J., Wendel, G.D., Burstain, J.M., McCracken, G.H. Jr., Radolf, J.D., Norgard, M.V.: Use of polymerase chain reaction and rabbit infectivity testing to detect *Treponema pallidum* in amniotic fluid, fetal and neonatal sera, and cerebrospinal fluid. J. Clin. Microbiol. 29:1711–1718 (1991).

68. Wicher, K., Noordhoek, G.T., Abbruscato, F., Wicher, V.: Detection of *Treponema pallidum* in early syphilis by DNA amplification. J. Clin. Microbiol. 30:497–500 (1992).

69. Hay, P.E., Clarke, J.R., Strugnell, R.A., Taylor-Robinson, D., Goldmeier, D.: Use of the polymerase chain reaction to detect DNA sequences specific to pathogenic treponemes in cerebrospinal fluid. FEMS Microbiol. Lett. 68:233–238 (1990).

70. Noordhoek, G.T., Wolters, E.C., De Jonge, M.E.J., van Embden, J.D.A.: Detection by polymerase chain reaction of *Treponema pallidum* DNA in cerebrospinal fluid from neurosyphilis patients before and after antibiotic treatment. J. Clin. Microbiol. 29:1976–1984 (1991).

71. Sanchez, P.J., Wendel, G.D. Jr., Grimprel, E., Goldberg, M., Hall, M., Arencibia-Mireles, O., Radolf, J.D., Norgard, M.V.: Evaluation of molecular methodologies and rabbit infectivity testing for the diagnosis of congenital syphilis and neonatal central nervous system invasion by *Treponema pallidum*. J. Infect. Dis. 167:148–157 (1993).

72. Lanar, D.E., McLaughlin, G.L., Wirth, D.F., Barker, R.J., Zolg, J.W., Chulay, J.D.: Comparison of thick films, *in vitro* culture and DNA hybridization probes for detecting *Plasmodium falciparum* malaria. Am. J. Trop. Med. Hyg. 40:3–6 (1989).

73. Barker, R.H., Jr., Brandling Bennett, A.D., Koech, D.K., Mugambi, M., Khan, B., David, R., David, J.R., Wirth, D.F.: *Plasmodium falciparum*: DNA probe diagnosis of malaria in Kenya. Exp. Parasitol. 69:226–233 (1989).

74. Selwyn, P.A., Hartel, D., Lewis, V.A., Scheonbaum, E.E., Vermund, S.H., Klein, R.S., Walker, A.T., Friedland, G.H.: A prospective study of the risk of tuberculosis among intravenous drug users with human immunodeficiency virus infection. N. Engl. J. Med. 320:545–550 (1989).

75. Barnes, P.F., Bloch, A.B., Davidson, P.T., Snider, D.E. Jr.: Tuberculosis in patients with human immunodeficiency virus infection. N. Engl. J. Med. 324:1644–1650 (1991).

76. Daley, C.L., Small, P.M., Schecter, G.F., Schoolnik, G.K., McAdam, R.A., Jacobs, W.R. Jr., Hopewell, P.C.: An outbreak of tuberculosis with accelerated progression among persons infected with the human immunodeficiency virus. N. Engl. J. Med. 326:231–235 (1992).

77. Fox, J.L.: Coalition reacts to surge of drug-resistant TB. ASM News 58:135–139 (1992).

78. Edlin, B.R., Tokars, J.I., Grieco, M.H., Crawford, J.T., Williams, J., Sodrillo, E.M., Ong, K.R., Kilburn, J.O., Dooley, S.W., Castro, K.G., Jarvis, W.R., Holmberg, S.D.: An outbreak of multidrug-resistant tuberculosis among hospitalized patients with the acquired immunodeficiency syndrome. N. Engl. J. Med. 326:1514–1521 (1992).

79. Nolte, F.S., Metchock, B.: *Mycobacterium*. In Murray, P.R. (ed.): Manual of Clinical Microbiology (6th ed.). Washington, D.C.: American Society for Microbiology, 1995.

80. Roberts, G.D., Goodman, N.L., Heifets, L., Larsh, H.W., Lindner, T.H., McClatchy, J.K., McGinnis, M.R., Siddiqi, S.H., Wright, P.: Evaluation of the BACTEC radiometric method for recovery of mycobacteria and drug susceptibility testing of *Mycobac-*

terium tuberculosis from acid-fast smear-positive specimens. J. Clin. Microbiol. 18:689–696 (1983).

81. Cousins, D.V., Wilton, S.D., Francis, B.R., Gow, B.L.: Use of polymerase chain reaction for rapid diagnosis of tuberculosis. J. Clin. Microbiol. 30:255–258 (1992).

82. Sritharan, V., Barker, R.H. Jr.: A simple method for diagnosing *M. tuberculosis* infection in clinical samples using PCR. Mol. Cell Probes 5:385–395 (1991).

83. Hermans, P.W., van Soolingen, D., Bik, E.M., de Haas, P.E., Dale, J.W., van Embden, J.D.: Insertion element IS*987* from *Mycobacterium bovis* BCG is located in a hot-spot integration region for insertion elements in *Mycobacterium tuberculosis* complex strains. Infect. Immun. 59:2695–2705 (1991).

84. De Wit, D., Steyn, L., Shoemaker, S., Sogin, M.: Direct detection of *Mycobacterium tuberculosis* in clinical specimens by DNA amplification. J. Clin. Microbiol. 28:2437–2441 (1990).

85. Buck, G.E., O'Hara, L.C., Summersgill, J.T.: Rapid, simple method for treating clinical specimens containing *Mycobacterium tuberculosis* to remove DNA for polymerase chain reaction. J. Clin. Microbiol. 30:1331–1334 (1992).

86. Savic, B., Sjobring, U., Alugupalli, S., Larsson, L., Miorner, H.: Evaluation of polymerase chain reaction, tuberculostearic acid analysis, and direct microscopy for the detection of *Mycobacterium tuberculosis* in sputum. J. Infect. Dis. 166:1177–1180 (1992).

87. Tomioka, H., Sato, K., Saito, H., Tasaka, H. Reactivities of various mycobacteria species against DNA probes (Gen-Probe Rapid Diagnostic System) specific to *Mycobacterium tuberculosis* complex, *Mycobacterium avium* and *Mycobacterium intracellulare*. Kekkaku 66:405–411 (1991).

88. Sherman, I., Harrington, N., Rothrock, A., George, H.: Use of a cutoff range in identifying mycobacteria by the Gen-Probe Rapid Diagnostic System. J. Clin. Microbiol. 27:241–244 (1989).

89. Ellner, P.D., Kiehn, T.E., Cammarata, R., Hosmer, M.: Rapid detection and identification of pathogenic mycobacteria by combining radiometric and nucleic acid probe methods 'see comments. J. Clin. Microbiol. 26:1349–1352 (1988).

90. Body, B.A., Warren, N.G., Spicer, A., Henderson, D., Chery, M.: Use of Gen-Probe and Bactec for rapid isolation and identification of mycobacteria. Correlation of probe results with growth index. Am. J. Clin. Pathol. 93:415–420 (1990).

91. Kusunoki, S., Ezaki, T., Tamesada, M., Hatanaka, Y., Asano, K., Hashimoto, Y., Yabuchi, E.: Application of colorimetric microdilution plate hybridization for rapid genetic identification of 22 *Mycobacterium* species. J. Clin. Microbiol. 29:1596–1603 (1991).

92. Telenti, A., Marchesi, F., Balz, M., Bally, F., Bottger, E.C., Bodmer, T.: Rapid identification of mycobacteria to the species level by polymerase chain reaction and restriction enzyme analysis. J. Clin. Microbiol. 31:175–178 (1993).

93. Schachter, J., Stamm, W.E.: *Chlamydia*. In Murray, P.R. (ed.): Manual of Clinical Microbiology (6th ed.). Washington, D.C.: American Society for Microbiology, 1995.

94. Deguchi, T., Yamamoto, H., Iwata, H., Yamamoto, N., Ito, Y., Ban, Y., Saito, I., Ezaki, T., Kawada, Y.: Detection of *Chlamydia trachomatis* by polymerase chain reaction. Kansenshogaku. Zasshi 65:1183–1187 (1991).

95. Watson, M.W., Lambden, P.R., Clarke, I.N.: Genetic diversity and identification of human infection by amplification of the chlamydial 60-kilodalton cysteine-rich outer membrane protein gene. J. Clin. Microbiol. 29:1188–1193 (1991).

96. Ostergaard, L., Birkelund, S., Christiansen, G.: Use of polymerase chain reaction for detection of *Chlamydia trachomatis*. J. Clin. Microbiol. 28:1254–1260 (1990).

97. Wahlberg, J., Lundeberg, J., Hultman, T., Uhlen, M.: General colorimetric method for DNA diagnostics allowing direct solid-phase genomic sequencing of the positive samples. Proc. Natl. Acad. Sci. U.S.A. 87:6569–6573 (1990).

98. Bauwens, J.E., Clark, A.M., Loeffelholz, M.J., Herman, S.A., Stamm, W.E.: Diagnosis of *Chlamydia trachomatis* urethritis in men by polymerase chain reaction assay of first catch urine. J. Clin. Microbiol. 31:3013–3016 (1993).

99. Bauwens, J.E., Clark, A.M., Stamm, W.E.: Diagnosis of *Chlamydia trachomatis* endocervical infections by a commercial polymerase chain reaction assay. J. Clin. Microbiol. 31:3023–3027 (1993).

100. Sweet, R.L., Weisenfeld, H.C., Uhrin, M., Dixon, B.: Comparison of EIA, culture, and polymerase chain reaction for *Chlamydia trachomatis* in a sexually transmitted disease clinic. J. Infect. Dis. 170:500–501 (1994).

101. Kristiansen, B.E., Ask, E., Jenkins, A., Fermer, C., Radstrom, P., Skold, O.: Rapid diagnosis of meningococcal meningitis by polymerase chain reaction. Lancet 337:1568–1569 (1991).

102. Gustaferro, C.A., Persing, D.H.: Chemiluminescent universal probe for bacterial ribotyping. J. Clin. Microbiol. 30:1039–1041 (1992).

103. Luft, B.J., Steinman, C.R., Neimark, H.C., Muralidhar, B., Rush, T., Finkel, M.F., Kunkel, M., Dattwyler, R.J.: Invasion of the central nervous system by *Borrelia burgdorferi* in acute disseminated infection. JAMA 267:1364–1367 (1992).

104. Barre-Sinoussi, F., Nuguyre, M., Dauguet, C., Vilmer, E., Griscelli, C., Brun-Vezinet, F., Rouzioux, C., Gluckman, J., Chermann, J., Montagnier, L.: Isolation of a T-lymphotropic retrovirus from a patient at risk for acquired immunodeficiency syndrome. Science 220:868–871 (1983).

105. Sloand, E.M., Pitt, E., Chiarello, R.J., Nemo, G.J.: HIV testing—State of the art. JAMA 266:2861–2866 (1991).

106. Barker, E., Barnett, S.W.: Human immunodeficiency viruses. In Murray, P.R. (ed.): Manual of Clinical Microbiology (6th ed.). Washington, D.C.: American Society for Microbiology, 1995.

107. Soriano, V., Hewlett, I., Tor, J., Esteve, M., Granada, I., Muga, R., Epstein, J., Medrano, L.: The evidence of human immunodeficiency virus infection in the seronegative subjects of high-risk groups. Med. Clin. 97:441–445 (1991).

108. Simonds, R.J., Holmberg, S.D., Hurwitz, R.L., Coleman, T.R., Bottenfield, S., Conley, L.J., Kohlenberg, S.H., Castro, K.G., Dahan, B.A., Schable, C.A., Rayfield, M.A., Rogers, M.F.: Transmission of human immunodeficiency virus type 1 from a seronegative organ and tissue donor. N. Engl. J. Med. 326:726–732 (1992).

109. Reesnick, H.W., Lelie, P.N., Huisman, J.G., Schassbert, W., Gonsalves, M., Aaij, C., Winkel, I.N., van der Does, J.A., Hekker, A.C., Desmyter, J.: Evaluation of six enzyme immunoassays for antibody against human immunodeficiency virus. Lancet 88:483–486 (1986).

110. Wang, Z., Kinard, S., Kung, K., Butcher, A., Spadoro, J.: Performance evaluation of a PCR-microtiter plate hybridization assay for the detection of HIV-1 DNA (abstract). Am. Soc. Microbiol. 92:418 (1992).

111. Whetsell, J., Drew, J.B., Milman, G., et al.: Comparison of three nonradioisotopic poly-

merase chain reaction-based methods for detection of human immunodeficiency virus type 1. J. Clin. Microbiol. 30:845–853 (1992).

112. Jackson, J.B.: Detection and quantitation of human immunodeficiency virus type 1 using molecular DNA/RNA technology. Arch. Pathol. Lab. Med. 117:473–477 (1993).

113. Menzo, S., Baganrelli, P., Giacca, M., Manzin, A., Varaldo, P.E., Clementi, M. Absolute quantitation of viremia in human immunodeficiency virus infection by competitive reverse transcription and polymerase chain reaction. J. Clin. Microbiol. 30:1752–1757 (1992).

114. Clementi, M., Bagnarelli, P., Manzin, A., Menzo, S.: Competitive polymerase chain reaction and analysis of viral activity at the molecular level. GATA 11:1–6 (1994).

115. Weiser, B., Nachman, S., Tropper, P., Viscosi, K.H., Grimson, R., Baxter, G., Fang, G., Reyelt, C., Hutcheon, N., Burger, H.: Quantitation of human immunodeficiency virus type 1 during pregnancy: Relationship of viral titer to mother-to-child transmission and stability of viral load. Proc. Natl. Acad. Sci. U.S.A. 91:8037–8041 (1994).

116. Connor, E.M., Sperling, R.S., Gelber, R., et al.: Reduction of maternal-infant transmission of human immunodeficiency virus type I with zidovudine treatment. N. Engl. J. Med. 331:1173–1180 (1994).

117. De Rossi, A., Ades, A.E., Mammano, F., Del Mistro, A., Amadori, A., Giaquinto, C., Chieco Bianchi, L.: Antigen detection, virus culture, polymerase chain reaction, and *in vitro* antibody production in the diagnosis of vertically transmitted HIV-1 infection. AIDS 5:15–20 (1991).

118. Laure, F., Courgnaud, V., Rouzioux, C., Blanche, S., Veber, F., Burgard, M., Jacomet, C., Griscelli, C., Brechot, C.: Detection of HIV-1 DNA in infants and children by means of the polymerase chain reaction. Lancet 2:538–541 (1988).

119. Kline, M.W., Lewis, D.E., Hollinger, F.B., Reuben, J.M., Hanson, I.C., Kozinetz, C.A., Dimitrov, D.H., Rosenblatt, H.M., Shearer, W.T.: A comparative study of human immunodeficiency virus culture, polymerase chain reaction and anti-human immunodeficiency virus immunoglobulin A antibody detection in the diagnosis during early infancy of vertically acquired human immunodeficiency virus infection. Pediatr. Infect. Dis. J. 13:90–94 (1994).

120. Petru, A., Dunphy, M.G., Azimi, P., Janner, D., Gallo, D., Hanson, C., Sohmer, P., Stanley, M.: Reliability of polymerase chain reaction in the detection of human immunodeficiency virus infection in children. Pediatr. Infect. Dis. J. 11:30–33 (1992).

121. Scarlatti, G., Lombardi, V., Plebani, A., Principi, N., Vegni, C., Ferraris, G., Bucceri, A., Fenyo, E.M., Wigzell, H., Rossi, P.: Polymerase chain reaction, virus isolation and antigen assay in HIV-1-antibody-positive mothers and their children. AIDS 5:1173–1178 (1991).

122. Weintrub, P.S., Ulrich, P.P., Edwards, J.R., Boucher, F., Levy, J.A., Cowan, M.J., Vyas, G.N.: Use of polymerase chain reaction for the early detection of HIV infection in the infants of HIV-seropositive women. AIDS 5:881–884 (1991).

123. Laure, F., Courgnaud, V., Brechot, C.: The polymerase chain reaction in the human immunodeficiency virus diagnosis. Ann. Biol. Clin. 48:413–415 (1990).

124. Chadwick, E.G., Yogev, R., Kwok, S., Sninsky, J.J., Kellogg, D.E., Wolinsky, S.M.: Enzymatic amplification of the human immunodeficiency virus in peripheral blood mononuclear cells from pediatric patients. J. Infect. Dis. 160:954–959 (1989).

125. Brandt, C.D., Rakusan, T.A., Sison, A.V., Josephs, S.H., Saxena, E.S., Herzog, K.D.,

Parrott, R.H., Sever, J.L.: Detection of human immunodeficiency virus type 1 infection in young pediatric patients by using polymerase chain reaction and biotinylated probes. J. Clin. Microbiol. 30:36–40.

126. Brandt, C.D., Rakusan, T.A., Sison, A.V., Saxena, E.S., Ellaurie, M., Sever, J.L.: Human immunodeficiency virus infection in infants during the first two months of life. Arch. Pediatr. Adolesc. Med. 148:250–254 (1994).

127. Candotti, D., Jung, M., Kerouedan, D., Rosenheim, M., Gentilini, M., M'Pele, P., Huraux, J.M., Agut, H.: Genetic variability affects the detection of HIV by polymerase chain reaction. AIDS 5:1003–1007 (1991).

128. Arvin, A.M., Prober, C.G.: Herpes simplex viruses. In Murray, P.R. (ed.): Manual of Clinical Microbiology (6th ed.). Washington, D.C.: American Society for Microbiology, 1995.

129. Hodinka, R.L., Friedman, H.M.: Human cytomegalovirus. In Murray, P.R. (ed.): Manual of Clinical Microbiology (6th ed.). Washington, D.C.: American Society for Microbiology, 1995.

130. Koutsky, L.A., Stevens, C.E., Holmes, K.K., Ashley, R.L., Kiviat, N.B., Critchlow, C.W., Corey, L.: Underdiagnosis of genital herpes by current clinical and viral-isolation procedures. N. Engl. J. Med. 326:1533–1539 (1992).

131. Demmler, G.J., Buffone, G.J., Schimbor, C.M., May, R.A.: Detection of cytomegalovirus in urine from newborns by using polymerase chain reaction DNA amplification. J. Infect. Dis. 158:1177–1184 (1988).

132. Smith, K.L., and Dunstan, R.A.: PCR detection of cytomegalovirus: A review. Br. J. Haematol. 84:187–190 (1993).

133. Boland, G.J., De Weger, R.A., Tilanus, M.G.J., Ververs, C., Bosboom-Kalsbeek, K., De Gast, G.C.: Detection of cytomegalovirus (CMV) in granulocytes by polymerase chain reaction compared with the CMV antigen test. J. Clin. Microbiol. 30:1763–1767 (1992).

134. Cagle, P.T., Buffone, G., Holland, V.A., Samo, T., Demmler, G.J., Noon, G.P., Lawrence, E.C.: Semiquantitative measurement of cytomegalovirus DNA in lung and heart-lung transplant patients by in vitro DNA amplification. Chest 101:93–96 (1992).

135. Patel, R., Smith, T.F., Espy, M., Wiesner, R.H., Krom, R.A., Portela, D., Paya, C.V. Detection of cytomegalovirus DNA in sera of liver transplant recipients. J. Clin Microbiol. 32:1431–1434 (1994).

136. Ruutu, P., Ruutu, T., Volin, L., Tukiainen, P., Ukkonen, P., Hovi, T.: Cytomegalovirus is frequently isolated in bronchoalveolar lavage fluid of bone marrow transplant recipients without pneumonia. Ann. Intern. Med. 112:913–916 (1990).

137. Miller, M.J., Bovey, S., Pado, K., Bruckner, D.A., Wagar, E.A. Application of PCR to multiple specimen types for diagnosis of cytomegalovirus infection: Comparison with cell culture and shell vial assay. J. Clin. Microbiol. 32:5–10 (1994).

138. Gerna, G., Zipeto, D., Parea, M., Grazia Revello, M., Silini, E., Percivalle, E., Zavattoni, M., Grossi, P., Milanesi, G.: Monitoring of human cytomegalovirus infections and ganciclovir treatment in heart transplant recipients by determination of viremia, antigenemia, and DNAemia. J. Infect. Dis. 164:488–498 (1991).

139. Einsele, H., Ehninger, G., Steidle, M., Vallbracht, A., Muller, M., Schmidt, H., Saal, J.G., Waller, H.D., Muller, C.A.: Polymerase chain reaction to evaluate antiviral therapy for cytomegalovirus disease. Lancet 338:1170–1172 (1991).

140. Bukh, J., Wantzin, P., Krogsgaard, K., Knudsen, F., Purcell, R.H., Miller, R.H., and the

Copenhagen Dialysis HCV Study Group.: High prevalence of hepatitis C virus (HCV) RNA in dialysis patients: Failure of commercially available antibody tests to identify a significant number of patients with HCV infection. J. Infect. Dis. 168:1343–1348 (1993).

141. Hagiwara, H., Hayashi, N., Mita, E., Takehara, T., Kasahara, A., Fusamoto, H., Kamada, T.: Quantitative analysis of hepatitis C virus RNA in serum during interferon alpha therapy. Gastroenterology 104:877–883 (1993).

142. Gretch, D., Corey, L., Wilson, J., dela Rosa, C., Wilson, R., Carithers, Jr., R., Busch, M., Hart, J., Sayers, M., Han., J.: Assessment of hepatitis C virus RNA levels by quantitative competitive RNA polymerase chain reaction: High-titer viremia correlates with advanced stage of disease. J. Infect. Dis. 169:1219–1225 (1994).

143. Kiviat, N.B., Koutsky, L.A.: Human papillomavirus. In Murray, P.R. (ed.): Manual of Clinical Microbiology (6th ed.). Washington, D.C.: American Society for Microbiology, 1995.

144. Schlegel, R., Phelps, W.C., Zhang, Y.-L., Barbosa, M.: Quantitative keratinocyte assay detects two biological activities of human papillomavirus DNA and identifies viral types associated with cervical carcinoma. EMBO J. 7:3181–3187 (1988).

145. Lungu, O., Sun, X.W., Felix, J., Richart, R.M., Silverstein, S., Wright, T.C.: Relationship of human papillomavirus type to grade of cervical intraepithelial neoplasia. JAMA 267:2493–2496 (1992).

146. Varma, V.A., Sanchez Lanier, M., Unger, E.R., Clark, C., Tickman, R., Hewan Lowe, K., Chenggis, M.L., Swan, D.C.: Association of human papillomavirus with penile carcinoma: A study using polymerase chain reaction and in situ hybridization. Hum. Pathol. 22:908–913 (1991).

147. Wiener, J.S., Effert, P.J., Humphrey, P.A., Yu, L., Liu, E.T., Walther, P.J.: Prevalence of human papillomavirus types 16 and 18 in squamous-cell carcinoma of the penis: A retrospective analysis of primary and metastatic lesions by differential polymerase chain reaction. Int. J. Cancer 50:694–701 (1992).

148. Cuzick, J., Terry, G., Ho, L., Hollingworth, T., Anderson, M.: Human papillomavirus type 16 DNA in cervical smears as predictor of high-grade cervical cancer. Lancet 339:959–960 (1992).

149. Johnson, T.L., Kim, W., Plieth, D.A., Sarkar, F.H.: Detection of HPV 16/18 DNA in cervical adenocarcinoma using polymerase chain reaction (PCR) methodology. Mod. Pathol. 5:35–40 (1992).

150. Rodu, B., Christian, C., Snyder, R.C., Ray, R., Miller, D.M.: Simplified PCR-based detection and typing strategy for human papillomaviruses utilizing a single oligonucleotide primer set. Biotechniques 10:632–637 (1991).

151. Nuovo, G.J., MacConnell, P., Forde, A., Delvenne, P.: Detection of human papillomavirus DNA in formalin-fixed tissues by in situ hybridization after amplification by polymerase chain reaction. Am J Pathol. 139:847–854 (1991).

152. Delvenne, P., Fontaine, M.-A., Delvenne, C., Nikkels, A., Boniver, J.: Detection of human papillomaviruses in paraffin-embedded biopsies of cervical intraepithelial lesions: Analysis of immunohistochemistry, in situ hybridization, and the polymerase chain reaction. Mod. Pathol. 7:113–119 (1994).

153. Warren, N.G., Hazen, K.C.: Candida, Cryptococcus, and other yeasts of medical importance. In Murray, P.R. (ed.): Manual of Clinical Microbiology (6th ed.). Washington, D.C.: American Society for Microbiology, 1995.

154. Dupont, B.: Clinical manifestations and management of candidosis in the compromised patient. In Warnock, D.W., Richardson, M.D. (eds.): Fungal Infection in the Compromised Patient. New York: John Wiley and Sons, Inc., 1990.

155. Ganesan, K., Banerjee, A., Datta, A.: Molecular cloning of the secretory acid proteinase gene from *Candida albicans* and its use as a species-specific probe. Infect. Immun. 59:2972–2977 (1991).

156. Niesters, H.G., Goessens, W.H.F., Meis, J.F.M.G., Quint, W.G.V.: Rapid polymerase chain reaction-based identification assays for *Candida* species. J. Clin. Microbiol. 31:904–910 (1993).

157. Miyakawa, Y., Mabuchi, T., Kagaya, K., Fukuzawa, Y.: Isolation and characterization of a species-specific DNA fragment for detection of *Candida albicans* by polymerase chain reaction. J. Clin. Microbiol. 30:894–900 (1992).

158. Holmes, A.R., Cannon, R.D., Shepherd, M.G., Jenkinson, H.F. Detection of *Candida albicans* and other yeasts in blood by PCR. J. Clin. Microbiol. 32:228–231 (1994).

159. Garcia, L.S., Sulzer, A.J., Healy, G.R., Grady, K.K., Bruckner, D.A.: Blood and tissue protozoa. In Murray, P.R. (ed.): Manual of Clinical Microbiology (6th ed.). Washington, D.C.: American Society for Microbiology, 1995.

160. Francis, V.S., Ayyanathan, K., Bhat, P., Srinivasa, H., Padmanaban, G.: Development of a DNA diagnostic probe for the detection of the human malarial parasite *Plasmodium falciparum*. Indian J. Biochem. Biophys. 25:537–541 (1988).

161. Relf, W.A., Boreham, R.E., Tapchaisri, P., Khusmith, S., Healey, A., Upcroft, P., Tharavanij, S., Kidson, C.: Diagnosis of *Plasmodium vivax* malaria using a specific deoxyribonucleic acid probe. Trans. R. Soc. Trop. Med. Hyg. 84:630–634 (1990).

162. Lupski, J.R.: Molecular epidemiology and its clinical application. JAMA 270:1363–1364 (1993).

163. Versalovic, J., Woods, C.R., Georghiou, P.R., Hamill, R.J., Lupski, J.R. DNA-based identification and epidemiologic typing of bacterial pathogens. Arch. Pathol. Lab. Med. 117:1088–1098 (1993).

164. Farmer, J.J. III: Conventional typing methods. J. Hosp. Infect. 11:309–314 (1988).

165. Garaizar, J., Kaufmann, M.E., Pitt, T.L.: Comparison of ribotyping with conventional methods for the type identification of *Enterobacter cloacae*. J. Clin. Microbiol. 29:1303–1307 (1991).

166. Blumberg, H.M., Rimland, D., Kiehlbauch, J.A., Terry, P.M., Wachsmuth, I.K.: Epidemiologic typing of *Staphylococcus aureus* by DNA restriction fragment length polymorphisms of rRNA genes: Elucidation of the clonal nature of a group of bacteriophage-nontypeable, ciprofloxacin-resistant, methicillin-susceptible *S. aureus* isolates. J. Clin. Microbiol. 30:362–369 (1992).

167. Porter, R.D. Conjugation. In Streips, U.N., Yasbin, R.E. (eds.): Modern Microbial Genetics. New York: Wiley-Liss, 1991.

168. Tompkins, L.S., Troup, N., Labigne-Roussel, A., Cohen, M.L.: Cloned, random chromosomal sequences as probes to identify *Salmonella* species. J. Infect. Dis. 154:156–162 (1986).

169. Ross, B.C., Raios, K., Jackson, K., Sievers, A., Dwyer, B.: Differentiation of *Mycobacterium tuberculosis* strains by use of a nonradioactive Southern blot hybridization method. J. Infect. Dis. 163:904–907 (1991).

170. Jones, K.F., Schneewind, O., Koomey, J.M., Fischetti, V.A.: Genetic diversity among the T protein genes of group A streptococci. Mol. Microbiol. 5:2947–2952 (1991).

171. Rodriguez, P., Vekris, A., De Barbeyrac, B., Dutilh, B., Bonnet, J., Bebear, C.: Typing of *Chlamydia trachomatis* by restriction endonuclease analysis of the amplified major outer membrane protein gene. J. Clin. Microbiol. 29:1132–1136 (1991).

172. Lupski, J.R., Weinstock, G.M.: Short, interspersed repetitive DNA sequences in prokaryotic genomes. J. Bacteriol. 174:4525–4529 (1992).

173. Grimont, F., Grimont, P.A.D.: Ribosomal ribonucleic acid gene restriction patterns as potential taxonomic tools. Ann. Inst. Pasteur 137B:165–175 (1986).

174. Schmidt, T.M.: Fingerprinting bacterial genomes using ribosomal RNA genes and operons. Methods Mol. Cell Biol. 5:3–12 (1994).

175. Bachmann, B.J.: Linkage map of *Escherichia coli* K-12, edition 8. Microbiol. Rev. 54:130–197 (1990).

176. Irino, K., Grimont, F., Casin, I., Grimont, P.A.D., The Brazilian Purpuric Fever Study Group: rRNA gene restriction patterns of *Haemophilus influenzae* biogroup aegyptius strains associated with Brazilian purpuric fever. J. Clin. Microbiol. 26:1535–1538 (1988).

177. Brenner, D.J., Mayer, L.W., Carlone, G.M., Harrison, L.H., Bibb, W.F., de Cunto Brandileone, M.C., Sottnek, F.O., Irino, K., Reeves, M.W., Swenson, J.M., Birkness, K.A., Weyant, R.S., Berkley, S.F., Woods, T.C., Steigerwalt, A.G., Grimont, P.A.D., McKinney, R.M., Fleming, D.W., Gheesling, L.L., Cooksey, R.C., Arko, R.J., Broome, C.V., The Brazilian Purpuric Fever Study Group: Biochemical, genetic, and epidemiologic characterization of *Haemophilus influenzae* biogroup aegyptius (*Haemophilus aegyptius*) strains associated with Brazilian purpuric fever. J. Clin. Microbiol. 26:1524–1534 (1988).

178. Steinbach, S., Sun, L., Ru-Zhang, J., Flume, P., Gilligan, P., Egan, T.M., Goldstein, R.: Transmissibility of *Pseudomonas cepacia* infection in clinic patients and lung-transplant recipients with cystic fibrosis. N. Engl. J. Med. 331:981–987 (1994).

179. Hermans, P.W.M., van Soolingen, D., Dale, J.W., Schuitema, A.R.J., McAdam, R.A., Catty, D., van Embden, J.D.A.: Insertion element IS*986* from *Mycobacterium tuberculosis*: A useful tool for diagnosis and epidemiology of tuberculosis. J. Clin. Microbiol. 28:2051–2058 (1990).

180. van Soolingen, D., de Haas, P.E.W., Hermans, P.W.M., Groenen, P.M.A., van Embden, J.D.A.: Comparison of various repetitive DNA elements as genetic markers for strain differentiation and epidemiology of *Mycobacterium tuberculosis*. J. Clin. Microbiol. 31:1987–1995 (1993).

181. van Embden, J.D.A., Cave, M.D., Crawford, J.T., Dale, J.W., Eisenach, K.D., Gicquel, B., Hermans, P., Martin, C., McAdam, R., Shinnick, T.M., Small, P.M.: Strain identification of *Mycobacterium tuberculosis* by DNA fingerprinting: Recommendations for a standardized methodology. J. Clin. Microbiol. 31:406–409 (1993).

182. van Soolingen, D., de Haas, P.E., Hermans, P.W.M., van Embden, J.D.A.: DNA fingerprinting of *Mycobacterium tuberculosis*. Methods Enzymol 235:196–205 (1994).

183. Sadhu, C., McEachern, M.J., Rustchenko Bulgac, E.P., Schmid, J., Soll, D.R., Hicks, J.B.: Telomeric and dispersed repeat sequences in *Candida* yeasts and their use in strain identification. J. Bacteriol. 173:842–850 (1991).

184. Schmid, J., Voss, E., Soll, D.R.: Computer-assisted methods for assessing strain relatedness in *Candida albicans* by fingerprinting with the moderately repetitive sequence Ca3. J. Clin. Microbiol. 28:1236–1243 (1990).

185. Versalovic, J., Koeuth, T., Lupski, J.R.: Distribution of repetitive DNA sequences in eu-

bacteria and application to fingerprinting of bacterial genomes. Nucleic acids Res. 19:6823–6831 (1991).

186. van Belkum, A.: DNA fingerprinting of medically important microorganisms by use of PCR. Clin. Microbiol. Rev. 7:174–184 (1994).

187. Versalovic, J., Schneider, M., de Bruijn, F.J., Lupski, J.R.: Genomic fingerprinting of bacteria using repetitive sequence based PCR (rep-PCR). *Methods Mol. Cell Biol.* 5:25–40 (1994).

188. Welsh, J., McClelland, M.: Genomic fingerprints produced by PCR with consensus tRNA gene primers. Nucleic Acids Res. 19:861–866 (1991).

189. McClelland, M., Petersen, C., Welsh, J.: Length polymorphisms in tRNA intergenic spacers detected by using the polymerase chain reaction can distinguish streptococcal strains and species. J. Clin. Microbiol. 30:1499–1504 (1992).

190. Welsh, J., McClelland, M.: PCR-amplified length polymorphisms in tRNA intergenic spacers for categorizing staphylococci. Mol. Microbiol. 6:1673–1680 (1992).

191. Stern, M.J., Ames, G.F.L., Smith, N.H., Robinson, E.C., Higgins, C.F.: Repetitive extragenic palindromic sequences: A major component of the bacterial genome. Cell 37:1015–1026 (1984).

192. Gilson, E., Clement, J.M., Brutlag, D., Hofnung, M.: A family of dispersed repetitive extragenic palindromic DNA sequences in *E. coli.* EMBO J 3:1417–1421 (1984).

193. Sharples, G.J., Lloyd, R.G.: A novel repeated DNA sequence located in the intergenic regions of bacterial chromosomes. Nucleic Acids Res. 18:6503–6508 (1990).

194. Hulton, C.S.J., Higgins, C.F., Sharp, P.M.: ERIC sequences: A novel family of repetitive elements in the genomes of *Escherichia coli, Salmonella typhimurium* and other enterobacteria. Mol. Microbiol. 5:825–834 (1991).

195. Versalovic, J., Koeuth, T., Zhang, Y.H., McCabe, E.R.B., Lupski, J.R.: Quality control for bacterial inhibition assays: DNA fingerprinting of microorganisms by rep-PCR. Screening 1:175–183 (1992).

196. de Bruijn, F.J.: Use of repetitive (repetitive extragenic element and enterobacterial repetitive intergenic consensus) sequences and the polymerase chain reaction to fingerprint the genomes of *Rhizobium meliloti* isolates and other soil bacteria. Appl. Environ. Microbiol. 58:2180–2187 (1992).

197. Woods, C. R., Versalovic, J., Koeuth, T., Lupski, J.R.: Analysis of relationships among isolates of *Citrobacter diversus* using DNA fingerprints generated by repetitive sequence-based primers in the polymerase chain reaction (rep-PCR). J. Clin. Microbiol. 30:2921–2929 (1992).

198. Georghiou, P.R., Doggett, A.M., Kielhofner, M.A., Stout, J.E., Watson, D.A., Lupski, J.R., Hamill, R.J.: Molecular fingerprinting of *Legionella* species by repetitive element PCR. J. Clin. Microbiol. 32:2989–2994 (1994).

199. Georghiou, P., Hamill, R.J., Wright, C.E., Versalovic, J., Koeuth, T., Watson, D.A., Lupski, J.R.: Molecular epidemiology of infections due to *Enterobacter aerogenes*: Identification of hospital outbreak-associated strains by molecular techniques. Clin. Infect. Dis. 20:84–94 (1995).

200. Versalovic, J., Kapur, V., Mason, E.O., Jr., Shah, U., Koeuth, T., Lupski, J.R., Musser, J.M.: Penicillin resistant *Streptococcus pneumoniae* strains recovered in Houston, Texas: Identification and molecular characterization of multiple clones. J. Infect. Dis. 167:850–856 (1993).

201. Sokol, D.M., Demmler, G.J., Buffone, G.J.: Rapid epidemiologic analysis of cytomegalovirus by using polymerase chain reaction amplification of the L-S junction region. J. Clin. Microbiol. 30:839–844 (1992).

202. Woods, C.R., Versalovic, J., Koeuth, T., Lupski, J.R.: Whole cell rep-PCR allows rapid assessment of clonal relationships of bacterial isolates. J. Clin. Microbiol. 31:1927–1931 (1993).

203. Versalovic, J., Kapur, V., Koeuth, T., Mazurek, G.H., Whittam, T.S., Musser, J.M., Lupski, J.R.: DNA fingerprinting of pathogenic bacteria by fluorophore-enhanced repetitive sequence based polymerase chain reaction. Arch. Pathol. Lab. Med. 119:23–29 (1995).

204. Williams, J.G.K., Kubelik, A.R., Livak, K.J., Rafalski, J.A., Tingey, S.V.: DNA polymorphisms amplified by arbitrary primers are useful as genetic markers. Nucleic Acids Res. 18:6531–6535 (1990).

205. Welsh, J., McClelland, M.: Fingerprinting genomes using PCR with arbitrary primers. Nucleic Acids Res. 18:7213–7218 (1990).

206. Groenen, P.M.A., Bunschoten, A.E., van Soolingen, D., van Embden, J.D.A.: Nature of DNA polymorphism in the direct repeat cluster of *Mycobacterium tuberculosis*: Application for strain differentiation by a novel typing method. Mol. Microbiol. 10:1057–1065 (1993).

207. Berg, D.E., Akopyantis, N.S., Kersulyte, D.: Fingerprinting microbial genomes using the RAPD or AP-PCR method. Methods Mol. Cell Biol. 5:13–24 (1994).

208. Brumfitt, W., Hamilton-Miller, J.: Methicillin-resistant *Staphylococcus aureus*. N. Engl. J. Med. 320:1188–1196 (1989).

209. Unal, S., Hoskins, J., Flokowitsch, J.E., Wu, C.Y.E., Preston, D.A., Skatrud, P.L.: Detection of methicillin-resistant staphylococci by using the polymerase chain reaction. J. Clin. Microbiol. 30:1685–1691 (1992).

210. Ubukata, K., Nakagami, S., Nitta, A., Yamane, A., Kawakami, S., Sugiura, M., Konno, M.: Rapid detection of the *mecA* gene in methicillin-resistant staphylococci by enzymatic detection of polymerase chain reaction products. J. Clin. Microbiol. 30:1728–1733 (1992).

211. Matsuhashi, M., Song, M.D., Ishino, F., Wachi, M., Doi, M., Inoue, M., Ubukata, K., Yamashita, N., Konno, M.: Molecular cloning of the gene of a penicillin-binding protein supposed to cause high resistance to β-lactam antibiotics in *Staphylococcus aureus*. J. Bacteriol. 167:975–980 (1986).

212. Song, M.D., Wachi, M., Doi, M., Ishino, F., Matsuhashi, M.: Evolution of an inducible penicillin-target protein in methicillin-resistant *Staphylococcus aureus* by gene fusion. FEBS Lett. 221:167–171 (1987).

213. Telenti, A., Imboden, P., Marchesi, F., Lowrie, D., Cole, S., Colston, M.J., Matter, L., Schopfer, K., Bodmer, T.: Detection of rifampicin-resistance mutations in *Mycobacterium tuberculosis*. Lancet 341:647–650 (1993).

214. Telenti, A., Imboden, P., Marchesi, F., Schidheini, T., Bodmer, T.: Direct, automated detection of rifampin-resistant *Mycobacterium tuberculosis* by polymerase chain reaction and single strand conformation polymorphism analysis. Antimicrob. Agents Chemother. 37:2054–2058 (1993).

215. Kapur, V., Li, L.-L., Iordanescu, S., Hamrick, M.R., Wanger, A., Kreiswirth, B.N., Musser, J.M.: Characterization by automated DNA sequencing of mutations in the gene (*rpoB*) encoding the RNA polymerase β subunit in rifampin-resistant *Mycobacterium*

tuberculosis strains from New York City and Texas. J. Clin. Microbiol. 32:1095–1098 (1994).

216. Meier, A., Kirschner, P., Springer, B., Steingrube, V.A., Brown, B.A., Wallace, R.J. Jr., Bottger, E.C. Identification of mutations in 23S rRNA gene of clarithromycin-resistant *Mycobacterium intracellulare*. Antimicrob. Agents Chemother. 38:381–384 (1994).

217. Greiner, T.C.: Polymerase chain reaction—Uses and potential applications in cytology. Diagn. Cytopathol. 8:61–64 (1992).

218. Long, A.A., Mueller, J., Andre-Schwartz, J., Barrett, K.J., Schwartz, R., Wolfe, H.: High-specificity *in-situ* hybridization—Methods and application. Diagn. Mol. Pathol. 1:45–57 (1992).

219. Hanson, C.A., Holbrook, E.A., Sheldon, S., Schnitzer, B., Roth, M.S.: Detection of Philadelphia chromosome-positive cells from glass slide smears using the polymerase chain reaction. Am. J. Pathol. 137:1–6 (1990).

220. Ledbetter, D.H., Ballabio, A.: Molecular cytogenetics of contiguous gene syndromes: Mechanisms and consequences of gene dosage imbalance. In Ledbetter, D.H., Ballabio, A.: The Metabolic and Molecular Bases of Inherited Disease (7th ed., Volume I). New York: McGraw-Hill, 1995.

221. Koenig, M., Hoffman, E.P., Bertelson, C.J., Monaco, A.P., Feener, C., Kunkel, L.M.: Complete cloning of the Duchenne muscular dystrophy (DMD) cDNA and preliminary genomic organization of the DMD gene in normal and affected individuals. Cell 50:509–517 (1987).

222. Lupski, J.R., de Oca Luna, R.M., Slaugenhaupt, S., Pentao, L., Guzzetta, V., Trask, B.J., Saucedo Cardenas, O., Barker, D.F., Killian, J.M., Garcia, C.A., Chakravarti, A., Patel, P.I.: DNA duplication associated with Charcot-Marie-Tooth disease type 1A. Cell 66:219–232 (1991).

223. Lupski, J.R., Chance, P.F., Garcia, C.A.: Inherited primary peripheral neuropathies— Molecular genetics and clinical implications of CMT1A and HNPP. JAMA 270:2326–2330 (1993).

224. Rowley, J.D.: Molecular cytogenetics: Rosetta stone for understanding cancer. Cancer Res. 50:3816–3825 (1990).

225. Kuwano, A., Ledbetter, S.A., Dobyns, W.B., Emanuel, B.S., Ledbetter, D.H.: Detection of deletions and cryptic translocations in Miller-Dieker syndrome by *in situ* hybridization. Am. J. Hum. Genet. 49:707–714 (1991).

226. Li, Y.S., Le Beau, M.M., Mick, R., Rowley, J.D.: The proportion of abnormal karyotypes in acute leukemia samples related to method of preparation. Cancer Genet. Cytogenet. 52:93–100 (1991).

227. Foster, C.S., Mostofi, F.K.: Prostate cancer: Present status. Hum. Pathol. 23:209–210 (1992).

228. Trask, B.J.: Fluorescence *in situ* hybridization: Applications in cytogenetics and gene mapping. Trends. Genet. 7:149–154 (1991).

229. Verma, R.S., Luke, S., Conte, R.A.: FISH technique—What's all the fuss about? GATA 11:106–109 (1994).

230. Hindkjaer, J., Koch, J., Terkelson, C., Brandt, C.A., Kolvraa, S., Bolund, L.: Fast, sensitive multicolor detection of nucleic acids by Primed In Situ labeling (PRINS). Cytogenet. Cell Genet. 66:152–154 (1994).

231. Hulten, M.A., Gould, C.P., Goldman, A.S., Waters, J.J.: Chromosome *in situ* suppression hybridisation in clinical cytogenetics. J. Med. Genet. 28:577–582 (1991).

232. Dauwerse, J.G., Wiegant, J., Raap, A.K., Breuning, M.H., van Ommen, G.J.B.: Multiple colors by fluorescence *in situ* hybridization using radiolabelled DNA probes create a molecular karyotype. Hum. Mol. Genet. 1:593–598 (1992).

233. Lawrence, J.B., Carter, K.C., Gerdes, M.: Extending the capabilities of interphase chromatin mapping. Nature Genet. 2:171–172 (1992).

234. Lengauer, C., Riethman, H., Cremer, T.: Painting of human chromosomes with probes generated from hybrid cell lines by PCR with *Alu* and L1 primers. Hum. Genet. 86:1–6 (1990).

235. Parrish, J.E., Nelson, D.L.: Practical aspects of fingerprinting human DNA using *Alu* polymerase chain reaction. Methods Mol. Cell. Biol. 5:71–77 (1994).

236. Lengauer, C., Riethman, H.C., Speicher, M.R., Taniwaki, M., Konecki, M., Konecki, D., Green, E.D., Becher, R., Olson, M.V., Cremer, T.: Metaphase and interphase cytogenetics with *Alu*-PCR-amplified yeast artificial chromosome clones containing the *BCR* gene and the protooncogenes c-*raf*-1, c-*fms*, and c-*erbB-2*. Cancer Res. 52:2590–2596 (1992).

237. Ried, T., Baldini, A., Rand, T.C., Ward, D.C.: Simultaneous visualization of seven different DNA probes by *in situ* hybridization using combinatorial fluorescence and digital imaging microscopy. Proc. Natl. Acad. Sci. U.S.A. 89:1388–1392 (1992).

238. van Dekken, H., Pizzolo, J.G., Kelsen, D.P., Melamed, M.R.: Targeted cytogenetic analysis of gastric tumors by *in situ* hybridization with a set of chromosome-specific DNA probes. Cancer 66:491–497 (1990).

239. van Dekken, H., Pizzolo, J.G., Reuter, V.E., Melamed, M.R.: Cytogenetic analysis of human solid tumors by *in situ* hybridization with a set of 12 chromosome-specific DNA probes. Cytogenet. Cell Genet. 54:103–107 (1990).

240. Balazs, M., Mayall, B.H., Waldman, F.M.: Interphase cytogenetics of a male breast cancer. Cancer Genet. Cytogenet. 55:243–247 (1991).

241. Anastasi, J., Le Beau, M.M., Vardiman, J.W., Fernald, A.A., Larson, R.A., Rowley, J.D.: Detection of trisomy 12 in chronic lymphocytic leukemia by fluorescence *in situ* hybridization to interphase cells: A simple and sensitive method. Blood 79:1796–1801 (1992).

242. Lengauer, C., Eckelt, A., Weith, A., Endlich, N., Ponelies, N., Lichter, P., Greulich, K.O., Cremer, T.: Painting of defined chromosomal regions by *in situ* suppression hybridization of libraries from laser-microdissected chromosomes. Cytogenet. Cell Genet. 56:27–30 (1991).

243. Cotran, R.S., Kumar, V., Robbins, S.L.: Neoplasia. In Cotran, R.S., Kumar, V., Robbins, S.L. (eds.): Robbins Pathologic Basis of Disease. Philadelphia: W.B. Saunders Co., 1989.

244. Groffen, J., Stephenson, J.R., Heisterkamp, N.: Philadelphia chromosomal breakpoints are clustered within a limited region *bcr* on chromosome 22. Cell 36:93 (1984).

245. Shtivelman, E., Lifshitz, B., Gali, R.P.: Fused transcript of *abl* and *bcr* genes in chronic myelogenous leukemia. Nature 315:550 (1985).

246. Heisterkamp, N., Jenster, G., ten Hoeve, J., Zovich, D., Pattengale, P.K., Groffen, J.: Acute leukaemia in *bcr/abl* transgenic mice. Nature 344:251–253 (1990).

247. Nowell, P.C., Hungerford, D.A.: A minute chromosome in human chronic granulocytic leukemia. Science 132:1497 (1960).

248. Rowley, J.D.: A new consistent chromosomal abnormality in chronic myelogenous leukaemia identified by quinacrine fluorescence and Giemsa staining. Nature 243:290–293 (1973).

249. Blennerhassett, G.T., Furth, M.E., Anderson, A., Burns, J.P., Chaganti, R.S., Blick, M., Talpaz, M., Dev, V.G., Chan, L.C., Wiedemann, L.M.: Clinical evaluation of a DNA probe assay for the Philadelphia (Ph1) translocation in chronic myelogenous leukemia. Leukemia 2:648–657 (1988).

250. Tkachuk, D.C., Westbrook, C.A., Andreeff, M., Donlon, T.A., Cleary, M.L., Surya-narayan, K., Homge, M., Redner, A., Gray, J., Pinkel, D.: Detection of *bcr-abl* fusion in chronic myelogenous leukemia by *in situ* hybridization. Science 250:559–562 (1990).

251. Amiel, A., Yarkoni, S., Slavin, S., Or, R., Lorberboum-Galski, H., Fejgin, M., Nagler, A.: Detection of minimal residual disease state in chronic myelogenous leukemia patients using fluorescence *in situ* hybridization. Cancer Genet. Cytogenet. 76:59–64 (1994).

252. Thirman, M.J., Gill, H.J., Burnett, R.C., Mbangkollo, D., McCabe, N.R., Kobayashi, H., Ziemin-van-der Poel, S., Kaneko, Y., Morgan, R., Sandberg, A.A., Chaganti, R.S.K., Larson, R.A., Le Beau, M.M., O, Diaz, M., Rowley, J.: Rearrangement of the *MLL* gene in acute lymphoblastic and acute myeloid leukemias with 11q23 chromosomal translocations. N. Engl. J. Med. 329:909–914 (1993).

253. DeLattre, O., Zucman, J., Melot, T., Sastre Garau, X., Zucker, J.-M., Lenoir, G.M., Ambros, P.F., Sheer, D., Turc-Carel, C., Triche, T.J., Aurias, A., Thomas, G. The Ewing family of tumors—A subgroup of small round cell tumors defined by specific chimer ic transcripts. N. Engl. J. Med. 331:294–299 (1994).

254. O'Leary, T.J., Stetler-Stevenson, M.: Diagnosis of t(14:18) by polymerase chain reaction. Arch. Pathol. Lab. Med. 118:789–790 (1994).

255. Stehelin, D., Varmus, H.E., Bishop, J.M., Vogt, P.K.: DNA related to the transforming gene(s) of avian sarcoma viruses is present in normal avian DNA. Nature 260:170–173 (1976).

256. Shih, C., Padhy, L.C., Murray, M., Weinberg, R.A.: Transforming genes of carcinomas and neuroblastomas introduced into mouse fibroblasts. Nature 290:261–264 (1981).

257. Parada, L.F., Tabin, C.J., Shih, C., Weinberg, R.A.: Human EJ bladder carcinoma oncogene is homologue of Harvey sarcoma virus *ras* gene. Nature 297:474–478 (1982).

258. Tabin, C.J., Bradley, S.M., Bargmann, C.I., Weinberg, R.A., Papageorge, A.G., Scolnick, E.M., Dhar, R., Lowy, D.R., Chang, E.H.: Mechanism of activation of a human oncogene. Nature 300:143–149 (1982).

259. Harvey, J.J.: An unidentified virus which causes the rapid production of tumors in mice. Nature 204:1104–1105 (1964).

260. Kirsten, W.H. Mayer, L.A.: Morphologic responses to a murine erythroblastosis virus. J. Natl. Cancer Inst. 39:311–334 (1967).

261. Deng, G.R., Liu, X.H., Wang, J.R.: Correlation of mutations of oncogene C-Ha-ras at codon 12 with metastasis and survival of gastric cancer patients. Oncogene Res. 6:33–38 (1991).

262. Namba, H., Rubin, S.A., Fagin, J.A.: Point mutations of *ras* oncogenes are an early event in thyroid tumorigenesis. Mol. Endocrinol. 4:1474–1479 (1990).

263. Motojima, K., Tsunoda, T., Kanematsu, T., Nagata, Y., Urano, T., Shiku, H.: Distinguishing pancreatic carcinoma from other periampullary carcinomas by analysis of mutations in the Kirsten-*ras* oncogene. Ann. Surg. 214:657–662 (1991).

264. Motojima, K., Urano, T., Nagata, Y., Shiku, H., Tsunoda, T., Kanematsu, T.: Mutations in the Kirsten-*ras* oncogene are common but lack correlation with prognosis and tumor stage in human pancreatic carcinoma. Am. J. Gastroenterol. 86:1784–1788 (1991).

265. Shen, C., Chang, J.G., Lee, L.S., Yang, M.J., Chen, T.C., Lin, K.Y., Lee, M.D., Chen, P.H.: Analysis of *ras* gene mutations in gastrointestinal cancers. Taiwan. I. Hsueh. Hui. Tsa. Chih. 90:1149–1154 (1991).

266. Stork, P., Loda, M., Bosari, S., Wiley, B., Popoenhusen, K., Wolfe, H.: Detection of K-*ras* mutations in pancreatic and hepatic neoplasms by non-isotopic mismatched polymerase chain reaction. Oncogene 6:857–862 (1991).

267. Cline, M.J.: The molecular basis of leukemia. N. Engl. J. Med. 330:328–336 (1994).

268. Zhang, Y., Coyne, M.Y., Will, S.G., Levenson, C.H., Kawasaki, E.S.: Single-base mutational analysis of cancer and genetic diseases using membrane bound modified oligonucleotides. Nucleic Acids Res. 19:3929–3933 (1991).

269. Lips, C.J.M., Landsvater, R.M., Hoppener, J.W.M., Geerdink, R.A., Blijham, G., Jansen-Schillhorn van Veen, J.M., van Gils, A.P.G., de Wit, M.J., Zewald, R.A., Berends, M.J.H., Beemer, F.A., Jansen, R.P.M., Brouwers-Smalbraak, J., Ploos van Amstel H.K., van Vroonhoven, T.J.M.V., Vroom, T.M.: Clinical screening as compared with DNA analysis in families with multiple endocrine neoplasia type 2A. N. Engl. J. Med. 331:828–835 (1994).

270. Slamon, D.J., Clark, G.M., Wong, S.G., Levin, W.J., Ulrich, A., McGuire, W.L.: Human breast cancer: Corrleation of relapse and survival with amplification of the *HER-2/neu* oncogene. Science 235:177–182 (1987).

271. Sellami, M., Gamoudi, M., Krichen, K., Kharrat, A., Ben Romdhane, K., Maalej, M.: Incidence of amplification of the C-erb B2/Her-2/neu gene in human breast cancer. Arch. Inst. Pasteur Tunis 68:33–41 (1991).

272. Zhou, D., Battifora, H., Yokota, J., Yamamoto, T., Cline, M.J.: Association of multiple copies of the c-*erbB*-2 oncogene with spread of breast cancer. Cancer Res. 47:6123–6125 (1987).

273. Brodeur, G.M.: Clinical significance of genetic rearrangements in human neuroblastomas. Clin. Chem. 35:B38–B42 (1989).

274. Frye, R.A., Benz, C.C., Liu, E.: Detection of amplified oncogenes by differential polymerase chain reaction. Oncogene 4:1153–1157 (1989).

275. Levine, A.J.: The p53 tumor-suppressor gene. N. Engl. J. Med. 326:1350–1351 (1992).

276. Donehower, L.A., Harvey, M., Slagle, B.L., McArthur, M.J., Montgomery, J., Butel, J.S., Bradley, A.: Mice deficient for p53 are developmentally normal but susceptible to spontaneous tumours. Nature 356:215–221 (1992).

277. Hollstein, M., Sidransky, D., Vogelstein, B., Harris, C.C.: p53 mutations in human cancers. Science 253:49–53 (1991).

278. Osborne, R.J., Merlo, G.R., Mitsudomi, T., Venesio, T., Liscia, D.S., Cappa, A.P., Chiba, I., Takahashi, T., Nau, M.M., Callahan, R.: Mutations in the p53 gene in primary human breast cancers. Cancer Res. 51:6194–6198 (1991).

279. Runnebaum, I.B., Nagarajan, M., Bowman, M., Soto, D., Sukumar, S.: Mutations in p53 as potential molecular markers for human breast cancer. Proc. Natl. Acad. Sci. U.S.A. 88:10657–10661 (1991).

280. de Vos, S., Wilczynski, S.P., Fleischhacker, M., Koeffler, P.: Alterations in uterine leiomyosarcomas versus leiomyomas. Gynecol. Oncol. 54:205–208 (1994).

281. Sameshima, Y., Tsunematsu, Y., Watanabe, S., Tsukamoto, T., Kawa-ha, K., Hirata, Y., Mizoguchi, H., Sugimura, T., Terada, M., Yokota, J.: Detection of novel germ-line p53 mutations in diverse-cancer-prone families identified by selecting patients with childhood adrenocortical carcinoma. J. Natl. Cancer Inst. 84:703–707 (1992).

282. Toguchida, J., Yamaguchi, T., Dayton, S.H., Beauchamp, R.L., Herrera, G.E., Ishizaki, K., Yamamuro, T., Meyers, P.A., Little, J.B., Sasaki, M.S., Weichselbaum, R.R., Yandell, D.W.: Prevalence and spectrum of germline mutations of the p53 gene among patients with sarcoma. N. Engl. J. Med. 326:1301–1308 (1992).

283. Malkin, D., Jolly, K.W., Barbier, N., Look, A.T., Friend, S.H., Gebhardt, M.C., Andersen, T.I., Borresen, A.-L., Li, F.P., Garber, J., Strong, L.C.: Germline mutations of the p53 tumor-suppressor gene in children and young adults with second malignant neoplasms. N. Engl. J. Med. 326:1309–1315 (1992).

284. Sidransky, D., Mikkelsen, T., Schwechheimer, K., Rosenblum, M.L., Cavenee, W., Vogelstein, B.: Clonal expansion of p53 mutant cells is associated with brain tumour progression. Nature 355:846–847 (1992).

285. Sameshima, Y., Matsuno, Y., Hirohashi, S., Shimosato, Y., Mizoguchi, H., Sugimura, T., Terada, M., Yokota, J.: Alterations of the p53 gene are common and critical events for the maintenance of malignant phenotypes in small-cell lung carcinoma. Oncogene 7:451–457 (1992).

286. Cavenee, W.K., Scrable, H.J., James, C.D.: Molecular genetics of human cancer predisposition and progression. Mutat. Res. 247:199–202 (1991).

287. Hansen, M.F., Cavenee, W.K.: Retinoblastoma and the progression of tumor genetics. Trends Genet. 4:125–128 (1988).

288. Hansen, M.F., Morgan, R., Sandberg, A.A., Cavenee, W.K.: Structural alterations at the putative retinoblastoma locus in some human leukemias and preleukemia. Cancer Genet. Cytogenet 49:15–23 (1990).

289. Fearon, E.R., Vogelstein, B.: A genetic model for colorectal tumorigenesis. Cell 61:759–767 (1990).

290. Bodmer, W., Bishop, T., Karran, P.: Genetic steps in colorectal cancer. Nature Genet. 6:217–219 (1994).

291. Peltomaki, P.T.: Genetic basis of hereditary nonpolyposis colorectal carcinoma (HNPCC). Ann. Med. 26:215–219 (1994).

292. Papadopoulos, N., Nicolaides, N.C., Wei, Y.-F., Ruben, S.M., Carter, K.C., Rosen, C.A., Haseltine, W.A., Fleischmann, R.D., Fraser, C.M., Adams, M.D., Venter, J.C., Hamilton, S.R., Petersen, G.M., Watson, P., Lynch, H.T., Peltomaki, P., Mecklin, J.-P., de la Chapelle, A., Kinzler, K.W., Vogelstein, B.: Mutation of a *mutL* homolog in hereditary colon cancer. Science 263:1625–1629 (1994).

293. Powell, S.M., Petersen, G.M., Krush, A.J., Booker, S., Jen, J., Giardiello, F.M., Hamilton, S.R., Vogelstein, B., Kinzler, K.W.: Molecular diagnosis of familial adenomatous polyposis. N. Engl. J. Med. 329:1982–1987 (1993).

294. Futreal, P.A., Liu, Q., Shattuck-Eidens, D., Cochran, C., Harshman, K., Tavtigan, S., Bennett, L.M., Haugen-Strano, A., Swensen, J., Miki, Y., Eddington, K., McClure, M., Frye, C., Weaver-Feldhaus, J., Ding, W., Gholami, Z., Soderkvist, P., Terry, L., Jhanwar, S., Berchuck, A., Igelhart, J.D., Marks, J., Ballinger, D.G., Barrett, J.C., Skolnick, M.H., Kamb, A., Wiseman, R.: BRCA1 mutations in primary breast and ovarian carcinomas. Science 266:120–122 (1994).

295. Wooster, R., Newhausen, S.L., Mangion, J., Quirk, Y., Ford, D., Collins, N. Nguyen, K., Seal, S., Tran, T., Averill, D., et al.: Localization of a breast cancer susceptibility gene, BRCA2, to chromosome 13q12–13. Science 265:2088–2090 (1994).

296. Weber, B.L.: Susceptibility genes for breast cancer. N. Engl. J. Med. 331:1523–1524 (1994).

297. Steeg, P.S., Cohn, K.H., Leone, A.: Tumor metastasis and *nm23*: Current concepts. Cancer Cells 3:257–262 (1991).

298. Leone, A., McBride, O.W., Weston, A., Wang, M.G., Anglard, P., Cropp, C.S., Goepel, J.R., Lidereau, R., Callahan, R., Linehan, W.M.: Somatic allelic deletion of *nm23* in human cancer. Cancer Res. 51:2490–2493 (1991).

299. Cohn, K.H., Wang, F.S., Desoto LaPaix, F., Solomon, W.B., Patterson, L.G., Arnold, M.R., Weimar, J., Feldman, J.G., Levy, A.T., Leone, A.: Association of *nm23*-H1 allelic deletions with distant metastases in colorectal carcinoma. Lancet 338:722–724 (1991).

300. Ling, V.: P-glycoprotein and resistance to anticancer drugs. Cancer 69:2603–2609 (1992).

301. Ng, W.F., Sarangi, F., Zastawny, R.L., Veinot Drebot, L., Ling, V.: Identification of members of the P-glycoprotein multigene family. Mol. Cell Biol. 9:1224–1232 (1989).

302. Chan, H.S., Thorner, P.S., Haddad, G., Ling, V.: Immunochemical detection of P-glycoprotein: Prognostic correlation in soft tissue sarcoma of childhood. J. Clin. Oncol. 8:689–704 (1990).

303. Klein, E.A., Allen, G., Fair, W.R., Reuter, V., Chaganti, R.S. Absence of structural alterations of the multidrug resistance genes in transitional cell carcinoma. Urol. Res. 18:281–286 (1990).

304. Ito, Y., Tanimoto, M., Kumazawa, T., Okumura, M., Morishima, Y., Ohno, R., Saito, H.: Increased P-glycoprotein expression and multidrug-resistant gene (*mdr1*) amplification are infrequently found in fresh acute leukemia cells. Sequential analysis of 15 cases at initial presentation and relapsed stage. Cancer 63:1534–1538 (1989).

305. Fugger, L., Tisch, R., Libau, R., van Endert, P., McDevitt, H.O.: The role of human major histocompatibility complex (HLA) genes in disease. In Scriver, C.R., Beaudet, A.L., Sly, W.S., Valle, D. (eds.): The Molecular and Metabolic Bases of Inherited Disease (7th ed., Volume I). New York: McGraw-Hill, 1995.

306. Watkins, D.I., McAdam, S.N., Liu, X., Strang, C.R., Milford, E.L., Levine, C.G., Garber, T.L., Dogon, A.L., Lord, C.I., Ghim, S.H., Troup, G.M., Hughes, A.L., Letvin, N.L.: New recombinant HLA-B alleles in a tribe of South American Amerindians indicate rapid evolution of MHC class I loci. Nature 357:329–333 (1992).

307. Gyllensten, U., Allen, M.: PCR-based HLA class II typing. PCR Methods Applications 1:91–98 (1991).

308. Doherty, D.G., Donaldson, P.T.: HLA-DRB and DQB typing by a combination of serol-

ogy, restriction fragment length polymorphism analysis and oligonucleotide probing. Eur. J. Immunogenet. 18:111–124 (1991).

309. Vijverberg, K., Schreuder, G.M., Kenter, M.J., Naipal, A.M., van Rood, J.J., Giphart, M.J.: Applicability of HLA-DQB oligonucleotide typing for the TA10 and 2B3 specificities in routine HLA typing. Tissue Antigens 35:165–171 (1990).

310. Peter, J.B., Hawkins, B.R.: The new HLA. Arch. Pathol. Lab. Med. 116:11–15.

311. Martin, M., Carrington, M., Mann, D.: A method for using serum or plasma as a source of DNA for HLA typing. Hum. Immunol. 33:108–113 (1992).

312. Bidwell, J.: Advances in DNA-based HLA-typing methods. Immunol. Today 15:303–307 (1994).

313. Erlich, H., Bugawan, T., Begovich, A.B., Scharf, S., Griffith, R., Saiki, R., Higuchi, R., Walsh, P.S.: HLA-DR, DQ and DP typing using PCR amplification and immobilized probes. Eur. J. Immunogenet. 18:33–55 (1991).

314. Fugger, L., Morling, N., Ryder, L.P., Odum, N., Svejgaard, A.: Technical aspects of typing for HLA-DP alleles using allele-specific DNA *in vitro* amplification and sequence-specific oligonucleotide probes. Detection of single base mismatches. J. Immunol. Methods 129:175–185 (1990).

315. Erlich, H., Bugawan, T., Scharf, S.: Analysis of HLA class II polymorphism using polymerase chain reaction. Arch. Pathol. Lab. Med. 117:482–485 (1993).

316. Sengar, D.P.S., Gidlstein, R.: Comprehensive typing of DQB1 alleles by PCR-RFLP. Tissue Antigens 43:242–248 (1994).

317. Breur-Vriesendorp, B.S., Dekker-Saeys, A.J., Ivanyi, P.: Distribution of HLA-B27 subtypes in patients with ankylosing spondylitis: The disease is associated with a common determinant of the various B27 molecules. Ann. Rheum. Dis. 46:353–356 (1987).

318. Dorman, J.S., LaPorte, R.E., Stone, R.A., Trucco, M.: Worldwide differences in the incidence of type I diabetes are associated with amino acid variation at position 57 of the HLA-DQβ chain. Proc. Natl. Acad. Sci. U.S.A. 87:7370–7374 (1990).

319. Khalil, I., d'Auriol, L., Gobet, M., Morin, L., Lepage, V., Deschamps, I., Park, M., Degos, L., Galibert, F., Hors, J.: A combination of HLA DQB Asp 57-negative and HLA DQA Arg 52 confers susceptibility to insulin-dependent diabetes mellitus. J. Clin. Invest. 85:1315–1319 (1990).

320. Tosi, G., Brunelli, S. Mantero, G., Magalini, A.R., Soffiati, M., Pinelli, L., Tridente, G., Accolla, R.S.: The complex interplay of the DQB1 and DQA1 loci in the generation of the susceptible and protective phenotype for insulin-dependent diabetes mellitus. Mol. Immunol. 31:429–437 (1994).

321. Mahran, M.Z., Ross, D.G., Sadeghi, N.A., Rabson, A.R.: Use of the polymerase chain reaction mismatch technique to identify the HLA-DQw8 allele in patients with insulin-dependent diabetes mellitus. Am. J. Clin. Pathol. 97:29–33 (1992).

322. Kadowaki, T., Kadowaki, H., Mori, Y., Tobe, K., Sakuta, R., Suzuki, Y., Tanabe, Y., Sakura, H., Awata, T., Goto, Y.-I., Hayakawa, T., Matsuoka, K., Kawamori, R., Kamada, T., Horai, S., Nonaka, I., Hagura, R., Akanuma, Y., Yazaki, Y.: A subtype of diabetes mellitus associated with a mutation of mitochondrial DNA. N. Engl. J. Med. 330:962–968 (1994).

CHAPTER 9

DEVELOPMENTS IN FOOD TECHNOLOGY: APPLICATIONS AND REGULATORY AND ECONOMIC CONSIDERATIONS[1]

PETER FENG, KEITH A. LAMPEL, and WALTER E. HILL
Center for Food Safety and Applied Nutrition, FDA,
Washington, D.C. 20204, (P. F., K. A. L.);
Seafood Products Research Center, FDA,
Bothell, WA 98041-3012 (W. E. H.)

9.1. INTRODUCTION

National regulatory programs for food safety did not exist in the United States before the 1900s. At that time, there were no great demands for testing foods for contamination, and consumers relied solely on manufacturers to produce foods that were unadulterated and safe for consumption. However, with the rapid growth in the economy and the increased variety of food produced, rising public concern for food safety made the testing of foods a necessity. Growth of the dairy industry in response to consumer demands exceeded the handling capacity of many farms, and milk became a dangerous source of infection for diptheria, typhoid, and scarlet fever [1]. At the same time, unscrupulous producers, seeking to increase profits while holding down production costs, began to adulterate many foods. Dilution of milk with water and addition of charcoal dust or sawdust to ground coffee and cocoa became frequent practices. Also, to prolong the shelf life of perishable foods but

[1]References to specific products or companies in this chapter do not represent an endorsement by the authors or by the U.S. Food and Drug Administration.

Nucleic Acid Analysis: Principles and Bioapplications, pages 203–229
© 1996 Wiley-Liss, Inc.

without regard for consumer safety, producers sometimes used borax and formaldehyde to preserve butter and milk.

In response to public demands for safe foods, the U.S. Congress passed the Food
and Drug Act of 1906, making it illegal for foods in interstate commerce to be adul-
terated or misbranded or to contain substances injurious to health. However, the
burden of proof of food adulteration or contamination rested with the federal government and not with the food industry. Therefore, interest in developing analytical
methods for food contaminants was pursued actively only by federal regulatory
agencies.

That situation changed in 1938 with the passing of the Federal Food, Drug, and
Cosmetic Act, which revised the authority of the federal government and better defined the requirements for food safety. In effect, the Act established the manufacturers' responsibility not only to know the law but also to comply with the requirements. The producers now had to ensure that the ingredients and the foods produced
were safe for consumption. As a result, the demand for test methods increased, and
research to develop better analytical methods was stimulated.

In recent years, with the advent of biotechnology, food-testing technology has itself grown to become an industry. Numerous companies are devoted to developing
assays for detecting microbial contaminants in foods. Using technologies such as
monoclonal antibodies, diagnostic nucleic acid probes (DNA probes), and others,
these assays are specific, faster, and often more sensitive than traditional microbiological methods. The new techniques may enable manufacturers and regulatory
agencies to monitor food quality better and to ensure the distribution of safe food
products to the consumers.

9.2. PRESENT APPLICATIONS

The U.S. Food and Drug Administration (FDA) has established six general categories of food-borne hazards, of which microbiological contamination appears to be
the most common and serious to humans [1,2]. Microbiological food-borne diseases are typically caused by the ingestion of live microbes or viruses (infections) or
by the consumption of preformed toxins (intoxications). Of the confirmed food-
borne outbreaks reported to the Centers for Disease Control and Prevention between 1983 and 1987, 66% were caused by bacteria, with salmonellosis being the
most prevalent disease [3]. It is not surprising, therefore, that the first nucleic acid
probe assays developed for testing foods were to detect bacterial pathogens, particularly *Salmonella*. Over the past 15 years, specific probes have been developed for
most of the traditional as well as newly emerging food-borne bacterial pathogens
and viruses.

9.2.1. Identification of Food-Borne Pathogens

Many of the DNA probes used for detecting pathogens in foods were initially developed for analysis of clinical specimens. It is not possible to discuss all the existing

probes or those currently being developed; therefore, only probes that have been tested in food analysis are discussed here. Selected probes and their respective target genes are listed in Table 9.1. A more comprehensive discussion and listing of DNA probe assays for detecting food-borne pathogens is reported elsewhere [4]. In addition, many probes are now being developed for use in conjunction with polymerase chain reaction (PCR) assays (see Section 9.5.1.). PCR uses short segments of DNA as primers to amplify a specific section of the bacterial or viral genome; therefore, the initial PCR step is essentially a DNA probe assay, where primers are hybridized to the specific target sequences. Although discussions on PCR are not within the scope of this chapter, selected PCR assays that have been used in food analysis are briefly mentioned. The applications of PCR assays in foods is reviewed elsewhere [5,6].

Campylobacter. The incidence of food-borne illness caused by *Campylobacter* spp. is steadily increasing in the United States [3]. *C. jejuni* is now recognized as one of the most common causes of acute gastroenteritis in humans, and infection is often associated with the consumption of raw milk or undercooked poultry [7]. The mechanism of *Campylobacter* virulence is not clear. Cytotoxins, endotoxins, and a cholera-like enterotoxin are produced by pathogenic isolates of *Campylobacter*; however, the role of these toxins in pathogenicity is uncertain [7]. Conventional isolation methods for *Campylobacter* spp. from foods require cultivation with selective media under reduced oxygen tension and elevated temperatures. The organisms are difficult to maintain in the laboratory.

Most of the DNA probes developed for the identification of *Campylobacter* are not targeted to specific virulence markers. Several genomic probes as well as probes specific for the 16S ribosomal RNA (rRNA) sequences have been used to identify these pathogens in clinical specimens. Some probes are species specific, genus specific, or targeted at specific groups, such as the thermophilic campylobacters [8].

A colorimetric DNA probe assay for the genus *Campylobacter* is available commercially from GENE-TRAK Systems (Framingham, MA). This probe, specific for the rRNA sequences, was evaluated for the detection of *Campylobacter* spp. in poultry. The recovery efficiency of the probe assays was similar to that of culture methods; however, several false-positive and false-negative reactions were also observed [9]. A few PCR assays have been evaluated for detecting *Campylobacter* spp. in chicken products [10] and *C. coli* and *C. jejuni* in raw milk and diary products [11]. These assays were much faster and showed comparable if not better detection sensitivity than conventional culture methods.

Escherichia coli. Most strains of *E. coli* are common harmless commensals in the human gut. However, several groups within the species can cause diarrheal disease in humans. These pathogenic strains, categorized on the basis of epidemiology, serology, and distinct virulence properties, include enterotoxigenic (ETEC), enteropathogenic (EPEC), enterohemorrhagic (EHEC), enteroinvasive (EIEC), and enteroadherent (EAEC) *E. coli*. Some of these pathogenic groups differ in physio-

TABLE 9.1. Selected DNA Probes Used for Identifying Food-Borne Pathogenic Bacteria

Microorganism	Target	References
Campylobacter spp.	Ribosomal RNA	Romaniuk and Trust [113]
		Stern and Mozola [9]
Escherichia coli		
Enterotoxigenic	Stable and labile toxin	Hill et al. [15]
		Hill and Payne [18]
		Hill et al. [16]
		Ferreira et al. [17]
		Cryan [19]
Enterohemorrhagic	Shiga-like toxins	Samadpour et al. [25]
		Smith et al. [27]
		Samadpour et al. [26]
O157:H7	*uid*A gene	Feng [29]
Enteropathogenic	Adhesion	Nataro et al. [22]
Enteroinvasive and *Shigella* spp.	Invasiveness	Jagow and Lampel [34]
Salmonella spp.	Mixed fragments	Fitts et al. [58]
	Unknown	Tsen et al. [67]
	Ribosomal RNA	Curiale et al. [63]
	IS*200*	Cano et al. [68]
Vibrio spp.		
V. parahaemolyticus	Hemolysin	Nishibushi et al. [90]
		Lee et al. [92]
V. vulnificus	Cytolysin	Morris et al. [91]
		Wright et al. [93]
V. cholerae	Cholera toxin	Yoh et al. [94]
Yersinia spp.		
Y. enterocolitica	Virulence plasmid	Jagow and Hill [75]
		Miliotis et al. [76]
		Kapperud et al. [77]
		Nesbakken et al. [79]
	Invasion gene	Feng [80]
		Goverde et al. [81]
Clostridium spp.		
C. perfringens	Enterotoxin	Van Damme et al. [87]
Listeria spp.	Ribosomal RNA	King et al. [45]
		Ninet et al. [46]
L. monocytogenes	Invasion gene	Datta et al. [41]
		Kim et al. [42]
	Listeriolysin	Datta et al. [44]
Staphylococcus spp.		
S. aureus	Enterotoxin	Notermans et al. [84]
		Neill et al. [85]

logical requirements and in some biochemical reactions from typical *E. coli* strains; therefore, conventional *E. coli* isolation and identification procedures may not be useful to identify these pathogens [12]. Furthermore, cultural enrichment procedures may cause the loss of plasmids [13] which encode for virulence properties of many of these isolates.

ETEC. Recognized as the most frequent cause of travelers' diarrhea, ETEC is also a major cause of infant diarrhea in third world nations. ETEC produces plasmid-encoded, heat-labile (LT) and heat-stable (ST) enterotoxins and factors essential for adherence and colonization of the intestinal mucosa [12]. The toxigenic strains or the toxins may be identified by tissue culture assays, serology, enzyme-linked immunosorbent assays (ELISA), and genetic probes [14]. Volunteer feeding studies have shown that a fairly large infective dose (10^5–10^6 cells) is required for ETEC to cause illness [12]. Therefore, in addition to detection, the number of ETEC in foods must be quantified in the assessment of food safety and risk.

Some of the first DNA probes used in food microbiology were targeted to the LT or ST genes of ETEC. An 850 base pair plasmid fragment specific for the LT gene was used to analyze artificially contaminated foods [15]. Colony hybridization assays showed that the probe assay was able to detect 100 cells per gram of food plated onto selective media [16]. The probe was also used effectively to detect ETEC in various types of seafood samples [17]. Comparisons of DNA probes, immunoassays, and animal models generally showed that these methods were equally efficient for identifying ETEC [15,18]. A DNA probe assay called SNAP is now commercially available, and it uses a probe specific for the ST gene of ETEC [19]. Primers specific for the LT genes have also been used in PCR assays to identify ETEC efficiently in minced meat samples [20].

EPEC. Gastroenteritis caused by EPEC is most prevalent in infants. Although the incidence of such outbreaks is now less frequent, they still occur occasionally. Isolates of EPEC do not produce toxins but elaborate an enteroadherence factor that is plasmid encoded and enables the pathogen to adhere to HEp-2 cells *in vitro* [12]. The *E. coli* methodology described in the FDA *Bacteriological Analytical Manual* [21] may be used to isolate EPEC. However, there are no reliable methods to detect this organism directly in foods. A plasmid fragment that encodes for an adherence factor has been used as a probe [22]; however, it has been tested only with clinical samples [14].

EHEC. EHEC causes a clinically distinct gastrointestinal disorder called *hemorrhagic colitis*, which may progress to the more severe hemolytic uremic syndrome. This organism has been implicated in numerous food-borne outbreaks in 1993, including the largest outbreak, which occurred in Washington state and was traced to the consumption of contaminated hamburger. The mechanism of EHEC pathogenesis has not been elucidated fully. Although it does not appear to be invasive, the pathogen can adhere to the intestines, where it produces several shiga-like toxins (SLT), also known as verotoxins, that are encoded by lysogenic bacteriophages [12].

EHEC does not grow well at elevated temperatures; hence, conventional methods for *E. coli* are not useful for isolating this pathogen from foods [23]. Although many serogroups of EHEC produce SLT, the most common serotype associated with food-borne diseases is O157:H7. This serotype has been found in 2% of beef and poultry samples, and it is identified serologically by the O157 and H7 antigens.

Most of the probe assays for EHEC are directed to the SLT genes [24]. Although many probes have been developed for the analysis of clinical specimens, only a few have been evaluated for testing foods. One set of SLT-specific probes was able to detect as few as 1.3 colony-forming units (CFU) of O157:H7 initially seeded per gram of ground beef and oyster samples [25], and was also used to identify EHEC isolates in retail raw goat milk, fruits, and surimi-based products. These probes have also been used to survey retail fresh seafood, beef, lamb, and pork [26] and chicken and sausage samples for the presence of SLT-producing EHEC [27] and to study the epidemiology and profile of O157:H7 isolates implicated in the 1993 outbreak in Washington state [28]. There are also many SLT gene-specific PCR assays that have been used to test clinical specimens or to determine the toxigenic potential of EHEC isolates from foods [5]. However, few PCR assays have been used directly to detect EHEC in food samples.

SLT-specific probes or PCR assays will identify all SLT-producing serotypes of EHEC; therefore, these tests are not specific for O157:H7, the principal pathogenic serotype. However, a serotype-specific DNA probe was recently reported that identifies only isolates of O157:H7 [29]. The probe is directed to a unique sequence in the *uid*A gene that is highly conserved within the O157:H7 serogroup. Analysis of isolates from the recent food-borne outbreaks showed that the probe distinguished O157:H7 from other SLT-producing serotypes. A modification of this probe is now being used as a primer in conjunction with SLT gene-specific primers in a multiplex PCR assay, where the serotype and SLT type of EHEC isolates may be determined simultaneously [30].

EIEC and Shigella. Bacillary dysentery, the human infection caused by EIEC and *Shigella* spp., accounted for almost 20% of food-borne illnesses reported in the United States between 1983 and 1987 [3]. Dysentery is caused by the bacterial penetration of the colonic mucosa followed by intracellular multiplication and destruction of tissue. The ability of these pathogens to invade and penetrate tissue cells is encoded by genes located on the chromosome and on a large virulence plasmid in *Shigella* spp. and EIEC [12]. Although conventional microbiological and serological methods may be used to isolate and identify EIEC and shigellae, these procedures are cumbersome and protracted. Furthermore, all isolates must be tested using animal models to confirm virulence.

Most of the probes developed to identify shigellae and EIEC are cloned DNA fragments targeted to the virulence genes on the invasion plasmid [31]. Colony hybridization analyses of clinical isolates showed the probe to be highly specific for Shigella and EIEC. Synthetic oligonucleotides directed to the invasion plasmid have also been tested as probes [32], or as PCR primers [33] for the analysis of shigellae in clinical specimens. Comparisons of animal assays for virulence versus reactivity with virulence gene-specific DNA probes generally showed good correlation between these methods [32].

Another oligonucleotide probe specific for the virulence plasmid was tested for its ability to identify EIEC and shigellae in artificially contaminated foods [34]. Probe sensitivity was highly susceptible to the level of normal flora present in the foods, but inclusion of selective enrichment improved probe performance. These probes have also been tested as primers in PCR assays and were able efficiently to identify shigellae in lettuce [35], raw milk [36], and various seafoods, greens, and dairy products [37].

E. coli species. GENE-TRAK manufactures a commercially available DNA probe assay for identifying *E. coli*. The assay consists of a species-specific colorimetric probe, and it can be used to screen samples for the presence of *E. coli* to monitor sanitary conditions or for breakdowns in quality control. The probe, however, will not distinguish between pathogenic and nonpathogenic *E. coli* or differentiate the various pathogenic serogroups. This DNA probe assay has not been evaluated extensively for the analysis of food samples.

Listeria. Outbreaks of listeriosis in the United States have been associated mostly with the consumption of contaminated unprocessed or underprocessed dairy products [2]. Although the overall incidence of food-borne illness caused by *Listeria monocytogenes* may be low, the high abortion and mortality rates associated with listeriosis, especially in immunocompromised individuals, have made *L. monocytogenes* one of the important food-borne pathogens of the past decade. Like *Salmonella*, all isolates of *L. monocytogenes* are considered to be virulent; therefore, no detectable levels are permitted in foods. Pathogenic *Listeria* spp. produce a hemolysin also known as *listeriolysin O*, which appears to be essential for virulence [38]. The ability to invade phagocytic cells is also a critical pathogenic factor for *Listeria* [39]. Although virulent strains may be identified by biochemical and hemolytic reactions, isolation procedures for *Listeria* require lengthy cold enrichment steps before plating on selective medium [40].

Several DNA probes directed to the virulence genes in *Listeria* spp. have been used to identify *Listeria* in artificially contaminated dairy products. These probes, once thought to be specific for the hemolysin gene [41], were later shown to be directed to a region encoding for an invasion-associated protein [39]. These oligonucleotide probes were able to detect 10–50 *L. monocytogenes* cells initially seeded per gram of milk or 100–500 bacterial cells per gram of cheese after enrichment. Nonisotopically labeled synthetic probes have been used to confirm food isolates of *Listeria* rapidly [42], and a probe specific for the listeriolysin O gene [43] was tested with seeded Brie cheese, lettuce, and shrimp samples. The probe identified *L. monocytogenes* cells in the presence of 10^6–10^8 CFU per gram of background flora [44]. A commercially available nonisotopic probe (GENE-TRAK) specific for the 16S rRNA of the *Listeria* genus has been tested in artificially contaminated foods. Comparative analysis with culture methods showed an overall agreement of 96% [45]. The Accuprobe for *L. monocytogenes* (GEN-PROBE) is another nonisotopic test that uses a chemiluminescent detection system. Though species specific, this test has a detection limit of 10^5 CFU and is used only for confirmation following isolation [46]. Furthermore, some of the enrichment culture media used for *L. monocytogenes* have been found to interfere with the Accuprobe test [47].

A few PCR assays, using primers specific for the hemolysin gene, have been tested in food analysis and found to detect low levels of *L. monocytogenes* seeded into various meat and dairy products [48,49]. Another PCR assay coupled with a selective capture procedure using immunomagnetic beads has been used to detect *L. monocytogenes* in cheese [50].

Salmonella.   Salmonellosis continues to be the most important and prevalent form of food-borne disease, with an estimated 2–4 million cases in the United States annually [3,51,52]. *Salmonella* infection typically results in gastroenteritis, and, although it is not generally fatal, deaths do occur, especially among those who are debilitated or immunodeficient. For example, individuals infected with the human immunodeficient virus (HIV) are estimated to be about 20 times more susceptible to salmonellosis than uninfected persons [53]. About 2,000 serovars of *Salmonella* have been serologically classified on the basis of somatic (O) and flagellar (H) antigens, and all species in the genus are considered pathogenic. Virulence factors of *Salmonella* include toxins and the ability to invade tissue cells [54]. Conventional culture methods for *Salmonella* spp. require pre-enrichment and selective enrichment before plating for isolation. Biochemical reactions and serology are used for identification. Recently, genetic methods such as pulsed-field gel electrophoresis and plasmid fingerprinting have become valuable tools in tracing the epidemiology of *Salmonella* outbreaks [55–57].

The first commercially available DNA probe assay for testing microbial contamination in foods was used to identify *Salmonella* spp. The prototype assay was a mixture of cloned DNA fragments that hybridized to more than 300 isolates of salmonellae [58]. Salmonellae seeded into foods were collected onto nitrocellulose membranes by filtration and tested with radioactively labeled probes by using a colony hybridization procedure. The sensitivity of the assay was unaffected by natural flora. The resulting commercial DNA probe kit for salmonellae (GENE-TRAK) was compared extensively with standard methods for the identification of *Salmonella* spp. in a wide variety of foods. In all cases, the probe correctly identified salmonellae with efficiencies equal to or better than the standard culture methods [59–61]. Compared with the other commercially available rapid methods, the sensitivity of the DNA probe method in identifying salmonellae in foods was equivalent to some enzyme-linked immunosorbent assays [62].

The first commercial DNA probe for *Salmonella* relied on radioactivity for detection, which precluded the widespread implementation of the assay. To overcome this drawback, GENE-TRAK introduced another probe-based assay that used an enzymatic detection system with a colorimetric substrate [63]. A collaborative study showed more than 97% agreement between standard methods and the nonisotopic probe assay in identifying *Salmonella* in foods. The assay also proved to be effective in detecting salmonellae in meat and poultry products [64]. The method was adopted as official first action by the Association of Official Analytical Chemists (AOAC) for screening *Salmonella* spp. in foods [65]; however, it has since undergone further modifications [66]. The GENE-TRAK probe assay is used only for presumptive screening of samples; hence, probe-positive results must still be confirmed by standard methods.

Recently, several additional probes, PCR, or combinations of these assays have been developed for detecting *Salmonella* spp. in foods. Some probes use chromogenic or fluorogenic markers and have been tested for the detection of *Salmonella* in seeded minced meat and fish [67] and in chicken and hamburger samples [68]. The detection sensitivity reported was about 10^3–10^4 CFU/g of food. A few combination probe/PCR assays have also been introduced for detecting *Salmonella* spp. in ground pork, beef, chicken, and turkey samples [69] as well as in oysters [70]. The latter assay was found to be highly specific and detected less than 40 *Salmonella* cells per gram of oyster meat without the need for cultural enrichment. Other PCR assays have utilized anti-salmonella monoclonal antibodies coupled to immunomagnetic beads to capture *Salmonella* cells specifically prior to PCR [71]. An initial inoculum of 0.1 CFU of salmonellae per g of chicken meat was detectable by this assay following enrichment.

Yersinia. Although not a frequent cause of gastroenteritis, *Y. enterocolitica* serotype O:8 has been implicated in several food-borne outbreaks in North America, and infections caused by serotypes O:3 and O:9 are common in Europe [72]. In Japan, enteritis caused by *Y. pseudotuberculosis* appears to be more prevalent. Although the pathogenic mechanisms of *Yersinia* spp. have not been fully elucidated, it is known that the production of enterotoxins and the ability to invade tissue culture cells are chromosomally encoded [73], and most other virulence properties reside on a plasmid that is conserved within the pathogenic *Yersinia* species [74]. Conventional methods to isolate *Yersinia* require a lengthy cold enrichment followed by selective plating. Furthermore, because most yersiniae are avirulent, all isolates must also be tested for pathogenicity using animal assays [72].

Several cloned DNA and oligonucleotide probes directed to the virulence plasmid have been used to identify pathogenic yersiniae in foods [75–77,107]. Probe detection sensitivities varied greatly depending on the type of food matrix and the level of normal flora present [78]. In general, probes were able to identify yersiniae in the presence of 10^4–10^5 CFU/ml of background flora [75,76]. Inclusion of an enrichment step in the procedure reduced background interference and greatly improved detection sensitivities [78]. Another plasmid-specific oligonucleotide probe was used to identify *Y. enterocolitica* serotype O:3 in naturally contaminated pork products [79]. Comparative analyses showed that the probe was more efficient than cultural procedures in identifying this serotype in foods.

Several cloned DNA and oligonucleotide probes specific for the chromosomally encoded invasion genes in *Yersinia* spp. have also been developed and evaluated [73,80,81]. Unlike plasmids that can be lost readily during cultivation, chromosomal genes are stably inherited; therefore, probes specific for chromosomal targets are not affected by decreased sensitivity due to plasmid loss. Although these invasion-specific probes for yersiniae have not been tested extensively in foods, a non-radioactive digoxigenin-labeled probe has been used to identify invasive *Y. enterocolitica* in naturally contaminated pig samples from a slaughterhouse [81].

A colorimetric probe, specific for the rRNA sequences of *Y. enterocolitica*, is commercially available from GENE-TRAK. However, the test is species specific and does not discriminate between virulent and avirulent isolates of *Y. enterocolitica*.

Recently, a PCR assay coupled with a nonisotopic probe was developed for identifying *Y. enterocolitica* isolates from pig and pork products [82]. Another PCR assay that can detect all common pathogenic serotypes of *Y. enterocolitica* uses immunomagnetic particles first to isolate yersiniae selectively from enrichment media. The PCR assay was able to detect 10–30 CFU/g of *Yersinia* in various meat samples containing in excess of 10^7 CFU/g of flora [83].

Other Food-Borne Pathogens. A number of other bacterial species cause food-borne illnesses. Intoxication by *Staphylococcus aureus* enterotoxins ranks second in frequency to *Salmonella* infections as the cause of food-borne gastroenteritis [52]. At least five immunologically distinct but genetically related enterotoxins can be produced by *S. aureus* strains. Although several probes directed to the toxin genes have been developed [84,85] very few have been tested for identifying toxigenic *S. aureus* in foods. However, a PCR assay using primers to the enterotoxin genes has been tested for detecting *S. aureus* in dried skimmed milk samples [86]. A colorimetric DNA probe assay is also commercially available from GENE-TRAK, but the probe will only identify *S. aureus* cells and will not determine their toxigenic potential.

Other food-borne intoxications are caused by the neurotoxins of *Clostridium botulinum* and the enterotoxins of *C. perfringens*. Several of these clostridial toxin genes have been cloned and sequenced [87,88], but only a few have been developed for use as specific probes. One of these, directed to the enterotoxin gene, has been used to identify toxigenic strains of *C. perfringens* [87]; but it has not been tested in the analysis of food samples. Similarly, very few DNA probes are reported for toxigenic *Bacillus cereus*, a common spoilage bacteria of dairy products and also associated with food-borne illness. However, genetic methods such as analysis of plasmid profiles have been used to study the epidemiology of *B. cereus* outbreaks [89].

Gastroenteritis caused by *Vibrio* spp. is most commonly associated with the consumption of seafood. These pathogens produce several enterotoxins, cytotoxins, and hemolysins, and several cloned DNA and oligonucleotide probes directed to these toxin genes have been developed [90,91]. An oligonucleotide probe specific for the hemolysin gene of *V. parahaemolyticus* was used to identify this species in seeded oysters; however, about 10^6 CFU was needed for a positive reaction [92]. Probes specific for the cytolysin gene of *V. vulnificus* [93] and the cholera toxin of *V. cholerae* [94] have also been tested for the identification of these species in oysters and environmental samples. Similarly, primers directed to these genes have been used effectively in PCR assays to identify *V. vulnificus* in oysters [95] and *V. cholerae* O1 serotype from seeded oysters, crabmeat, shrimp, and lettuce samples [96].

In addition to bacterial pathogens and toxins, enteroviruses are also commonly encountered in foods and can cause human illness. Transmission of hepatitis A via food and water is well known, and the virus is especially prevalent in undercooked shellfish [97]. Because hepatitis A is highly infectious, the presence of a few viral particles in food is of concern. Several probes have been developed to identify hepatitis A, but most of these have been tested only with clinical and environmental specimens [98–100]. There are also large numbers of food-borne viral gastroenteritis cases caused by the Norwalk agent and rotaviruses [101,102]. The Norwalk agent has long been difficult to propagate *in vitro*; however, recent studies have provided genetic sequences that are being used to develop probes or primers for PCR analysis. There are also probes against rotaviruses, but these have been tested only with clinical specimens [101].

9.2.2. Difficulties of Using DNA Probes in Food Analysis

DNA probes are already used in clinical laboratories for identifying bacteria and viruses in patient specimens [103]. It is difficult, however, for the food microbiologist to apply these same probe techniques directly to the analysis of foods. The following impediments, although discussed in relation to DNA probes, applies to almost all methods used for food analysis.

High Backgrounds. Clinical specimens such as blood or cerebrospinal fluid are usually sterile; hence, identification of infectious agents in these samples is relatively straightforward and not hampered by background organisms. Such circumstances happen rarely in food analysis, as many foods normally contain high levels of indigenous microflora, which interfere with the task of detecting specific bacterial agents that are present in much smaller numbers. This is especially critical when dealing with pathogens such as *Shigella* spp. and *E. coli* serotype O157:H7, which have low infectious doses. Assay interference by background flora may be reduced by selective treatments or cultural enrichments to favor the growth of target organisms. For instance, *Y. enterocolitica* is fairly resistant to alkali; hence a brief treatment of food samples homogenates with potassium hydroxide can reduce the numbers of competitors and increase the detection sensitivity of assays and DNA probes [75]. However, these procedures and enrichments extend the time required to complete the assay.

Time and Temperature Factors. In food-borne outbreaks, the incriminated food may not always be available for analysis because it may have already been consumed or discarded. In cases where a food sample is available, the analysis for bacterial pathogens present may be further complicated by time and temperature factors. For instance, when an outbreak is identified, usually a few days have elapsed by the time the suspected food is collected for analysis. This delay can significantly alter the dynamics of the microbial population in a food. Competing microflora may have overgrown, inhibited, or even inactivated the target pathogens. The number of viable pathogenic cells may also have decreased due to detrimental substances in the food itself or because of additives such as processing aids and bacteriocidal or bacteriostatic preservatives. In addition, some of these substances may cause physiological changes to bacterial cells during processing or storage, rendering them nonculturable, though still viable [104]. Because cultural enrichment to amplify target bacteria is still essential for most assays, including DNA probes, nonculturable pathogens in foods will not be easily detected. Processing and storage conditions may also affect the dynamics of bacterial populations in a food, making it susceptible to contamination. For example, heat treatments may reduce the level of indigenous microflora in a sample by several orders of magnitude. In the absence of competing flora, heat-resistant pathogens or bacteria introduced by postprocessing contamination can flourish easily. All of these time and temperature factors can have an effect on the sensitivity and the accuracy of assay systems to detect bacterial pathogens in foods.

Complex Matrices. The complexity of food matrices makes food analysis a highly challenging task. Although some foods can be blended to a reproducible consistency, many foods, especially those high in fat content or viscosity, are difficult to homogenize, which can result in significant sampling errors. Furthermore, compositional variations in foods may hamper the isolation and extraction of target nucleic acids or interfere with the performance of a DNA probe. Therefore, before a DNA probe can be effectively used for food analysis, considerable efforts must be expended to develop efficient and cost-effective extraction protocols for each food type. This may require input from food chemists, microbiologists, and molecular biologists—a situation that has not been realized often in the past.

Controls. The heterogeneity of food compositions also complicates the implementation and interpretation of control experiments. For example, to establish lower detectable limits of an assay system, foods are usually seeded with known numbers of target bacterial cells and analyzed to determine recovery efficiency. However, variations in the physical (fat content, pH, water activity) and microbiological (level and species of indigenous microflora) make-ups of various foods may require alterations in preparation procedures, which can affect the efficiency of a assay system from one food type to another or even from one sample to another. These factors can inject a degree of uncertainty into the analysis as to whether results obtained for a food are accurate or whether they are false reactions due to the nature of the food or because the preparation or sampling techniques are inadequate.

All of these aforementioned factors have contributed to the difficulties of adapting various probe assays to food analysis. Furthermore, these factors continue to plague even those assays that have been validated for food testing, as they can be used only for screening and presumptive testing, where all probe-positive reactions must still be confirmed by conventional, often cultural, microbiological procedures.

9.3. FACTORS IN DEVELOPING PROBES FOR FOOD ANALYSIS

A primary objective of many food testing methods is to assess the microbiological safety of food samples. To develop a DNA probe assay to achieve this objective requires expert scientific judgment and careful planning. Prior to development, however, extensive practical and economic considerations must also be addressed, such as the cost of the technology to the food industry, the way tests will be applied, and the potential for the assay to be marketed commercially. A more important factor to consider is the receptiveness to using DNA probes in food analysis by the industry and, in the case of public health laboratories and regulatory agencies, the acceptance of DNA probe results by the judicial systems.

9.3.1. Scientific Aspects for Probe Development

The actual scientific process for developing a gene probe, though complex, is fairly well understood and straightforward. A genetic target is selected, probes are made or synthesized, and an assay system is designed, developed, and tested. Often the most difficult aspects in this process are the selections of the specific food-borne pathogen and the genetic target sequences to be used for probe development.

Logistically and economically, it is impractical to test all foods for all known pathogens and toxins. It is also impossible to predict accurately which pathogenic microbes will be found in foods. Therefore, target selection priorities must be set on the basis of past outbreaks to determine the incidences of particular pathogens and the types of foods implicated. This deductive process, however, is now complex and challenging due to the changing nature of the food supply, processing methods, and distribution and handling systems. Accordingly, bacteria are evolving, adjusting, and expanding to other habitats in response to their environments. For example, in the past few decades, *Yersinia, Campylobacter, Listeria,* and enterohemorrhagic *E. coli* of O157:H7 serotype have emerged as food-borne pathogens of significant concern. Even established food-borne pathogens have adapted to new niches, e.g., *Clostridium botulinum* in smoked fish and cooked, chopped garlic; *S. aureus* enterotoxins in canned Chinese mushrooms; and *Salmonella enteritidis* in fresh in-shell chicken eggs. Thus, microbiologists must be careful in selecting targets for developing new tests and resist the temptation to respond to every food-borne disease outbreak or to every newly emerged pathogen.

Once the target pathogen has been identified, a probe is designed on the basis of its intended application. For instance, if all the species are considered pathogenic, as with *Salmonella*, a genus-specific probe would be preferred. On the other hand, if only selected isolates are pathogenic, as in the case of *Y. enterocolitica*, then the probe should target the species-specific virulence markers rather than the entire genus.

In the earlier development of genus-specific probes, e.g., *Salmonella*, random DNA restriction enzyme fragments were screened individually for specificity against a large number of *Salmonella* species [58]. More sophisticated approaches are now available from the extensive taxonomic studies on the genetic evolution of bacteria. For example, nucleotide sequence data of the ribosomal RNA genes have revealed variable and conserved genetic regions that are highly suitable as probe targets for specific taxonomic groups [105]. This approach has already been used to develop genus-specific probes for *Listeria* [45], *Salmonella* [63], and other pathogens.

Development of probes specific only for the pathogenic strains within a genus or species is more complex and requires knowledge of the mechanisms of pathogenicity and the genes involved. This information is usually obtained through published laboratory reports on the many aspects of bacterial virulence, including expression and regulation, nucleotide sequences of the virulence genes, physiological factors, and environmental effects. Even so, the actual mechanism(s) by which many pathogens cause disease may still remain unclear.

One of the first virulence factors of food-borne enteric pathogens to be studied genetically was the heat-labile enterotoxin elaborated by ETEC. The genetic region that encoded for this toxin was identified after extensive phenotypic testing by observing the physiological responses stimulated when the toxin was applied to tissue culture cells or injected into animals [106]. These genetic regions were then cloned to produce DNA fragments that were tested as gene probes to identify ETEC in food [15]. Later, advances in nucleic acid sequencing and synthesis made it possible to substitute cloned DNA fragments with toxin gene-specific oligonucleotide probes [16]. Synthetic probes are inexpensive, easier to make and purify, and are highly specific, capable of differentiating single base pair differences in the sequence [29].

Advances in computer technologies have also facilitated the design and development of probes. DNA sequence data banks can now be easily searched to identify unique gene sequences to which specific probes can be synthesized and evaluated. Despite these conveniences, the economics of probe development remains complex and risky. It takes a major financial commitment to develop probe assays, yet it is unpredictable whether that probe will be accepted by the food industry and be of market value.

9.3.2. Economics of Probe Use and Development

The economic aspects of gene probes in the food industry are examined here from two perspectives: that of food manufacturers who may wish to use the probes to monitor the quality of their product and that of companies that develop the probe assays for commercialization.

By law, food manufacturers must ensure that their products are in compliance with quality assurance guidelines. This monitoring process for food safety has traditionally relied on time-consuming conventional microbiological methods. Probe assays can provide faster and more sensitive analyses and thereby potentially reduce inventory costs and the loss of perishable products. However, comparative studies have also shown that the performance of any assay may vary greatly from one food type to another; therefore, manufacturers must carefully identify their needs and evaluate the feasibility of using DNA probes to meet those needs. An important factor in these decisions will be the comparative studies of DNA probes with standard methods to determine whether the advantages are sufficient to warrant changes. The most obvious economic factors to consider in this assessment are the costs of labor, materials, equipment, and training, analysis time, availability of reagents and suppliers, and the need for additional confirmation if the probe tests are approved for use only as screening assays. But, regardless of these economic factors, the decision by the food industry to implement probe assays may ultimately be hinged on the "legal" acceptance of probe data. Results of DNA probe analyses of foods have not yet been used by regulatory agencies and courts in compliance cases; hence, the judicial validity of DNA probes has not been established. This lack of experience or precedance in using DNA probe data in legal compliance violation proceedings has not instilled confidence in the food industry nor prodded anyone toward using DNA probes for testing foods.

The economic considerations for companies developing DNA probe assays for the food industry are entirely different. Although business concerns are usually not important for setting basic research priorities, the commercial development of a DNA probe test is based entirely on market demand, which is constantly fluctuating. Therefore, information surveys are often essential in identifying existing markets or forecasting new markets. Occasionally however, a market is well defined and assays are developed to meet specific needs. For example, a food-processing firm wishing to monitor the presence of a specific bacterium commonly associated with its products may contract for the development of a specific DNA probe assay to monitor the microbiologial quality of raw ingredients, the integrity of processing procedures, the overall sanitation of the plant, and the adequacy of storage conditions. DNA probes may also be tailored to meet the needs of regulatory agencies. For instance, not all species of *Yersinia* are pathogenic; therefore, to use DNA probes specific for the *Yersinia* virulence genes to monitor foods will be more significant in terms of public health and legal compliance actions.

As expected, prototype DNA probe kits are generally more expensive than subsequent products, because they must bear most of the research and development costs, including laboratory set up, equipment, and supplies. In addition, the complexities of food samples necessitate the development of suitable microbiological enrichment schemes and extraction protocols to complement the DNA probe assay. Only then is the assay ready for validation.

The process to obtain official approval for a method for use in food analysis is complex and requires considerable amounts of time and economic resources. The stepwise process usually begins with an intralaboratory study to obtain preliminary data on the effectiveness of the test. A proposal is then formulated and submitted to the Association of Official Analytical Chemists International (AOAC), requesting a formal interlaboratory collaborative study. Collaborators are usually selected from laboratories associated with all facets of the food industry, including academia, regulatory, and commercial sectors. Results of the collaborative study are then analyzed, submitted for official review, and published. Generally, several years are required for the entire process, which could be longer if the assay requires further modifications.

Collaborative studies provide data on the performance of a DNA probe assay in comparison with standard methods for the analysis of foods. Several probe assays for *Salmonella* were evaluated and received official first action approval by the AOAC [59,65]. These studies also provide the assay manufacturers with other useful information, e.g., clarity of instructions, ease of use, reagent preparations and assay manipulations, so that any deficiencies can be modified. Although a test kit should be designed to be performed easily by laboratory personnel, some degree of training is usually essential, particularly with DNA probe assays, which are more complex technologically. Therefore, the extent of user training required may be another economic factor to consider in marketing a DNA probe assay.

Finally, commercialization of new diagnostic technologies, such as DNA probes, always involves an amount of economic risk. In this age of biotechnology, by the time an assay has been developed, validated, and is ready for market, a new technology may have resulted in even better and faster alternatives.

9.4. REGULATORY IMPLICATIONS

The microbiological quality of foods is an important factor in ensuring food safety and is of great concern to public health and regulatory agencies. To comply with government regulations and guidelines, the food industry often uses official microbiological methods to evaluate all stages of food processing to ensure quality control. The highly specific nature of probes may make DNA probe testing a useful tool in monitoring the microbiological content of foods. However, the implementation of probes could also significantly alter current food testing practices; for, as test sensitivity improves, existing microbiological specifications for foods may be affected.

9.4.1. Applications of Probes in Food Analysis

After a food product is released to the retailer, it may be sampled by state or federal public health agencies to determine if it is in compliance with safety regulations. However, a food is usually not extensively tested unless it has been implicated in a food-borne outbreak. Most of the microbiological testing of foods, therefore, takes place during production, through analysis of the raw materials, in-line sampling, and testing of the finished products. Although some foods are tested for specific pathogens, many quality control programs, implemented at the discretion of the manufacturer, rely on total bacterial counts as indicators of general sanitary conditions. Bacteria such as coliforms or *E. coli* are also used often as indicators of fecal contamination or enteric pathogens such as *Salmonella* or *Shigella* which may be present in the sample. However, with the advent of highly specific assays such as DNA probes, the concept of using indicators may become obsolete since it will be possible to test directly for specific pathogens at all stages of food processing.

Analysis of end products is generally regarded as an unreliable and ineffective means of ensuring the absence of bacterial pathogens or toxins in foods. Therefore, the federal government has recommended that the food industry implement Hazard Analysis and Critical Control Point (HACCP) programs to ensure that the products are safe for consumption. The concept of HACCP is to analyze systematically and identify all critical points of food processing, then implement controls to ensure that contaminants (microbiological, chemical, or others) are eliminated or kept at a safe level. DNA probe assays may be very useful in the analysis and implementation of HACCP programs. For example, time and temperature of pasteurization are two critical control points in milk processing and are specified based on the reduction in the numbers of *Coxiella burnetii*. In the processing of beef and egg products, the absence of recoverable *Salmonella* indicates adequate pasteurization [108]. In both of these instances, the greater sensitivity of DNA probes over standard methods may enable the establishment of more accurate and probably safer processing parameters for pasteurization.

The production of safe foods alone, however, is not sufficient, as processed foods can be easily contaminated before consumption. Food poisonings are often associated with food services, such as restaurants and cafeterias, because the volume of food and the excessive amount of handling required to serve large numbers of people increases the risk of contamination and abuse. This becomes especially critical in hospitals and nursing homes, where the exposure of immunocompromised patients to food-borne pathogens could be life threatening. The responsibility for monitoring and regulating local food services and retail manufacturers resides with the State public health departments. Federal agencies also have district laboratories to carry out compliance programs to analyze foods for safety. DNA probe testing may also be incorporated into these programs to enable rapid detection of contaminated foods and to monitor foods more effectively that are prone to contamination.

9.4.2. Effect of Probe Assays on Regulations

Since the enactment of the Food and Drug Act of 1906, much progress has been made in the area of food safety. Although food-borne illnesses still occur, the incidence of reported outbreaks is relatively low [109]. This may be attributed in part to the government compliance programs and to the efforts of the food industry to establish and maintain sound quality control programs. Also, as analysis methods improve, the level of sensitivity for detecting pathogenic bacteria in foods also increases. Although this may be good news for the consumer, the greater detection sensitivity creates challenges for both the food industry and regulatory agencies.

Current specifications for the production of certain foods require that pathogens such as *Salmonella* be absent—the "zero tolerance" policy. The term *absent*, however, is inaccurate and misleading because the absence of a pathogen in a food is dependent on the sensitivity of the methods used. The word *absent*, therefore, actually means *not found*. The problem of greater sensitivity arises in that some foods previously analyzed by culture procedures and determined to meet the requirement of "not found" may no longer meet the same specifications when a more sensitive method such as a DNA probe is used. Such circumstances may raise many interesting questions. For example, what course of action will regulatory agencies take regarding enforcement of compliance laws? Will food manufacturers need to upgrade their processing methods and quality control programs to comply with the greater test sensitivity? What will be the ultimate limit or "zero tolerance" allowed by government regulations for each pathogen in foods, and will the food industry need to implement changes every time a more sensitive method of analysis is introduced?

Under the Food, Drug, and Cosmetic Act, Section 406, the FDA has the authority to establish tolerance levels for poisonous or deleterious substances in foods. Such standards, however, are difficult to set for bacteria and generally do not exist because the number of bacteria tolerated in foods will depend on the type of food, the pathogen, the sensitivity of the analytical methods, and other variables. Furthermore, the effects of pathogens on humans also varies greatly depending on the health of the individual. These epidemiological data are not available for all bacteria, especially for emerging pathogens such as *L. monocytogenes* and enterohemorrhagic *E. coli* O157:H7.

The implication of using more sensitive assays in regulatory programs may also raise questions about the existing infectious doses established for selected pathogens. Such data are usually obtained from volunteer feeding studies and based on the susceptibility of healthy individuals. If a food tested with a more sensitive assay is found to contain pathogen levels lower than the known infectious dosages, will that food be considered a health risk to immunocompromised individuals and require regulatory action? Are regulatory agencies compelled by law to impose stricter guidelines in response to the improved sensitivities of assays such as DNA probes or PCR? Will future guidelines become method-dependent?

The responsibility for risk management to ensure the safety of food products resides with individual manufacturers. However, any costs for food safety expended by the producer are eventually passed on in the form of higher prices. The consumer therefore ultimately bears the financial burden of risk–benefit analyses. If safety specifications become more stringent as a result of better tests and, thus, more costly, will the consumer be willing or able to bear such costs? The regulatory agencies and the food industry must therefore establish a delicate balance between food safety and cost, bearing in mind that the concept of a food with zero microbiological risk or "absolute safety" is probably not realistic at any price.

9.5. FUTURE APPLICATIONS

Some pathogens such as *Shigella* have infectious dosages of 1–10 organisms. Current DNA probe assays lack the sensitivity to detect directly such low levels of pathogens in foods. This has necessitated the inclusion of time-consuming growth or enrichment steps to increase the numbers of target bacteria. Probe assays may become more applicable to food analysis if the present levels of sensitivity can be increased without increases in assay time. Technological developments in target and signal amplification may provide possible solutions.

9.5.1. Target Amplification

Several techniques are now available that can rapidly amplify the number of target genes *in vitro*. The polymerase chain reaction (PCR), which uses a thermostable

DNA polymerase and automated thermal cycling, can theoretically amplify a single gene copy several million-fold in just a few hours [110]. PCR is widely used in research and in forensic laboratories, and its potential applications for identifying bacterial pathogens in foods are also being actively explored [5,6]. Many PCR assays are also being developed in conjunction with DNA probes to enable specific detection of amplified products. PCR assays may be most useful in the analysis for fastidious food-borne organisms or viruses that are difficult to cultivate *in vitro* [111].

The Q-β replicase uses RNA polymerase to amplify an RNA probe enzymatically after it has bound to the target gene [112]. Q-β amplification is faster than PCR, but it is susceptible to contaminating enzymes and may produce high background signals. The self-sustained sequence replication (3SR), also known as *nucleic acid sequence–based amplification* (NASBA), is a two-cycle system that uses DNA synthesis and RNA transcription [112]. The amplification proceeds at a single temperature, and hence does not require thermal cycling as does PCR. Although 3SR is faster than PCR, it uses multiple enzymes; hence, it has a lower specificity and is affected by contaminating enzymes. Neither of these two systems has been explored for use in food analysis.

Although amplification systems are not yet sufficiently developed for routine identification of pathogens in foods, their potential is evident. However, *in vitro* amplification systems often will not discriminate between nucleic acid templates derived from viable and nonviable organisms; hence, the assay results should be interpreted with caution. Furthermore, the efficiencies of these systems in food analysis will depend on effective extraction procedures to provide suitable templates for amplification. Thus, they are plagued by the same factors that affect DNA probe assays in food analysis.

Finally, users should be aware that a positive test with DNA probe and PCR assays merely indicates that particular genetic sequences are present. Hence, a toxin gene-specific assay, for instance, will only show that the organisms has the potential to produce the toxin; it does not indicate that the toxin is actually produced or that it has other genes required for pathogenicity.

9.5.2. Signal Amplification

The use of isotopically labeled probes for food analysis is illogical and counterproductive. However, isotopic markers have traditionally been used in hybridization reactions; therefore, they were also used in the early development of probe assays. Despite the intrinsic sensitivity of isotopes, the hazards of handling them have greatly discouraged the implementation of DNA probe assays by the food industry and promoted extensive research into the development of nonisotopic markers. Current nonisotopic systems use various types of fluorescent and chemiluminescent labels and also use signal amplification techniques to attain sufficient detection sensitivities. Some nonisotopically labeled probes have shown sensitivities comparable to isotopic probes. Other nonisotopic systems use a combination of DNA probes and enzyme immunoassays to generate an amplified signal. Using horseradish peroxidase enzyme and a colorimetric substrate, these assay signals may be quantitated with a spectrophotometer. This design has been used to develop commercially available probe assays for the identification of various food-borne pathogens, including *Listeria*, *Salmonella*, *Yersinia*, *Campylobacter*, and *E. coli*.

9.6. CONCLUSIONS

The use of DNA probes in food analysis has not gained the wide acceptance that many originally predicted. As shown in Table 9.1, probes have been developed to detect most common food-borne bacterial pathogens; yet few are actually being used routinely for food analysis. Also, unlike antibody assays for detecting food-borne pathogens, which are readily available commercially, only a few DNA probe assays are marketed. The initial use of isotopic labels, the inability to detect directly pathogenic bacteria in foods, and the continued reliance on cultural confirmation have hampered its acceptance.

But, in relation to how long antibody-based methods have existed before they were widely utilized in food diagnostics, we may infer that the use of DNA probes in food analysis is probably still in its infancy. Even so, in the context of evolving analytical technologies, probe assays are already changing rapidly. The inherent specificities of probes are being used increasingly in combination with other technologies such as enzyme immunoassays, PCR, or as PCR primers to enable specific and sensitive detection of food-borne pathogens. Thus, regardless of how widely probe assays will be used in the future to detect pathogens in foods, the DNA probe technology has already made a significant impact on food microbiology and has firmly established itself as an integral part of the analytical methods for detecting food-borne pathogens and their toxins.

REFERENCES

1. Foster, E.M.: Perennial issues in food safety. In Cliver, D.O. (ed.): Foodborne Diseases. New York: Academic Press, 1990.
2. Archer, D., Young, F.E.: Contemporary issues: Diseases with a food vector. Clin. Microbiol. Rev. 1:377–398 (1988).
3. Bean, N.H., Griffin, P.M., Goulding, J.S., Ivey, C.B.: Foodborne disease outbreaks, 5 year summary, 1983–1987. J. Food Prot. 53:711–728 (1990).
4. Lampel, K.A., Feng, P., Hill, W.E.: Gene probes used in food microbiology. In Bhatnagar, D., Cleveland, T.E.: Molecular Approaches to Improving Food Safety. New York: Van Nostrand Reinhold, 1992.
5. Hill, W.E., Olsvik, Ø.: Detection and identification of foodborne microbial pathogens by the polymerase chain reactions: food safety applications. In Patel, PD: Rapid Analysis Techniques in Food Microbiology. London, Glasgow, Weinheim, New York, Tokyo, Melbourne, Madras: Blackie Academic and Professional, 1994.
6. Jones, D.D., Bej, A.K.: Detection of foodborne microbial pathogens using polymerase

chain reaction methods. In Griffin, H.G., Griffin, A.M. (eds.): PCR Technology: Current Innovations. Boca Raton: CRC Press, 1994.

7. Stern, N.J., Kazmi, S.U.: *Campylobacter jejuni.* In Doyle, M.P. (ed.): Foodborne Bacterial Pathogens. Basel: Marcel Dekker, 1989.

8. Rashtchian, A., Curiale, M.S.: DNA probe assays for detection of *Campylobacter* and *Salmonella.* In Swaminathan, B., Prakash, G. (eds.): Nucleic Acid and Monoclonal Antibody Probes. Basel: Marcel Dekker, 1989.

9. Stern, N.J., Mozola, M.A.: Methods for selective enrichment of *Campylobacter* spp. from poultry for use in conjunction with DNA hybridization. J. Food. Prot. 55:767–770 (1992).

10. Giesendorf, B.A., Quint, W.G.V., Henkens, M.H.C., Stegeman, H., Huf, F.A., Niesters, H.G.M.: Rapid and sensitive detection of *Campylobacter* spp. in chicken products by using the polymerase chain reaction. Appl. Environ. Microbiol. 58:3804–3808 (1992).

11. Wegmuller, B., Luthy, J., Candrian, U.: Direct polymerase chain reaction detection of *Campylobacter jejuni* and *Campylobacter coli* in raw milk and dairy products. Appl. Environ. Microbiol. 59:2161–2165 (1993).

12. Doyle, M.P., Padhye, V.V.: *Escherichia coli.* In Doyle, M.P. (ed.): Foodborne Bacterial Pathogens. New York, Basel: Marcel Dekker, 1989.

13. Hill, W.E., Carlisle, C.L.: Loss of plasmids during enrichment for *Escherichia coli.* Appl. Environ. Microbiol. 41:1046–1048 (1981).

14. Echeverria, P., Taylor, D.N., Seriwatana, J., Brown, J.E., Lexomboon, U.: Examination of colonies and stool blots for detection of enteropathogens by DNA hybridization with eight DNA probes. J. Clin. Microbiol. 27:331–334 (1989).

15. Hill, W.E., Madden, J.M., McCardell, B.A., Shah, D.B., Jagow, J.A., Payne, W.L., Boutin, B.K.: Foodborne enterotoxigenic *Escherichia coli:* Detection and enumeration by DNA colony hybridization. Appl. Environ. Microbiol. 45:1324–1330 (1983).

16. Hill, W.E., Payne, W.L., Zon, G., Moseley, S.L.: Synthetic oligodeoxyribonucleotide probes for detecting heat-stable enterotoxin-producing *Escherichia coli* by DNA colony hybridization. Appl. Environ. Microbiol. 50:1187–1191 (1985).

17. Ferreira, J.L., Hill, W.E., Hamdy, M.K., Zapatka, F.A., McCay, S.G.: Detection of enterotoxigenic *Escherichia coli* in foods by DNA colony hybridization. J. Food Sci. 51:665–667 (1986).

18. Hill, W.E., Payne, W.L.: Genetic methods for the detection of microbial pathogens. Identification of enterotoxigenic *Escherichia coli* by DNA colony hybridization: Collaborative study. J. Assoc. Off. Anal. Chem. 67:801–807 (1984).

19. Cryan, B.: Comparison of three assay systems for detection of enterotoxigenic *Escherichia coli* heat-stable enterotoxin. J. Clin. Microbiol. 28:792–794 (1990).

20. Wernars, K., Delfgou, E., Soentoro, P.S., Notermans, S.: Successful approach for detection of low numbers of enterotoxigenic *Escherichia coli* in minced meat by using polymerase chain reaction. Appl. Environ. Microbiol. 57:1914–1919 (1991).

21. Food and Drug Administration: Bacteriological Analytical Manual, 7th ed. Arlington, VA: Association of Official Analytical Chemists International, 1992.

22. Nataro, J.P., Baldini, M.M., Kaper, J.B., Black, R.E., Bravo, N., Levine, M.M.: Detection of an adherence factor of enteropathogenic *Escherichia coli* with a DNA probe. J. Infect. Dis. 152:560–565 (1985).

23. Doyle, M.P., Schoeni, J.L.: Survival and growth characteristic of *Escherichia coli* associated with hemorrhagic colitis. Appl. Environ. Microbiol. 48:855–856 (1984).

24. Newland, J.W., Neill, R.J.: DNA probes for Shiga-like toxins I and II and for toxin converting bacteriophages. J. Clin. Microbiol. 26:1292–1297 (1988).

25. Samadpour, M., Liston, J., Ongerth, J.E., Tarr, P.I.: Evaluation of DNA probes for detection of Shiga-like toxin-producing *Escherichia coli* in food and calf fecal samples. Appl. Environ. Microbiol. 56:1212–1215 (1990).

26. Samadpour, M., Ongerth, J.E., Liston, J., Tran, N., Nguyen, D., Whittam, T.S., Wilson, R.A., Tarr, P.I.: Occurrence of Shiga-like toxin-producing *Escherichia coli* in retail fresh seafood, beef, lamb, pork, and poultry from grocery stores in Seattle, Washington. Appl. Environ. Microbiol. 60:1038–1040 (1994).

27. Smith, H.R., Cheasty, T., Roberts, D., Thomas, A., Rowe, B.: Examination of retail chicken and sausages in Britain for vero cytotoxin-producing *Escherichia coli*. Appl. Environ. Microbiol. 57:2091–2093 (1991).

28. O'Brien, A.D., Melton, A.R., Schmitt, C.K., McKee, M.L., Batts, M.L., Griffin, D.E.: Profile of *Escherichia coli* O157:H7 pathogen responsible for hamburger-borne outbreak of hemorrhagic colitis and hemolytic uremic syndrome in Washington. J. Clin. Microbiol. 31:2799–2801 (1993).

29. Feng, P.: Identification of *Escherichia coli* O157:H7 by DNA probe specific for an allele of *uidA* gene. Mol. Cell. Probes 7:151–154 (1993).

30. Cebula, T.A., Payne, W.L., Feng, P.: Simultaneous identification of *Escherichia coli* of the O157:H7 serotype and their Shiga-like toxin type by MAMA/multiplex PCR. J. Clin. Microbiol. 33:248–250 (1995).

31. Vankatesan, M., Buysse, J.M., Vandendries, E., Kopecko, D.J.: Development and testing of invasion-associated DNA probes for detection of *Shigella* spp. and enteroinvasive *Escherichia coli*. J. Clin. Microbiol. 27:261–266 (1988).

32. Panda, C.S., Riley, L.W., Kunmari, S.N., Khanna, K.K., Prakash, K.: Comparison of alkaline phosphatase-conjugated oligonucleotide DNA probe with the sereny test for identification of *Shigella* strains. J. Clin. Microbiol. 28:2122–2124 (1990).

33. Frankel, G., Riley, L., Giron, J., Valmassoi, J., Friedmann, A., Strockbine, N., Falkow, S., Schoolnik, G.K.: Detection of *Shigella* in feces using DNA amplification. J. Infect. Dis. 161:1252–1256 (1990).

34. Jagow, J.A., Lampel, K.A.: Detecting enteroinvasive *Shigella* in food using a DNA probe. Abstr. Am. Soc. Microbiol. P-14:321 (1989).

35. Lampel, K.A., Jagow, J.A., Trucksess, M., Hill, W.E.: Polymerase chain reaction for detection invasive *Shigella flexneri* in foods. Appl. Environ. Microbiol. 56:1536–1540 (1990).

36. Keasler, S.P., Hill, W.E.: Polymerase chain reaction identification of enteroinvasive *Escherichia coli* seeded into raw milk. J. Food Prot. 55:382–384 (1992).

37. Andersen, M.R., Omiecinski, C.J.: Direct extraction of bacterial plasmid from food for polymerase chain reaction. Appl. Environ. Microbiol. 58:4080–4082 (1992).

38. Cossart, P., Vincente, M.F., Megaud, J., Baquero, F., Perez-Dias, J.C., Berche, P.: Listeriolysin O is essential for virulence of *Listeria monocytogenes*: Direct evidence obtained by gene complementation. Infect. Immun. 57:3629–3936 (1989).

39. Kohler, S., Leimeister-Wachter, M., Chakraborty, T., Lottspeich, F., Goebel, W.: The

gene coding for protein p60 of *Listeria monocytogenes* and its use as a specific probe for *Listeria monocytogenes.* Infect. Immun. 58:1943–1950 (1990).

40. Lovett, J.: *Listeria monocytogenes.* In Doyle, M.P. (ed.): Foodborne Bacterial Pathogens. Basel: Marcel Dekker, 1989.

41. Datta, A.R., Wentz, B.A., Shook, D., Trucksess, M.W.: Synthetic oligodeoxyribonu-cleotide probes for detection of *Listeria monocytogenes.* Appl. Environ. Microbiol. 54:2933–2937 (1988).

42. Kim, C., Swaminathan, B., Cassaday, P.K., Mayer, L.W., Holloway, B.P.: Rapid confir-mation of *Listeria monocytogenes* isolated from foods by a colony blot assay using a digoxigenin-labeled synthetic oligonucleotide probe. Appl. Environ. Microbiol. 57:1609–1614 (1991).

43. Datta, A.R., Wentz, B.A., Russell, J.: Cloning of listeriolysin O gene and development of specific gene probes for *Listeria monocytogenes.* Appl. Environ. Microbiol. 56:3874–3877 (1990).

44. Datta, A.R., Moore, M.A., Wentz, B.A., Lane, J.: Identification and enumeration of *Listeria monocytogenes* by nonradioactive DNA probe colony hybridization. Appl. Environ. Microbiol. 59:144–149 (1993).

45. King, W., Raposa, S., Warshaw, J., Johnson, A., Halbert, D., Klinger, J.D.: A new colori-metric nucleic acid hybridization assay for *Listeria* in foods. Int. J. Food Microbiol. 8:225–232 (1989).

46. Ninet, B., Bannerman, E., Billie, J.: Assessment of the Accuprobe *Listeria monocyto-genes* culture identification reagent kit for rapid colony confirmation and its applica-tion in various enrichment broths. Appl. Environ. Microbiol. 58:4055–4059 (1992).

47. Partis, L., Newton, K., Murby, J., Wells, R.J.: Inhibitory effects of enrichment media on the Accuprobe test for *Listeria monocytogenes.* Appl. Environ. Microbiol. 60:1693–1694 (1994).

48. Furrer, B., Candrian, U., Hofelein, C., Luthy, J.: Detection and identification of *Listeria monocytogenes* in cooked sausage products and in milk by *in vitro* amplification of he-molysin gene fragments. J. Appl. Bacteriol. 70:372–379 (1991).

49. Niederhauser, C., Candrian, U., Hofelein, C., Jermini, M., Buhler, H.-P., Luthy, J.: Use of polymerase chain reaction for detection of *Listeria monocytogenes* in food. Appl. Environ. Microbiol. 58:1564–1568 (1992).

50. Fluit, A.C., Toresma, R., Visser, M.J.C., Aarsman, C.J.M., Poppelier, M.J.J.G., Keller, B.H.I., Klapwijk, P., Verhoef, J.: Detection of *Listeria monocytogenes* in cheese with the magnetic immunopolymerase chain reaction assay. Appl. Environ. Microbiol. 59:1289–1293 (1993).

51. Roberts, T.: Human illness cost of foodborne bacteria. Am. J. Agric. Econ. 71:468–474 (1989).

52. Todd, E.C.D.: Preliminary estimates of costs of foodborne disease in the United States. J. Food Prot. 52:595–601 (1989).

53. Celum, C.L., Chaisson, R.E., Rutherford, G.W., Barnhart, J.L., Echenberg, D.F.: Incidence of salmonellosis in patients with AIDS. J. Infect. Dis. 156:998–1002 (1987).

54. D'Aoust, J.-Y.: *Salmonella.* In Doyle, M.P. (ed.): Foodborne Bacterial Pathogens. New York, Basel: Marcel Dekker, 1989.

55. O'Brien, T.F., Hopkins, J.D., Gilleece, E.S., Medeiros, A.A., Kent, R.L., Blackburn,

B.O, Holmes, M.B., Reardon, J.P., Vergeront, J.M., Schell, W.L., Christenson, E., Bissett, M.L., Morse, E.V.: Molecular epidemiology of antibiotic resistance in *Salmonella* from animals and human beings in the United States. N. Engl. J. Med. 307:1–6 (1982).

56. Owen, R.J.: Chromsomal DNA fingerprinting—A new method of species and strain identification applicable to microbial pathogens. J. Med. Microbiol. 30:89–99 (1989).

57. Earnshaw, R., Gidley, J.: Molecular methods for typing bacterial food pathogens. Trends Food Sci. Technol. 3:39–43 (1992).

58. Fitts, R., Diamond, M., Hamilton, C., Neri, M.: DNA–DNA hybridization assay for detection of *Salmonella* spp. in foods. Appl. Environ. Microbiol. 46:1146–1151 (1983).

59. Flowers, R.S., Klatt, M.J., Mozola, M.A., Curiale, M.S., Gabis, D.A., Silliker, J.H.: DNA hybridization assay for detection of *Salmonella* in foods: Collaborative study. J. Assoc. Off. Anal. Chem. 70:521–529 (1987).

60. Sall, B.S., Lombardo, M., Sheridan, B., Parsons, G.H.: Performance of a DNA probe-based *Salmonella* test in the AACC check sample program. J. Food Prot. 51:579–580 (1988).

61. Izat, A.L., Driggers, C.D., Colberg, M., Reiber, M.A., Adams, M.H.: Comparison of the DNA probe to culture methods for the detection of *Salmonella* on poultry carcasses and processing waters. J. Food Prot. 52:564–570 (1989).

62. St. Clair, V.J., Klenk, M.M.: Performance of three methods for the rapid identification of *Salmonella* in naturally contaminated foods and feed. J. Food Prot. 53:961–964 (1990).

63. Curiale, M.S., Klatt, M.J., Mozola, M.A.: Colorimetric deoxyribonucleic acid hybridization assay for rapid screening of *Salmonella* in food: Collaborative study. J. Assoc. Off. Anal. Chem. 73:248–256 (1990).

64. Rose, B.E., Llabres, C.M., Bennett, B.: Evaluation of colorimetric DNA hybridization test for detection of salmonellae in meat and poultry products. J. Food Prot. 54:127–130 (1991).

65. Chan, S.W., Wilson, S.G., Vera-Garcia, M., Whippie, K., Ottaviani, M., Whilby, A., Shah, A., Johnson, A., Mozola, M.A., Halbert, D.N.: Comparative study of DNA hybridization method and conventional culture procedure for detection of *Salmonella* in foods. J. Assoc. Off. Anal. Chem. 73:419–424 (1990).

66. Foster, K., Garramone, S., Ferraro, K., Groody, E.P.: Modified colorimetric DNA hybridization method and conventional culture method for detection of *Salmonella* in foods: Comparison of methods. J. Assoc. Off. Anal. Chem. Int. 75:685–692 (1992).

67. Tsen, H.-Y., Wang, S.-J., Green, S.S.: *Salmonella* detection in meat and fish by membrane hybridization with chromogenic/phosphatase biotin DNA probe. J. Food Sci. 56:1519–1523 (1991).

68. Cano, R.J., Torres, M.J., Klem, R.E., Palomares, J.C., Casadesus, J.: Detection of salmonellas by DNA hybridization with a fluorescent alkaline phosphatase substrate. J. Appl. Bacteriol. 72: 393–399 (1992).

69. Cano, R.J., Rasmussen, S.R., Sanchez Fraga, G., Palomares, J.C.: Fluorescent detection-polymerase chain reaction (FD-PCR) assay on microwell plates as a screening test for salmonellas in foods. J. Appl. Bacteriol. 75:247–253 (1993).

70. Jones, D.D., Law, R., Bej, A.K.: Detection of *Salmonella* spp. in oysters using polymerase chain reaction (PCR) and gene probes. J. Food Sci. 58:1191–1193 (1993).

71. Fluit, A.C., Widjojoatmodjo, M.N., Box, A.T.A., Torensma, R., Verhoef, J.: Rapid detection of salmonellae in poultry with the magnetic immuno-polymerase chain reaction assay. Appl. Environ. Microbiol. 59:1342–1346 (1993).

72. Schiemann, D.A.: *Yersinia enterocolitica* and *Yersinia pseudotuberculosis*. In Doyle, M.P. (ed.): Foodborne Bacterial Pathogens. Basel: Marcel Dekker, 1989.

73. Miller, V.L., Farmer, J.J., Hill, W.E., Falkow, S.: The *ail* loci is found uniquely in *Yersinia enterocolitica* serotypes commonly associated with disease. Infect. Immun. 57:121–131 (1989).

74. Portnoy, D.A., Moseley, S.A., Falkow, S.: Characterization of plasmid and plasmid-associated determinants of *Yersinia enterocolitica* pathogenesis. Infect. Immun. 31:775–782 (1981).

75. Jagow, J.A., Hill, W.E.: Enumeration by DNA colony hybridization of virulent *Yersinia enterocolitica* colonies in artificially contaminated food. Appl. Environ. Microbiol. 51:441–443 (1986).

76. Miliotis, M.D., Galen, J.E., Kaper, J.B., Morris, J.G.: Development and testing of a synthetic oligonucleotide probe for the detection of pathogenic *Yersinia* strains. J. Clin. Microbiol. 27:1667–1670 (1989).

77. Kapperud, G., Dommarsnes, K., Skurnik, M., Hornes, E.: A synthetic oligonucleotide probe and a cloned polynucleotide probe based on the *yopA* gene for detection and enumeration of virulent *Yersinia enterocolitica*. Appl. Environ. Microbiol. 56:17–23 (1990).

78. Jagow, J.A., Hill, W.E.: Enumeration of virulent *Yersinia enterocolitica* colonies by DNA colony hybridization using alkaline treatment and paper filters. Mol. Cell. Probes 2:189–195 (1988).

79. Nesbakken, T., Kapperud, G., Dommarsnes, K., Skurnik, M., Hornes, E.: Comparative study of DNA hybridization method and two isolation procedures for detection of *Yersinia enterocolitica* O:3 in naturally contaminated pork products. Appl. Environ. Microbiol. 57:389–394 (1991).

80. Feng, P.: Identification of invasive *Yersinia* species using oligonucleutode probes. Mol. Cell. Probes 6:291–297 (1992).

81. Goverde, R.L.J., Jansen, W.E., Bruning, H.A., Huis in't Veld, J.H.J., Mooi, F.R.: Digoxigenin-labelled *inv*- and *ail*-probes for detection and identification of pathogenic *Yersinia enterocolitica* in clinical specimens and naturally contaminated pig samples. J. Appl. Bacteriol. 74:301–313 (1993).

82. Kwaga, J., Iversen, J.O., Misra, V.: Detection of pathogenic *Yersinia enterocolitica* by polymerase chain reaction and digoxigenin-labeled polynucleotide probes. J. Clin. Microbiol. 30:2668–2673 (1992).

83. Kapperud, G., Vardund, T., Skjerve, E., Hornes, E., Michaelsen, T.E.: Detection of pathogenic *Yersinia enterocolitica* in foods and water by immunomagnetic separation, nested polymerase chain reaction, and colorimetric detection of amplified DNA. Appl. Environ. Microbiol. 59:2938–2944 (1993).

84. Notermans, S., Heuvelman, K.J., Wernars, K.: Synthetic enterotoxin B DNA probes for detection of enterotoxigenic *Staphylococcus aureus* strains. Appl. Environ. Microbiol. 54:531–533 (1988).

85. Neill, R.J., Fanning, G.R., Delahoz, F., Wolff, R., Gemski, P.: Oligonucleotide probe for detection and differentiation of *Staphylococcus aureus* strains containing genes for en-

terotoxin A, B, and C and toxic shock syndrome toxin 1. J. Clin. Microbiol. 28:1514–1518 (1990).

86. Wilson, I.G., Cooper, J.E., Gilmour, A.: Detection of enterotoxigenic *Staphylococcus aureus* in dried skimmed milk: Use of the polymerase chain reaction for amplification and detection of staphylococcal enterotoxin genes *entB* and *entC* and the thermonuclease gene *nuc*. Appl. Environ. Microbiol. 57:1793–1798 (1991).

87. Van Damme-Jongsten, M., Rodhouse, J., Gilbert, R.J., Notermans, S.: Synthetic DNA probes for detection of enterotoxigenic *Clostridium perfringens* strains isolated from outbreak of food poisoning. J. Clin. Microbiol. 28:131–133 (1990).

88. Thompson, D.E., Brehm, J.K., Oultram, J.D., Swinfield, T.-J., Shone, C.C., Atkinson, T., Melling, J., Minton, N.P.: The complete amino acid sequence of the *Clostridium botulinum* type A neurotoxin, derived by nucleotide sequence analysis of the encoding gene. Eur. J. Biochem. 189:73–81 (1990).

89. DeBuono, B.A., Brondum, J., Kramer, J.M., Gilbert, R.J., Opal, S.M.: Plasmid, serotypic and enterotoxin analysis of *Bacillus cereus* in an outbreak setting. J. Clin. Microbiol. 26:1571–1574 (1988).

90. Nishibushi, M., Hill, W.E., Zon, G., Payne, W.L., Kaper, J.B.: Synthetic oligodeoxyribonucleotide probes to detect Kanagawa phenomenon-positive *Vibrio parahaemolyticus*. J. Clin. Microbiol. 23:1091–1095 (1986).

91. Morris, J.G., Wright, A.C., Roberts, D.M., Wood, P.K., Simpson, L.M., Oliver, J.D.: Identification of environmental *Vibrio vulnificus* isolates with a DNA probe for the cytotoxin-hemolysin gene. Appl. Environ. Microbiol. 53:193–195 (1987).

92. Lee, C., Chen, L.-H., Liu, M.-L., Su, Y.-C.: Use of oligonucleotide probe to detect *Vibrio parahaemolyticus* in artificially contaminated oysters. Appl. Environ. Microbiol. 58:3419–3422 (1992).

93. Wright, A.C., Miceli, G.A., Landry, W.L., Christy, J.B., Watkins, W.D., Morris, J.G.: Rapid identification of *Vibrio vulnificus* on nonselective media with an alkaline phosphatase-labeled oligonucleotide probe. Appl. Environ. Microbiol. 59:541–546 (1993).

94. Yoh, M., Miyagi, K., Matsumoto, Y., Hayashi, K., Takarada, Y., Yamamoto, K., Honda, T.: Development of an enzyme-labeled oligonucleotide probe for the Cholera toxin gene. J. Clin. Microbiol. 31:1312 1314 (1993).

95. Hill, W.E., Keasler, S.P., Trucksess, M.W., Feng, P., Kaysner, C.A., Lampel, K.L.: Polymerase chain reaction identification of *Vibrio vulnificus* in artificially contaminated oysters. Appl. Environ. Microbiol. 57:707–711 (1991).

96. Koch, W.H., Payne, W.L., Wentz, B.A., Cebula, T.A.: Rapid polymerase chain reaction method for detection of *Vibrio cholerae* in foods. Appl. Environ. Microbiol. 59:556–560 (1993).

97. Cliver, D.O.: Epidemiology of viral foodborne disease. J. Food Prot. 57:263–266 (1994).

98. Jiang, X., Estes, M.K., Metcalf, T.G.: Detection of hepatitis A virus by hybridization with single stranded RNA probes. Appl. Environ. Microbiol. 53:2487–2495 (1987).

99. Gerba, P., Margolin, B., Trumper, E.: Enterovirus detection in water with gene probes. Z. Ges. Hyg. 34:518–519 (1988).

100. Petitjean, J., Quibriac, M., Freymuth, F., Fuchs, F., Laconche, N., Aymard, M., Kopecka, H.: Specific detection of enteroviruses in clinical samples by molecular hybridization using poliovirus subgenomic riboprobes. J. Clin. Microbiol. 28:307–311 (1990).

101. Cukor, G., Blacklow, N.R.: Human viral gastroenteritis. Microbiol. Rev. 48:157–179 (1984).

102. Larkin, E.P.: Detection, quantitation, and public health significance of foodborne viruses. In Pierson, M.D., Stern, N.J. (eds.): Foodborne Microorganisms and Their Toxins: Developing Methodologies. Basel: Marcel Dekker, 1986.

103. Tenover, F.C.: Diagnostic deoxyribonucleic acid probes for infectious diseases. Clin. Microbiol. Rev. 1:82–101 (1988).

104. Roszak, D.B., Colwell, R.R.: Survival strategies of bacteria in the natural environment. Microbiol. Rev. 51:365–379 (1987).

105. Hogan, J.J.: DNA probes to ribosomal RNA. In Swaminathan, B., Prakash, G. (eds.): Nucleic Acid and Monoclonal Antibody Probes. Basel: Marcel Dekker, 1989.

106. Dallas, W.S., Gill, D.M., Falkow, S.: Cistrons encoding *Escherichia coli* heat-labile toxin. J. Bacteriol. 139:850–858 (1979).

107. Hill, W.E., Payne, W.L., Aulisio, C.C.G.: Detection and enumeration of virulent *Yersinia enterocolitica* in foods by DNA colony hybridization. Appl. Environ. Microbiol. 46:636–641 (1983).

108. Snyder, O.P.: HACCP in the retail food industry. Dairy Food Environ. Sanit. 11:73–81 (1991).

109. Cliver, D.O.: Transmission of diseases via foods. In Cliver, D.O. (ed.): Foodborne Diseases. New York: Academic Press, 1990.

110. Saiki, R.K., Gelfand, D.H., Stoffel, S., Scharf, S.J., Higuchi, R., Horn, G.T., Mullis, K.B., Erlich, H.A.: Primer-directed enzymatic amplification of DNA with a thermostable DNA polymerase. Science 239:487–491 (1988).

111. Gouvea, V., Glass, R.I., Woods, P., Taniguchi, K., Clark, J.F., Forrester, B., Fang, Z.-Y.: Polymerase chain reaction amplification and typing of rotavirus nucleic acid from stool specimens. J. Clin. Microbiol. 28:276–282 (1990).

112. Wolcott, M.J.: DNA-based rapid methods for the detection of foodborne pathogens. J. Food Prot. 54:387–401 (1991).

113. Romaniuk, P.J., Trust, T.J.: Rapid identification of *Campylobacter* species using oligonucleotide probes to 16S ribosomal RNA. Mol. Cell. Probes 3:133–142 (1989).

CHAPTER 10

CURRENT DEVELOPMENT AND APPLICATIONS OF NUCLEIC ACID TECHNOLOGY IN THE ENVIRONMENTAL SCIENCES

ASIM K. BEJ and MEENA H. MAHBUBANI
University of Alabama at Birmingham,
Birmingham, AL 35294 (A. K. B.); Miles College,
Birmingham, AL 35208 (M. H. M.)

10.1. INTRODUCTION

Microbial monitoring and the study of microbial activities in the environment are necessary to ensure the safety of public health and to understand the survival and interactions of microorganisms with other organisms. In some cases the microorganisms of environmental concern are identical to those of clinical importance, but in other cases monitoring and metabolic activities of nonpathogenic microorganisms must be studied to understand their interaction with other organisms and the constantly changing environmental parameters that directly or indirectly affect the physiology of the organisms. Also, the detection and isolation of a wide range of microorganisms from the environment are essential components of the study of microbial ecology. The development and application of molecular techniques for monitoring and understanding microbial activities in the environment has helped undertake appropriate measures against various microbial pathogens and enhanced our knowledge about their metabolic activities when they are present in their natural habitat. Modern molecular techniques have been effectively used to study genetic regulation of microbial pathogenicity, survival in different environmental condi-

Nucleic Acid Analysis: Principles and Bioapplications, pages 231–274
© 1996 Wiley-Liss, Inc.

tions, existence in viable but nonculturable states, interactions with other organisms including plants and crops, and other specific metabolic activities of microorganisms when they are present in the environment.

Isolation and identification of the target microbial population, which comprises a minor proportion of the total microbial community, from the complex environmental matrix requires either culturing of the microorganisms by conventional microbiological methods or recovering nucleic acid from the environmental sample, followed by purification of interfering factors so that molecular techniques can be successfully applied. The development and subsequent applications of molecular techniques for the extraction and purification of DNA and RNA from the complex environmental matrices and detection of a target microbial population by identifying their nucleic acids by gene probe or polymerase chain reaction (PCR) methods have provided us information in the areas of microbial ecology, public health indicators of microbial contamination, and bioremediation. Furthermore, genetic manipulations in many of the naturally occurring microorganisms has endowed them with the ability to perform the desired tasks in the environment such as enhanced food and agricultural production, metal and mineral leaching, biopesticide/biocontrol, and waste treatment undoubtedly will revolutionize the environmental microbial sciences in the near future.

This chapter discusses the development, applications, usefulness, and future directions of gene probes and PCR methodologies in environmental microbiology.

10.2. NUCLEIC ACID TECHNOLOGY IN ENVIRONMENTAL SCIENCES

From the environmental perspective, the primary goal for using the nucleic acid technology is to detect and isolate organisms, including pathogens, with a specific genotype and gather knowledge on the absolute composition and structure of the microbial communities and the dynamics of individual populations or gene within the community. This information has the potential to contribute to understanding the mechanisms that control the pathogenicity, population dynamics, and specific activities and processes in the environment that may be of great use to human health, agriculture, and removal of toxic products in the environment. Two major nucleic acid technologies—gene probes and PCR—have been developed and applied in recent years to detect microorganisms in the environment and study their roles and interactions with each other and with the environmental parameters.

10.3. GENE PROBES AND NUCLEIC ACID HYBRIDIZATIONS

The earliest applications of the gene probe technology in environmental microbiology centered on the development of gene probes for identification and isolation of specific microorganisms, including microbial pathogens with a specific genotype, that have demonstrated the possibility of having defined roles in various environmental processes. More recently, interest has been directed toward the increase in the sensitivity of the DNA detection assays of such microorganisms.

Nucleic acid hybridization utilize gene probes for the detection of target nucleic acid sequences by base-pairing of the nucleotide sequences with a homologous complementary probe sequence. The criteria, parameters, and considerations for gene probe hybridizations are elaborated in Chapter 1.

10.3.1. Liquid-Phase Gene Probe Hybridization

The first nucleic acid hybridization techniques were developed for solution hybridization with primary applications in comparing species relatedness using DNA hormology [1]. Changes in the melting temperature or hyperchromicity, separation by hydroxyapatite, or specific nuclease treatment can be applied to distinguishing double-stranded DNA (dsDNA) or RNA hybrid molecules from single-stranded DNA (ssDNA). Because the kinetics of hybridization are faster in solution and less DNA probe is required, solution hybridization has been used for determining genomic or microbial community complexity in environmental samples [2,3]. Although applications of gene probes in liquid phase hybridization in the environmental sciences are limited, there are several reports that describe the application of DNA reassociation kinetics to determine the diversity of the total pool of DNA within a biological community [4]. A high degree of correlation has been shown to exist between taxonomic diversity measurements based on phenotypic characterization of isolates and genetic diversity based on DNA reassociation kinetics. The greater the diversity of the DNA, the slower the reassociation, and the lesser the genetic diversity the faster the DNA reannealing kinetics [4]. Based on this observation, Atlas et al. [5] and Bej et al. [6] reported that the liquid phase DNA hybridization kinetics of the total DNA from a soil microcosm in which a strain of genetically engineered microorganisms (GEMs) that degrade 2,4,5-trichlorophenoxy acetic acid had been released showed that the DNA denaturation and reannealing kinetics can be used to determine the genetic diversity in the environment. A solution hybridization approach was developed for the detection of a genetically engineered 2,4,5-trichlorophenoxy acetic acid–degrading *Pseudomonas cepacia* in artificially contaminated sediment [7]. In this study, the detection method involved recovery of nucleic acids from the contaminated sediment samples followed by hybridization in solution with a radiolabeled RNA probe. Following hybridization, the RNA–DNA hybrids were separated by Sephadex or hydroxyapatite column chromatography and detected by a liquid scintillation counter. The sensitivity of detection by following this approach was 10^2–10^3 target microbes per gram of sediment. In a separate study, a 784 bp DNA probe that was derived from the total RNA–cDNA approach was used for the detection of *Listeria monocytogenes* by a solution hybridization approach called *heteroduplex nucleic acid* (HNA) enzyme-linked immunosorbent assay [8]. In this approach, the biotinylated DNA probe was hybridized in aqueous phase with target RNA molecules and then specific HNAs were captured by HNA-specific antibodies. Captured HNA molecules were treated with a streptavidin enzyme conjugate. In this study, the HNA probe was specific for the detection of *Listeria* spp. and the sensitivity of detection was 5×10^2 cells per 25 g of irradiated meat samples. A 16S rRNA-based hybridization probe has been developed for the detection of *Listeria monocytogenes* using crude RNA extract from the pure cell cultures [9].

10.3.2. Solid Phase Gene Probe Hybridization

The most current and popular gene probe hybridization methods use solid surfaces such as nylon or nitrocellulose membranes, or polystyrene, on which the target or the probe DNA molecules are immobilized to capture the complementary sequences during the hybridization process. One of the advantages of this approach is that the hybridization can be performed on purified nucleic acids or directly on the bacterial colonies/phage plaques. Various solid-phase nucleic acid hybridizations have been elaborated in Chapter 1.

10.3.3. Nucleic Acid Probes

Nucleic acid probes to detect marker genes can be designed to detect a particular genotype or to detect unique sequences inserted into the genome of the target microorganisms (such as a transposon or an oligonucleotide sequence). The probe itself can be double stranded, comprising either total genomic DNA [10] or specific sequences of genomic or plasmid origin [11]. Also, single-stranded oligonucleotide probes constructed *in vitro* have been used successfully to detect specific 16S rRNA sequences [12,13] and are routine in many hybridization experiments. Nucleic acid probes can be labeled using either radioisotopes or hapten molecules for nonisotopic detection of the hybridization signals (see Chapter 1 for details).

10.3.4. Polymerase Chain Reaction

The PCR is an *in vitro* method for replicating defined DNA sequences of specific organisms using their DNA or RNA as templates. The amount of target sequence is increased exponentially by using PCR. Whereas previously only minute amounts of a specific gene could be obtained from a cell, now even a single gene copy can be amplified to 1 million copies within a few hours by PCR. PCR consists of repetitive cycles of DNA denaturation to convert double-stranded DNA to single-stranded DNA, annealing of oligonucleotide primers to the target DNA, and extension of the DNA by nucleotide addition to the primers by the action of a thermostable DNA polymerase [14–16] (Fig. 10.1). Repeated cycles of denaturation of the target DNA, primer annealing to the template DNA, followed by primer extension across the template DNA by the action of thermostable DNA polymerase result in an exponential increase in the amount of the target DNA.

The application of the PCR DNA amplification method to environmental sciences has shown great potential to solve many of the problems that could not be overcome by conventional microbiological culture methods or by the traditional

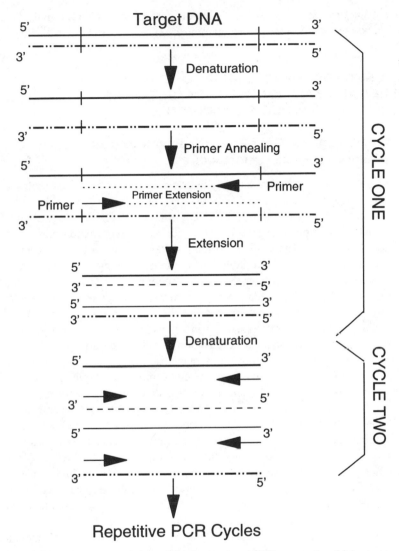

Fig. 10.1. Schematic representation of the conventional PCR approach, which uses a pair of oligonucleotide primers to amplify a targeted DNA segment or a gene.

gene probe DNA hybridization method alone [17–21]. The PCR DNA amplification method has many advantages over conventional methods for its applications in the environmental sciences. Some of the recent advances in PCR methodology that may be very useful in environmental applications are (1) "hot start," which increases the specificity of the reaction, (2) removal of PCR carryover contamination; and (3) thermostable DNA polymerase from various thermotolerant microorganisms, some of them with additional activities such as reverse transcriptase (*Tth*I DNA polymerase from *Thermus thermophilus*), which is particularly important in the study of *in situ* gene expression in the environment [20,22].

10.4. USE OF GENE PROBE NUCLEIC ACID HYBRIDIZATION APPROACH FOR DETECTION OF MICROORGANISMS IN THE ENVIRONMENT

10.4.1. Application of Gene Probes in Environmental Sciences Using Colony Hybridization Methods

For environmental microbiology, colony hybridization is perhaps the simplest application of nucleic acid–gene probe hybridization technology. Colony hybridization can be integrated with the conventional environmental microbiological sampling analysis that is traditionally performed on various selective and nonselective microbiological agar media. The original protocol developed by Grunstein and Hogness [23] was shown to be suitable for high density plate screening for pure cultures [24]. A single colony of the target microorganism *Pseudomonas putida* carrying catabolic plasmid (TOL) for the degradation of toluene was detected by using a whole TOL plasmid probe in a background of approximately 1 million *E. coli* colonies [25].

Gene probe can be hybridized with primary isolation from environmental samples [26–29] or from the secondary cultivation of already described strains [30–32]. The primary reasons for using the colony hybridization method on primary cultivation of the microorganisms grown from the environmental samples are (1) avoiding a cultivation bias encountered by selective media that may underestimate total abundance of a given genotype, (2) ensuring that a given genotype is represented in the population sampled even if the genes are poorly expressed or are poorly selected, (3) providing optimum growth conditions for organisms that are under stress due to the environmental factors and are not culturable on selective media, and (4) reducing the analysis time for cultivation, presumptive quantification, and conformation of a genotype and/or phenotype. Colony hybridization on secondary cultivation of pure cultures is usually used to confirm a specific genotype or to test a unique DNA sequence for gene probe development [11,33]. In the area of environmental sciences, the major applications of colony hybridization are to the detection, enumeration, and isolation of microorganisms with specific genotypes and/or phenotypes and for the development of gene probes. Fredrickson et al. [34] combined DNA hybridization with the most probable number (MPN) method, which permitted the detection of *Rhizobium* spp. and *P. putida* that had been genetically marked with a transposon *Tn5*, at approximately 10^2 cells per gram of soil. Jain et al. [35] used colony hybridization and gene probes of varying specificity to study the maintenance of catabolic and antibiotic resistance plasmids in groundwater aquifer microcosms. They found that introduced catabolic plasmids or organisms can be maintained in groundwater aquifers without selective pressure. In another study, colony hybridization with a naphthalene gene probe was used to correlate gene frequency with naphthalene degradation by bacteria that are capable of naphthalene degradation in activated sludge [36]. In this study, it was shown that the gene probe analysis for the catabolic genotype was nearly two orders of magnitude more sensitive than the standard plate assays.

The application of the colony hybridization method for the detection and enumeration of mercury-resistant bacteria in contaminated environmental samples [37,38] and naphthalene-degrading bacteria in aromatic hydrocarbon contaminated soils [11] has been reported. Although a single gene probe was used for the detection of each of the target microbes that are capable of degradation of mercury and naphthalene, respectively, later it was determined that in the case of mercury resistance there may be at least four degradation genes present in the mercury-stressed environment. Therefore, for accurate enumeration of the mercury-resistant bacteria a mixture of several gene probes needs to be used [39,40]. In general, the gene probe colony hybridization method for identification and enumeration of specific microorganisms in the environmental samples requires a large number of the target microorganisms so that at least one positive colony can be detected on an agar plate containing 100–1000 colonies. Additional sensitivity can be achieved by plating the isolated bacteria onto selective agar before colony hybridization. This approach has been described to increase the detection limit for *Listeria* from dairy products [30] and *E. coli* in seeded lake water [41]. More recently, a plasmid containing the cloned listeriolysin gene of *L. monocytogenes* was used to identify *Listeria* strains in artificially contaminated food (25 g of cheese, lettuce, and shrimp) by the colony hybridization method [42]. Pettigrew and Sayler [43] applied a plasmid gene probe carrying chlorobiphenyl-degrading genes to isolate 4-chlorobiphenyl–degrading organisms from contaminated lake water using the colony hybridization method. This approach has significantly reduced the cultivation and enrichment steps that are performed when using conventional microbiological methods. The colony hybridization approach was used to monitor genetically engineered microorganisms that are programmed to self-destroy by an environmental signal in soil microcosms [44–47].

10.4.2. Nucleic Acid Isolation and Purification for Direct Gene Probe Hybridization Analysis

The first step in the application and evaluating of PCR and gene probe–based hybridization methods in environmental samples is the recovery of DNA that can be utilized for such methods. Removal of various inhibitors such as humic materials, proteins, and heavy metals is essential for successful application of nucleic acid technology such as gene probe hybridizations and PCR DNA amplifications. Two approaches have been reported to recover DNA from environmental samples: (1) isolation of microbial cells followed by lysis and purification of the nucleic acids (cell extraction method) [48] and (2) direct lysis of the microbial cells in the environmental matrix followed by nucleic acid purification (direct lysis method) [49].

Purification of Nucleic Acids from Soil and Sediments. Purification of the released DNA either by direct lysis or by cell extraction methods is performed by applying a combination of the various standard purification methods such as phenol-chloroform extraction, treatment with ammonium acetate followed by ethanol precipitation, repeated polyvinylpolypyrolidone (PVPP) treatment or dialysis, hydroxylapatite or affinity chromatography, and multiple cesium chloride density gradient centrifugation [20,50]. In a separate approach, treatment with lysozyme followed by repeated freezing and thawing the samples to release nucleic acids from soil samples has been described by Tsai and Olson [51]. The released nucleic acids were then purified by standard phenol-chloroform extraction and chromatography methods. Using this approach, it was possible to detect <10 cells of *E. coli* per 1 g of seeded or unseeded soil samples [52–54].

In another study, bacterial cells were separated from soil colloids on the basis of their buoyant densities [55]. In this method, a modified sucrose gradient centrifugation protocol has been developed to separate most of the soil colloids from the bacterial cells in the sample for PCR amplification of the target DNA. However, since this approach retained minute quantities of inhibitory colloidal soil particles in the bacterial cell suspension, it was necessary to run an additional 25 cycles of PCR DNA amplification, with an aliquot of the amplification product of the first 25 cycles ("double PCR"), to detect 1–10 target microorganisms per gram of soil sample. In a separate study, DNA was extracted and purified from soil by the following sequential steps: cold lysozyme and SDS-assisted lysis with either freezing-thawing or bead beating, cold phenol extraction of the resulting soil suspension. CsCl and potassium acetate precipitation, and finally, spermine-HCl or glass milk purification [56]. The resulting DNA was pure enough for PCR amplification of up to 20 ng of soil-derived DNA from *Pseudomonas fluorescens* (RP4:*pat*) per 50 μl reaction mix. Recently, an effective method for removal of inhibitory humic substances in crude DNA extract from soil samples was described by using a Sephadex G-200 spun column (Pharmacia LKB Biotechnology, Inc.) saturated with Tris·Cl-EDTA (pH 8.0) buffer [5–5]. This extraction method resulted in the detection of <70 cells of *E. coli* by PCR DNA amplification from a 1/100 fraction of the purified sample using 16S rRNA as a target. Also, a simple and effective method of purification of PCR-amplifiable DNA from soil matrices has been described by separating crude DNA extracts by direct lysis method in a low-melting-point agarose gel (1.2% w/v) mixed with polyvinylpyrrolidone (PVP) (2% wt/vol) [57]. The water-soluble PVP forms hydrogen bonds with phenolic compounds in the DNA extracts and prevents comigration during agarose gel electrophoresis. Differentially migrated DNA bands were purified from the agarose gel, and PCR amplification yielded ~0.5 μg of amplified DNA from seeded *E. coli* cells using 16S rRNA as a target. In another approach, total DNA was purified from low quantities (100 mg) of soil samples seeded with bacterial species such as *Agrobacterium tumifaciens* and *Frankia* spp. In this extraction procedure, 100 mg of soil sample was resuspended in 0.5 ml of TENP (50 mM Tris·Cl [pH 8.0], 100 mM NaCl, 1% [wt/vol] PVPP) buffer and sonicated with a titanium microtip for 5 minutes at 15 W power. The sample was centrifuged and the pellet was heated at 900 W five times and suspended in 0.1 ml of TENP buffer. The sample was then "freeze-boiled" (liquid nitrogen and boiling water) three times for lysis and washed four successive times with TENP buffer. The final purification of the supernatant was performed in an Elutip-d-column (Schleicher & Schuell) for PCR amplification detection of the target microorganisms with the sensitivity of 10^7–10^3 cells of *A. tumifaciens* and 0.2×10^5 cells of indigenous Frankia spp. per gram of soil sample [58]. Another more tedious and long procedure for extraction of PCR-amplifiable DNA from soil, sediment, and sand samples has been described for the detection of native bacterial populations using 16S rRNA genes and mercury resistance gene (*mer*) [59,60]. This procedure required lysis of the cells by treatment with sodium dodecyl sulfate followed by concentration of the nucleic acids with polyethylene glycol and NaCl and precipitation overnight with ethanol. After CsCl density gradient centrifugation purification, the DNA was further purified by phenol–chloroform (vol/vol) treatment and precipitated overnight with sodium acetate and ethanol. The DNA was recovered by centrifugation and used for PCR DNA amplification. The entire procedure for the extraction and purification of DNA takes 3 days to complete. A slightly different approach has been described for purification of total DNA from humic acid–contaminated soil samples, and the effect of the presence of humic acids in the purified DNA on PCR and DNA-DNA hybridization has been evaluated [61]. In this approach, in addition to lysozyme, another cell lysing enzyme, lyticase (Boehringer Mannheim), was used. Following lysis, the sample was purified by phenol-chloroform-isoamyl alcohol, and the DNA was precipitated by ethanol. Additional brownish colored humic acids that are associated with the DNA sample were purified by passing the sample through an ion exchange column (Qiagen-Tip 500). The purified DNA was used for DNA-mediated genetic transformation of *E. coli,* slot blot hybridization, or PCR amplification of target genes.

Purification of Nucleic Acids from Microorganisms in Water. The isolation and purification of nucleic acids from microorganisms present in water is relatively easy and less complicated. A simple and rapid method for isolating nucleic acids from aquatic samples was described by Sommersville et al. [62]. In this approach, 300 to >1,000 ml of water sample was concentrated on a single cylindrical filter membrane (Sterivex-GS Filter unit type SVGS01015, Millipore Corporation, Bedford, MA) to harvest cells, followed by alkaline lysis and proteolysis performed within the filter housing. Crude high molecular weight nucleic acids were extruded from the filter unit, and further purification was carried out by conventional methods such as CsCl density gradient centrifugation or by treatment with NH_4OAc followed by ethanol precipitation. Using this approach, it was possible to extract sufficiently pure chromosomal DNA, plasmid DNA, or various species of RNAs (5S, 16S, and 23S rRNAs). This rapid and convenient approach of filtration, to concentrate relatively large volumes of environmental waters for recovery of dissolved and particulate DNA, can be used for PCR-based analysis of community diversity and microbial activity in the aquatic ecosystems. In various other studies, CsCl-EtBr density gradient centrifugation [63] for all environmental DNAs, multiple PVPP treatment for rRNA studies [64] and repeated phenol-chloroform extractions for total planktonic DNA [65,66] or cyanobacterial DNA [67] were found to be essential to yield sufficiently pure DNA for various molecular biological analyses. In another approach in the use of PCR methodology for the detection of microbial cells (both indicator and pathogens) in drinking water or relatively clear waters, cells are concentrated on a PCR-compatible membrane filter, lysed by a repeated freeze-thaw method, followed by PCR DNA amplification without removing the filter from the reaction tube [19,68]. Alternatively, filtered cells can be lysed directly on the polycarbonate filter by lysozyme, and the released DNA can be purified by phenol-choroform extraction followed by ethanol precipitation to yield adequately purified DNA for PCR amplification and analysis [69,70].

Although the fundamental steps for isolation and purification of DNA from soil and sediment can be summarized (Fig. 10.2), a standard protocol for removing all possible inhibitors including the humic acids in various types of soil and sediment samples that can be utilized for PCR amplification and gene probe analysis is yet to be described.

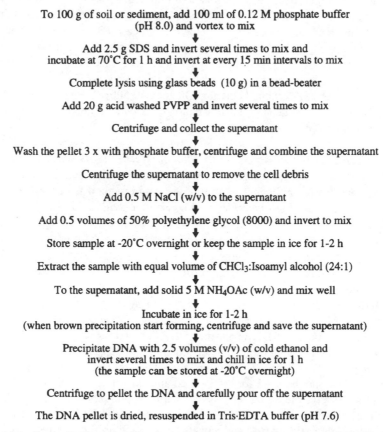

To 100 g of soil or sediment, add 100 ml of 0.12 M phosphate buffer (pH 8.0) and vortex to mix
↓
Add 2.5 g SDS and invert several times to mix and incubate at 70°C for 1 h and invert at every 15 min intervals to mix
↓
Complete lysis using glass beads (10 g) in a bead-beater
↓
Add 20 g acid washed PVPP and invert several times to mix
↓
Centrifuge and collect the supernatant
↓
Wash the pellet 3 x with phosphate buffer, centrifuge and combine the supernatant
↓
Centrifuge the supernatant to remove the cell debris
↓
Add 0.5 M NaCl (w/v) to the supernatant
↓
Add 0.5 volumes of 50% polyethylene glycol (8000) and invert to mix
↓
Store sample at -20°C overnight or keep the sample in ice for 1-2 h
↓
Extract the sample with equal volume of CHCl$_3$:Isoamyl alcohol (24:1)
↓
To the supernatant, add solid 5 M NH$_4$OAc (w/v) and mix well
↓
Incubate in ice for 1-2 h
(when brown precipitation start forming, centrifuge and save the supernatant)
↓
Precipitate DNA with 2.5 volumes (v/v) of cold ethanol and invert several times to mix and chill in ice for 1 h
(the sample can be stored at -20°C overnight)
↓
Centrifuge to pellet the DNA and carefully pour off the supernatant
↓
The DNA pellet is dried, resuspended in Tris·EDTA buffer (pH 7.6)

Fig. 10.2. Generalized protocol for the isolation and purification of total DNA from soil and sediment samples.

10.4.3. Application of Gene Probes for the Detection of Microbial Pathogens in the Environment

The detection of environmentally relevant microbial pathogens such as *E. coli* [6, 19–21,26,71–78], *Salmonella* [27,79–81], *Giardia* [82–88], *Legionella* [46,85,86, 89], and other gastrointestinal pathogens [21,75,76] by using gene probes or PCR gene probes have been reported. Use of species-specific probes can be more determinative in identifying target microbial pathogens from the complex environmental samples than whole genome probe hybridization. However, to apply such a probe one has to identify a unique DNA segment from the target microbial pathogens and extensively test against nontarget microbial strains to determine the specificity of the probe before applying to environmental samples. Another advantage is that the cloned fragment can be maintained on a plasmid in a host cell and can easily be recovered and used as a probe. Such species-specific probes have been developed and applied for *Lactobacillus* [43], *Arthrobacter* [90,91], *Thiobacillus ferrooxidans* [92], and *Bacteroides* [93].

Application of gene probes to distinguish the pathogenic microorganisms from the nonpathogenic strains can provide very useful information about the nature of a given environmental sample. In this approach, gene(s) associated with pathogenicity have been exploited to distinguish pathogenic versus nonpathogenic strains of *E. coli* [26], *Listeria* [32], and *Yersinia* [28].

Synthetic oligonucleotide probes have been applied extensively for identification of various microbial pathogens from the environmental samples. These short oligonucleotide probes have higher specificity and can be synthesized in large quantities in a single cycle of synthesis. Short oligonucleotide probes for the detection of specific microbial pathogens can be purchased commercially. Microbial pathogens such as *Legionella* [94] *Mycobacterium* [95] hepatitis A virus [96], *E. coli* and coliform bacteria [69], and *Giardia* [95] are some of the examples that have been detected by hybridization methods using oligonucleotide probes.

Oligonucleotide probes based on the rRNA genes have been developed that are kingdom, genus, species, and strain specific [97]. Use of 16S rRNA as a target is more advantageous because of its large size and availability of conserved regions for screening the same group of organisms [11]. Using oligonucleotide probes for 16S rRNA, eubacteria, archaebacteria, and the eukaryotes were distinguished at a single cell level [98]. The value of the culture-independent gene probe method was further emphasized by Ward et al. [99] who analyzed 16S rRNA sequences from the hot spring community to reveal a wider diversity of organisms than would be expected from culture techniques. The well known pink-filament community associated with the 84°–88°C outflow of Octopus Spring at the Yellowstone National Park was subjected to PCR gene probe characterization for its phylogenetic diversity [100]. In that study, the 16S-like rRNA genes (rDNA) of the mixed population DNA were amplified by PCR, cloned in pBluescript KS+ or KS– (Stratagene, La Jolla, CA) plasmid, sorted by the DNA–DNA hybridization method using an oligonucleotide gene probe specific for the 16S 4DNA, and sequenced. The sequence comparisons of these pink filaments with the related bacterial species showed that these microorganisms are closely related among cultured organisms to the hydrogen-oxidizing bacterium *Aquifex pyrophilus* and its close relative *Hydrogenobacter thermophilus*.

Oligonucleotide probes tagged with fluorochrome were used in detecting single cells of *Fibrobacter succinigenes* and *Methanosarcina acetivorans* in mixed ruminant bacteria [12]. Also, using this fluorochrome labeling approach, Amman et al. [13] demonstrated detection of lower than 3% of *Desulfovibrio gigas* from a total suspension of various nonspecific microorganisms. A fluorescent-labeled population-specific oligonucleotide probe based on the 16S rRNA of *Desulfovibrio vulgaris* was used to monitor and isolate sulfate-reducing microbial species in a multispecies bioreactor [101]. Similarly, eight group-specific 16S rRNA probes were used for the detection of phylogenetically defined groups of methanogens for environmental and determinative microbiology studies [102]. A DNA–DNA slot blot hybridization approach was used to detect various *Vibrio*-related strains in marine hatcheries and fish farms. An interesting observation was reported by Kopczynski et al. [103] on the significant differences of chimeric small-subunit rDNAs of uncultivated microorganisms during PCR amplification analysis of cyanobacterial mat on a hot spring. In a separate study, two 16S rRNA-based oligonucleotide probes were used to detect members of the *Archaea* kingdom using dot blot and whole cell (*in situ*) hybridization approaches [104]. In this study, it was shown that optimization of "cell fixation" and "hybridization solution" are essential for probe penetration and morphological integrity of the cells. Whole fixed bacterial cells from complex microbial communities in activated sludge samples were detected by using 200–300 nucleotide long polynucleotide probes that were generated by *in vitro* transcription of 23S rRNA. These polynucleotide probes had multiple reporter molecules, which allowed the simultaneous detection of three populations in the sludge samples following induction [105]. Although these long polynucleotide probes penetrated into the whole fixed cells and provided signals for differential identification of specific microbial species, the probe sensitivity and specificity were strongly influenced by the stringency of the hybridization reactions. Also, single-stranded RNA probes generated by the *in vitro* transcription method provided stronger hybridization signals [106]. A generalized protocol for whole cell hybridization is illustrated in Figure 10.3. The use of 16S rRNA oligonucleotide probes for hybridization detection of the iron (Fe^{3+}) and manganese (Mn^{4+}-reducing bacterium *Shewanella putrefaciens* in a water column and sediment samples [107] showed the importance of this bacterium in biogeochemical cycling of these two metal ions in the environment. Although this approach is useful to enumerate total target microbial cells in the environmental samples, this method does not distinguish between viable and nonviable cells.

10.4.4. PCR Gene Probe Hybridization for Genetically Engineered Microorganisms in the Environment

Use of gene probe hybridization technology has been applied to tracking GEMs in the environment. Although most of the studies on the fate of released GEMs were

STEP 1: Cell Fixation

Three volumes of paraformaldehyde solution (4% [w/v] in PBS, pH 7.2) is added directly to the
cell culture and fixed for 3 h at room temperature
↓
The cells were washed in PBS and stored in 1:1 mixure of PBS and 98% ethanol at -20°C
↓
The cells are spotted onto precleaned, gelatin-coated [0.1% gelatin, 0.01% $KCr(SO_4)_2$]
microscopic slides
↓
The cells are dried at 46°C for 30 min, and dehydrated in 50, 80, and 98% (v/v) ethanol (3 min
each)

STEP 2: *In situ* hybridization

Eight microliter hybridization solution (see text) + 50 ng oligonucleotide probe which are labeled
with rhodamine (tetramethylrhodamine-5-isothiocyanate) is added to the sample (in some probe
hybridizations, 20% formamide (w/v) may be required for optimum stringency of the hybridization
and the stringency of hybridization can be optimized by addition of higher concentrations of
formamide)
↓
The sample is incubated in the hybridization solution for 2 h at 45°C in a moisture chamber
equilibrated isotonically to the hybridization solution
↓
After hybridization the sample is washed in 2 ml washing solution (see text) by rinsing and then put
into 50 ml of washing solution at 48 °C for 20 min
↓
The slides are rinsed in distilled water, dried and mounted for microscopic examination

Fig. 10.3. Schematic representation of *in situ* whole cell hybridization using a rhodamine-labeled oligonucleotide probe. This procedure is adopted from Burggraf et al., 1994 (see reference).

conducted in the controlled environment, the application of gene probes has been shown to be extremely useful for monitoring them. The greatest advantage of using gene probes for tracking GEMs in the environment is that the molecule of interest can be assayed directly without culturing the microorganism. Using gene probe technology, the following can be achieved from GEMs released in the environment: (1) the maintenance and stability of the DNA of interest in the environment within the original host, (2) horizontal transfer of the marker gene between microorganisms by conjugation or by natural transformation, and (3) the effect of the recombinant DNA on community structure and dynamics due to its expression and transfer to other organisms. After addition of the marker gene–DNA sequences in the host microorganisms by cloning or other gene manipulation, application of the gene probe technology can be used to study the survival of the recombinant host microorganisms in the environment.

Gene probes for tracking polychlorinated biphenyl (PCBs)–degrading microorganisms in contaminated soil environments has been developed and applied [108]. In this study, using radiolabeled *cbpABCD* and pAW6194 probes and dot blot analysis, less than 1% of the garden top soil and >80% of bacteria isolated from PCB-contaminated soil showed positive hybridization. A bacterial haloacetate dehalogenase (H-l type enzyme) probe has been developed and used for identification of re-leased GEMs from various paddy soil environments. This gene seems to have unique nucleotide sequences and is uniformly distributed in nature. Therefore, this dehalogenase gene can be used as a marker for monitoring GEMs in the environment. The sensitivity of detection when using this probe was 4 cfu/g of soil. The efficiency of several gene probe methods was assessed in freshwater microcosms while monitoring the survival of a strain of *Pseudomonas cepacia* that was capable of degrading 4-chlorobiphenyl [109]. Transfer of plasmids from the released host microorganisms to the indigenous microbial population has been demonstrated in domestic wastewater treatment simulations by using gene probe methods [110]. In a soil microcosm study by Bentjen et al. [111], the transport and colonization of GEMs in plant roots has been studied using a gene probe assay on a mutant version of the *Tn5* gene that was incorporated in the host GEMs. Using the gene probe method, the mutant *Tn5*-containing bacteria could be separated from endogenous kanamycin-resistant bacteria. In a related study, a single copy of the transposon *Tn5* was transferred into the genomic DNA of *Rhizobium leguminosarum* released in the soil. These genetically altered microorganisms were detected by "double" PCR amplification using transposon *Tn5* as target, to a sensitivity of 1–10 cfu per gram of soil [55]. Although it is adequate to use *Tn5*, which contains an antibiotic resistance gene, as a model target for PCR detection, it may not be an appropriate marker for releasing GEMs in the environment because of its ability to be transferred into indigenous microorganisms, making them antibiotic resistant.

Several million-fold amplification of the engineered genes by PCR methodology from the released microorganisms followed by gene probe DNA–DNA hybridization methodology can potentially increase the sensitivity of detection of released GEMs in the environment. The application of PCR gene probe methodology was first applied to monitor genetically engineered microorganisms by amplification of a portion of the 1.3 kb repeat sequence from *Pseudomonas cepacia* AC1100, a herbicide (2,4,5-T) degrading bacterium, following release in the soil microcosm, to a sensitivity of 100 GEMs in 100 g of sediment against a background of 10^{11} diverse nontarget microorganisms. This sensitivity was at least 10^3-fold higher than nonamplified conventional dot blot hybridization detection [112]. A 0.3 kb unique DNA sequence from *Pennisetum purpureum* (napier grass) has been cloned into pRC10, a derivative of 2,4-dichlorophenoxyacetic degrading plasmid, and transferred into *E. coli* [113]. This genetically altered microbe was released into filter-sterilized lake and sewage water samples to a concentration of 10^4 cells per ml and detected by PCR at a sensitivity severalfold higher than the conventional plating technique, even after 10–14 days of incubation, using the unique cloned DNA sequence as a target. By using two highly conserved regions of the *lux*A gene as primers, PCR amplification and gene probe methods have been applied to detect and study the marine luminous bacteria [114]. This approach has enhanced the species specific identification of various luminescent *Vibrio* spp. in marine environments and has great taxonomic value.

In another study using a gene probe solution hybridization method, genetically engineered 2,4,5-T-degrading *Pseudomonas* was detected with a sensitivity of

100–1,000 cells per gram of sediment [7]. A plant growth promoting rhizosphere bacterium, *Pseudomonas fluorescens*, was genetically modified by transferring a reporter gene on the chromosome, the mannityl opine catabolism gene (*moc*), from *Agrobacerium tumificiens* [115]. Nucleic acid–based hybridization or PCR amplification of the *moc* gene was used for the detection of this genetically modified microorganism in the environment.

From these studies it can be concluded that the PCR method can be used for monitoring released GEMs in an environment consisting of a complex habitat of diverse microorganisms, where it may be tedious and time consuming to discriminate the GEMs from the indigenous microorganisms [21,47].

10.4.5. PCR Gene Probe Hybridization for Indigenous Microorganisms in the Environment

A number of microorganisms are present in the environment that are involved in the biodegradation of various pollutants and toxic wastes. It is possible to monitor such microorganisms that are involved in bioremediation of the polluted and toxic waste sites by application of the PCR gene probe methods using conserved regions of the genes that are involved in such function. In one study using the nucleotide sequence information of a chlorocatechol dioxygenase degrading gene (*tfd*C) from *Alcaligenes eutrophus* JMP134 (pJP4), it was possible to design oligonucleotide primers for the detection of various chloro-aromatic degrading bacteria by PCR amplification followed by gene probe hybridization [116]. PCR amplification followed by gene probe hybridization using such oligonucleotide primers gives information on the variations, similarities, and functional aspects of various pollutant degrading genes present in closely or distantly related microorganisms, in the environment, in a very short period of time. It allows the detection of the specific microorganism carrying the degrading gene from a complex mixed population in the environment. When such a polluted site is identified, it is important to investigate the possibility of the presence of various degrading microbes at that site. In addition, by using the mRNA reverse transcriptase cDNA PCR approach, it is possible to determine the degrading activities of these microorganisms in the contaminated sites. In another study, specific detection of the same species of the herbicide (2,4-dichlorophenoxyacetic acid) degrading bacterium *A. eutrophus* was achieved by PCR amplification of a region of *tfd*B gene from pJP4 and its derivative plasmid pRO103 [117]. In this study, by direct PCR amplified DNA analysis it was possible to detect approximately 3,000 cfu or 15.6 pg of plasmid DNA. The sensitivity of such detection was onefold higher when DNA–DNA hybridization was performed with an oligonucleotide probe internal to the amplified DNA. Using gene probe DNA–DNA hybridization technology, in one study the genetic and phenotypic diversity and in another study the identification of functionally dominant 2,4-Dichlorophenoxy acetic acid-degrading mciroorganisms in the soil and sediment samples were described [118, 119]. PCR DNA amplification of a portion of the *nah* gene followed by Southern blot DNA–DNA hybridization was used for identification of naphthalene-degrading microorganisms in soil and sediment samples with the sensitivity of 10^3 cells per gram of soil [120]. Native bacterial populations were detected from various samples of soil, sediment, and sand by PCR amplification of a conserved region of the 16S rDNA segment and the mercury resistance (*mer*) gene [59]. Since the rDNA target is present in the chromosome and the *mer* gene is plasmid borne, amplification of both targets simultaneously has the potential to serve as a model system to study the microbial interactions and gene transfer in the natural environment.

The ribulose biphosphate (*rbc*L) gene was used as a target to amplify planktonic DNA and analyses the microbial community in the aquatic environment. Using the same target, dissolved DNA associated with the phytoplankton in the aquatic environment was determined by PCR amplification of the extracellular DNA fraction [63]. The ribulose biphosphate is considered to be the most abundant protein in the environment. Therefore, by use of this conserved gene as a target for PCR amplification, it is possible to analyze important ecological functions in the environments.

Using such oligonucleotide primers and PCR amplification followed by gene probe hybridization, it is possible to detect the specific microorganisms carrying the genes from a complex mixed microbial population that are involved in the degradation of xenobiotic compounds in the environment. Formation of biofilms on various surfaces by microorganisms in the environment can be beneficial or detrimental. For example, microbial aggregation or attachment is required for various water treatments, while on the other hand extensive corrosion and biodeterioration can be caused due to the formation of such microbial biofilms. Characterization and ecology of microbial populations in biofilms has been hindered due to the available determinative techniques that require culture of microorganisms in selective media. These methods eliminate many of the important microbes from the biofilms since they survive only in a mixed culture and live on the cometabolism. Use of PCR amplification of specific targets makes it possible to identify a group of microbes in such a biofilm that may have been missed by the conventional technique. To determine the feasibility of such use, a sulfidogenic biofilm has been established in an anaerobic fixed-bed bioreactor, and PCR amplification was performed for the detection of the population architecture of all the gram-negative sulfate-reducing bacteria using a region of the 16S rRNA conserved in the resident sulfate-reducing bacteria [121]. A novel quantitative analysis of microbial communities in oil field production waters has been developed using the "reverse sample genome proving" method [122]. In this approach, the denatured chromosomal DNAs from bacteria obtained from the target environment (an oil field was used in this study) was spotted on a master hybridization filter. DNA isolated from the environmental samples was labeled and hybridized with the master filters to identify which of the bacterial genomes spotted on the master filter are most prevalent in the sample.

10.4.6. Purification of RNA from Environmental Samples and Use of Gene Probes in Detecting Microbial Activities in the Environment

For the isolation and purification of total RNA from microorganisms in the environmental samples, 10 g of soil sample seeded with *Pseudomonas aeruginosa* PU21 was treated with guanidium thiocyanate (4 M) mixed with sodium citrate (25 mM), sarcosyl (5% w/v), and 2-mercaptoethanol (0.1 M) to achieve lysis of cells, fixation of total cellular RNA, and hydrolysis of DNA. The treated sample was then purified with phenol-chloroform-isoamyl alcohol (24:24:1 v/v) followed by precipitation of the RNA with isopropanol [52]. This approach yielded 17 μg of total RNA and 0.16 μg of mRNA from 1 g of soil containing 8×10^8 *P. aeruginosa* PU21 cells. This extraction method can be completed within a few hours and has potential for the study of gene expression in various microorganisms in the environment by reverse transcription of the target mRNA coupled with PCR DNA amplification with specific primers sets. More recently, lysozyme–hot phenol treatment followed by gel filtration with Sephadex G-75 spun columns was used to recover total rRNA from microbial communities in sediment, soil, and water samples [123]. In this approach, the samples were treated with Tris-lysing buffer consisting of 50 mM Tris·Cl (pH 8.0), 25% sucrose, and 5 mg lysozyme per ml. Following centrifugation the pellet was resuspended in ACE buffer consisting of 10 mM NaOAc (pH 5.1), 10 mM NaCl, and 3 mM EDTA. The sample was then treated with ACE-buffered phenol at 65°C and purified with ACE-buffered phenol-chloroform-isoamyl alcohol. The aqueous phase was treated with NaOAc and the nucleic acids were precipitated with ethanol. The purified total RNA was further purified from the residual humic substances and concentrated by passing through Sephadex G-75 spun columns. The purified total RNA from various environmental samples have been shown to be useful for molecular procedures such as hybridization.

Purification of mRNA from water samples that is encoded by the *merA* gene in *P. aeruginosa* upon induction has been reported [124]. In this approach, the cells from the water samples were collected on a filter and the filter was frozen in dry ice until used. The filter containing the cells were boiled for 2–5 minutes in STE buffer (10 mM Tris·Cl [pH 8.0], 1 mM EDTA, 100 mM NaCl), which was mixed with 1% (w/v) SDS and 0.1% (v/v) diethylpyrocarbonate. The sample was chilled in ice for 15 minutes. The sample was then treated with guanidium isothiocyanate-sarkosyl solution (GIPS), phenol:chloroform, NaOAc, mixed, and centrifuged. The aqueous phase was transferred into a fresh tube and mixed with glycogen. The nucleic acid sample was precipitated with isopropanol. This approach of mRNA recovery from the environmental sample has been shown to be devoid of DNA and is particularly important to study the *in situ* gene expression in the environment.

Application of gene probe technology has special value in detecting RNA from target microorganisms in the environment. Information from genetic and physiological research suggests that many bacteria possess a wide variety of genes that are not expressed in the environment [125]. Moreover, many of the genes that are expressed in the laboratory may not do well in the environment due to stress conditions. Therefore, to understand the function and expression of a gene of interest, gene probe hybridization can be performed on the target mRNA rather than DNA. To determine the effect of 2-hydroxybenzoate on the rate of mineralization of ^{14}C-naphthalene, gene probes have been used against total mRNA isolated from the contaminated soil environment [126]. Extraction of rRNA–tRNA followed by Northern hybridization to determine the species composition in a marine sample

Collect cells on high capacity cartridge-type filters with the diameter of 142 mm and freeze in dry
ice (minimum 10⁹ cells may be required)
(0.22 μm pore size Durapore or Sterivex -GS cartridge, Millipore)

↓

Five ml of SDS in STE buffer which is preheated to 85°C was added to the filter and DEPC treated
water added to a final concentration of 0.1%

↓

The sample is boiled for 5 min with intermittent vortexing

↓

The liquid is transferred to a new tube and kept in ice

↓

Additional 5 ml of STE-SDS in DEPC treated water is added to the filter, vortexed and the liquid is
combined to the previous sample

↓

Ten ml of GIPS, 1 ml of 2 M NaOAc (pH 4.0) + 10 ml phenol and 2 ml CHCl₃ (49:1) is added to
the samples and mixed well

↓

Following centrifugation at 4°C, the supernatant is treated with glycogen and precipitate 2 x with
isopropanol

↓

The RNA pellet is washed with cold 70% alcohol

↓

The pellet is dried and resuspended in 1 mM EDTA

↓

The sample is treated with 0.1X volume of 2 M NaCl and 0.7X volume of isopropanol

↓

The pellet is washed 1X with cold 70% alcohol, dried and resuspended in DEPC-treated water to
Tris·EDTA (pH 8.0) buffer

Fig. 10.4. Generalized protocol for the extraction and purification of mRNA from environmental biomass using boiling lysis method. This protocol is adopted from Jeffrey et al., 1994
(see reference).

has been evaluated [127]. A generalized protocol for the extraction purification of
RNA from environmental samples is summarized in Figure 10.4. Although it is a
relatively new approach, the use of gene probes against the mRNA of the target
genes in the environment has the potential to provide us with information about the
gene activity, physiological state, and viability of microorganisms in the environment.

10.4.7. Detection of Indicator Microorganisms in Water

The bacteriological safety of water supplies is tested by monitoring coliform bacteria. The presence of coliform bacteria in water indicates potential human fecal contamination and the possibility of the presence of enteric pathogens. Coliform bacteria are traditionally detected by culturing on media such as MacConkey, m-Endo,
eosin methylene blue, or brilliant-green-lactose-bile media. These media are selective for gram-negative bacteria and differentially detect lactose-utilizing bacteria. At
an incubation temperature of 37°C, total coliform bacteria are enumerated, while at
44.5°C fecal coliforms, mainly *E. coli*, are enumerated. *E. coli* is primarily associated with human feces and is therefore a useful indicator of human fecal contamination.

The culture method for monitoring *E. coli* in environmental and potable waters has several problems associated with it. The conventional confirmative tests for the detection of *E. coli*, all of which require culturing of the organism, are time consuming and do not detect viable but nonculturable bacteria that may occur due to the chlorine injury during the process of water purification and treatment. Moreover, the cells may die between the time of collection and the test. A colorimetric test, Colilert, for the detection of *E. coli* is based on the detection of β-D-glucuronidase enzyme produced by the *uid*A gene. This method requires the culturing of bacteria. In addition, this method fails to detect β-D-glucuronidase–negative *E. coli*. Bej et al. [71,72,74] have developed a PCR gene probe–based method for the detection of coliform bacteria. Amplification of a portion of the *lacZ* gene detects *E. coli* and other coliform bacteria, including *Shigella* spp. Amplification of part of the *lam*B gene detects *E. coli, Salmonella*, and *Shigella* spp. In another study, Bej et al. [6] developed a method for the detection of *E. coli* and *Shigella* spp. using four different regions of the *uid*A gene, which codes for the β-D-glucuronidase enzyme and part of the *uid*R gene, which is the regulatory region of the *uid*A gene, as targets. Besides being less time consuming and having higher specificity and sensitivity, the most important advantage of this method over conventional and other commercially available methods is that *uid*A-negative *E. coli* can be detected that do not show a positive signal with the conventional tests because of the lack of the β-D-glucuronidase enzyme. The sensitivity of the method is 1–10 fg of genomic DNA and 1–5 viable *E. coli* cells. Similarly, Cleuziat and Baudouy-Robert [128] used a large region of the *uid* gene of *E. coli* as a target for PCR amplification and gene probe detection of *E. coli* and *Shigella* spp. This PCR gene probe–based method has the specificity and sensitivity required for monitoring coliforms as indicator organisms in environmental and potable waters. A field evaluation of the PCR application detection of enteric pathogens and indicator microorganisms has been reported using *uid*A and *lacZ* as targets [72].

Recently, the coding sequence of the *uid*A gene from *S. sonnei, S. flexneri, S. dysenteriae*, and *S. boydii* have been sequenced and compared with the *uid*A gene of *E. coli*. Oligonucleotide primers from the *E. coli uid*A gene sequence have been designed with the 3′ end mismatched nucleotides with all four *Shigella uid*A genes. PCR amplification using these oligonucleotide primers showed specific amplifications for the *E. coli* strains and no amplification for any *Shigella* strains [78] (Fig. 10.5). Therefore, these oligonucleotide primers can be used for the detection of indicator microorganism. *E. coli*, for water quality monitoring, as required by the U.S. Environmental Protection Agency (U.S. EPA). The application of gene probes and PCR amplification for detection of group-specific or species-specific microorganisms in water is rapid and shows great promise as a routine monitoring technique for microbial water quality monitoring.

10.4.8. Detection of Water-Borne Microbial Pathogens by PCR

Apart from the detection and monitoring of indicator microorganisms for water quality assessment, it is also important to detect various water-borne microbial

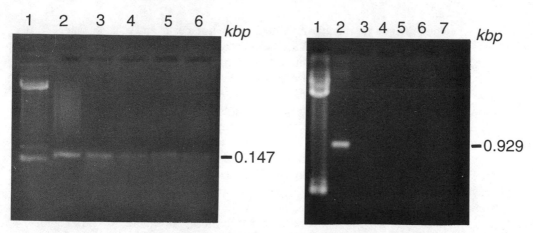

Fig. 10.5. Agarose gel electrophoresis analyses of the PCR amplified DNAs showing specific identification of *Escherichia coli* using *uid*A-specific oligonucleotide primers with mismatched nucleotides at their extreme 3'-ends as compared with the *uid*A gene of four species of shigellae. **Left** gel: **lane 1**, 123 bp DNA ladder as size standard; lanes 2–6, PCR amplification of 0.147 kbp DNA fragments using oligonucleotide primers that amplified *E. coli* (**lane 2**), and all four species of shigellae, *Shigella sonnei* (**lane 3**), *S. boydii* (*lane 4*), *S. dysenteriae* (**lane 5**), and *S. flexneri* (**lane 6**). **Right** gel: lanes 2–6, PCR amplification of *E. coli* (**lane 2**), *Shigella sonnei* (**lane 3**), *S. boydii* (**lane 4**), *S. dysenteriae* (**lane 5**), and *S. flexneri* (**lane 6**) using oligonucleotide primers consisting of 3'-end mismatched nucleotides showing amplification of a 0.929 kbp DNA fragment for *E. coli* only. **Lane 1**, 123 bp DNA ladder as a size standard; **lane 7**, PCR amplification of a sample in which no target DNA was added.

pathogens with high sensitivity and specificity. *Legionella* is a water-borne microbial pathogen and can cause Legionnaires' disease in humans via aerosol. Starnbach et al. [89] reported the detection of *Legionella pneumophila* by amplification of a fragment of DNA of unknown function from *Legionella* using PCR. Their sensitivity of detection was equivalent to 35 cfu detected by viable plating. Mahbubani et al. [129] have developed a method for the detection of *Legionella* in environmental water sources based on PCR and gene probes. All species of *Legionella*, including all 15 serogroups of *L. pneumophila*, were detected by PCR amplification of a 104 bp DNA sequence that codes for a region of 5S rRNA followed by radiolabeled oligoprobe hybridization to an internal region of the amplified DNA. Strains of *L. pneumophila* (all serogroups) were specifically detected based on amplification of a portion of the coding region of the macrophage infectivity potentiator (*mip*) gene. *Pseudomonas* spp. that exhibit antigenic cross-reactivity in serological detection methods did not produce positive signals in the PCR gene probe method using Southern blot analyses. Single-cell, single-gene *Legionella* detection was achieved with the PCR gene probe methods. PCR gene probe–based detection of *Salmonella* spp. has been developed using the *hns* [80] or *him*A gene [130], *Salmonella* plasmid

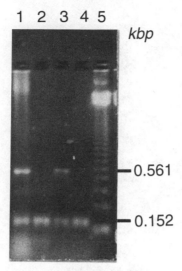

Fig. 10.6. Agarose gel electrophoresis analysis of multiplex PCR amplified DNAs showing differentiation between virulent and avirulent strains of *Salmonella typhimurium*. **Lanes 1** and **3**, simultaneous amplifications of a 0.561 kbp segment of the **spvB** (**Salmonella Plasmid Virulence**) gene on a large 96 kbp plasmid and a 0.152 kbp segment of the *hns* gene from *S. typhimurium* UK1 and ATCC 29629; **lanes 2** and **4**, PCR amplification of the 0.152-kbp DNA segments of the *hns* gene from *S. typhimurium* ATCC 19430 and 35664 indicating the absence of the virulence plasmid in these strains.

virulence gene (*spv*) (Fig. 10.6) or genes for fimbriae (*fim*) [81]. Also, species-specific detection of *Shigella* (*S. flexneri, S. sonnei, S. dysenteriae, S. boydii*) has been developed with a PCR gene probe approach (Mahbubani et al., in preparation).

Another protozoan pathogen, *Giardia lamblia* causes defined water-borne diarrhea in the United States and in many other parts of the world. Diagnosis of *G. lamblia* from environmental samples is performed by concentrating 100 gallons of water followed by microscopic examination using fluorescent dye. Using PCR amplification of different segments of the giardin gene of *G. lamblia*, it was possible to differentiate *G. lamblia* from *G. muris*. Also, a single *Giardia* cyst was detected by PCR amplification after separating the cyst by a micromanipulator. The use of an "immuno PCR" approach to selectively capture *G. lamblia* cysts in 100 gallon concentrated river water samples by immunomagnetic beads followed by PCR amplification showed a sensitivity of detection of as low as five cysts [87,88] (Fig. 10.7). Although the specificity and sensitivity of the detection of *Giardia* shows great promise for monitoring this pathogen in water rapidly, the reproducibility and reliability of this method for the detection of this pathogen for routine monitoring in 100 gallons of concentrated environmental waters from various geographical sources needs to be demonstrated.

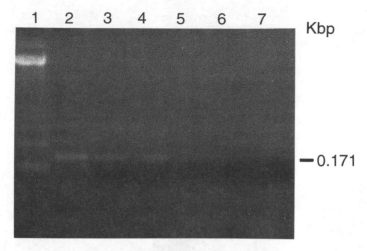

Fig. 10.7. Immuno-PCR amplification of a 0.171-kbp segment of the giardin gene from *Giardia lamblia* that were artificially contaminated in 100 gallons of concentrated river waters. **Lane 1**, 123-bp DNA ladder as size standard; **lane 2**, 10^4 **cysts;** lane 3, 10^3 **cysts;** lane 4, 10^2 **cysts;** lane 5, 10^1 **cysts;** lane 6, 10^0 **cyst; and** lane 7, no cyst was added to the water.

10.4.9. Multiplex PCR Amplification and Gene Probe Methods for Environmental Monitoring of Microorganisms

It is possible that the environmental samples and drinking waters may contain more than one type of microbial pathogen in addition to the indicator microorganism. Use of multiplex PCR for amplification and detection of more than one target in a single PCR reaction can be useful for monitoring multiple microbial pathogens in a single environmental or water sample. This method was first described by Chamberlain et al. [132] to detect human genes. A modification of this approach of simultaneous PCR amplification of multiple targets associated in different bacteria in the environmental samples has been demonstrated [70–74]. Multiplex amplification of two different *Legionella* genes, one specific for *L. pneumophila* (*mip*) and the other for the genus *Legionella* (5S rRNA), was achieved by staggered addition of two different sets of primers at two different concentrations [70]. By this method it is possible to detect *Legionella* and *L. pneumophila* should two different species of *Legionella* be present in one sample. In a field study of water quality monitoring, simultaneous PCR amplification was performed using *lac*Z and *uid*A as targets. In this study, it was possible to detect in one sample total coliform bacteria by amplification of the *lac*Z gene, the indicator microorganism *E. coli*, and a pathogen *Shigella* spp. by the amplification of the *uid*A gene [71,74]. Also, in this study the *lac*Z PCR detection gave results statistically equivalent to those of the conventional plate count and defined substrate methods accepted by the U.S. EPA for water quality monitoring. The *uid*A PCR method was more sensitive than the 4-methylumbelliferyl-β-D-glucuronide–based defined substrate test for the specific detection of *E. coli*. In another study multiplex amplification of five different targets in a single PCR reaction has been achieved for the detection of non-*pneumophila Legionella* spp. *L. pneumophila*, total coliforms, *E. coli*, *Shigella* spp., and total eubacterial species. A triplex PCR gene probe hybridization assay using heat-labile toxin (LT), shiga-like toxin I (SLT I), and shiga-like toxin II (SLT II) genes as target was used for identification of toxigenic strains of *E. coli* in water samples [77]. A multiplex PCR-based detection of another microbial pathogen, *Vibrio parahemolyticus*, which may be hemolytic (Kanagawa positive [K^+]) or non-hemolytic (Kanagawa negative [K^-]) form, has been developed in environmental samples by using two different target

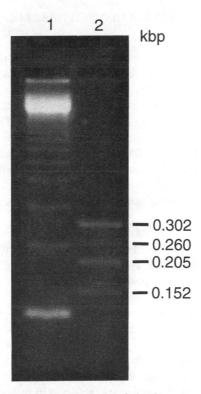

Fig. 10.8. Agarose gel electrophoresis analysis of simultaneous amplifications of multiple microbial pathogens in a single PCR reaction using the multiplex PCR approach using oligonucleotide primers that are specific for gene segments characteristic of each of these microbial pathogens. **Lane 1**, 123-bp DNA ladder as size standard; **lane 2**, amplification of 0.152-kbp segment of the *hns* gene for detection of total salmonellae; amplification of 0.205-kbp segment of the hemolysin gene for the detection of *Vibrio vulnificus*; amplification of 0.260-kbp segment of the *spvB* gene for the detection of total virulent strains of *S. typhimurium*; amplification of 0.302-kbp segment of the cytotoxin hemolysin gene for the detection of *V. cholerae*.

genes, thermostable direct hemolysin (*tdh*) for K$^+$ and thermolabile (*tl*) for K$^-$ strains [133] (Fig. 10.8). In future studies, it may be desirable to group certain microbial pathogens and indicators in the environmental samples and design the primers for specific targets. For example, one can group all the environmental and water-borne respiratory pathogens and PCR amplify all the specific target genes in a single reaction for their detection. When several genetically engineered microorganisms are released together for the degradation of complex hazardous wastes and pollutants, they can be monitored together, possibly both qualitatively and quantitatively, in a single PCR reaction by amplifying a unique segment of the DNA of each of the GEMs and together by amplifying a common segment of all the GEMs not present in other eubacterial species.

10.4.10. Detection of Viable but Nonculturable Microorganisms in the Environment

There are several reports on the existence of many microorganisms including human pathogens in the environment in a viable but nonculturable, i.e., dormant stage [73, 85, 86, 134–138]. These microbial pathogens are shown to be potentially infectious when suitable conditions prevail [135]. One obvious difficulty in elucidating this potential hazard is the inability of detecting these viable but nonculturable cells in the environment since the routine microbiological methods will not allow them to grow (on agar media) or will not distinguish them from the dead cells (by microscopic technique). The terms *alive* and *viable* are subject to different definitions, but a reasonably acceptable definition would be that the live cells are considered those capable of cell division, metabolism (respiration), or gene transcription (mRNA production) [136].

For the detection of those microbial cells that are in a viable but nonculturable state in the environment, it is desirable to target mRNA rather than the DNA for cDNA synthesis followed by PCR amplification. The potential problem of this approach would be that most of the prokaryotic mRNAs have half-lives of only a few minutes. Mahbubani et al. [85,86] have shown that the mRNA of the *mip* gene of *L. pneumophila* can be stabilized simply by growing the cells for 10–15 minutes in the presence of chloramphenicol before harvesting. They have shown that the PCR amplification of the *mip* mRNA could be a potential means for the detection of metabolically active *L. pneumophila* cells. Use of chloramphenicol for increasing stability of bacterial mRNA is yet to be tested in other microorganisms. Another perplexing issue that may create additional problems in such an approach is the efficiency of gene expression of these dormant microbiol pathogens. It is possible that the transcription or regulatory systems of the target genes in these microbial pathogens are inhibited by various environmental factors and inhibitors when they are present in the natural environment. Therefore, in this situation the quantity of the target mRNA level may be so low that it may remain undetected even by the most sophisticated method like PCR. However, it has been shown by Bej et al. [71, 74] and Brauns et al. [134] that targeting DNA for PCR amplification may be sufficient for the detection of culturable and nonculturable microbial pathogens. Both viable culturable and viable nonculturable cells of *L. pneumophila*, formed during exposure to hypochlorite, showed positive PCR amplification, whereas nonviable cells did not. Field verification of this approach for the detection of metabolically active (viable vs. dead) *L. pneumophila* from contaminated environmental samples is yet to be done [71–74]. Besides *L. pneumophila*, another important marine water-borne microbial pathogen, *Vibrio vulnificus*, which can cause fatal infections in humans when ingested with contaminated raw oyster, has been found to enter a viable but nonculturable state during the colder months and resuscitate from the nonculturable state when a suitable environment prevails [134]. Using PCR amplification of the hemolysin gene, Brauns et al. [134] detected 72 pg of DNA from culturable and 31 ng from nonculturable cells. Although, the decreased sensitivity of detection of nonculturable cells by the PCR method is not well understood at this time, several possible explanations have been described [134]. Among these possibilities, the important criteria that may be of concern in applying the PCR methodology for the detection of viable but nonculturable microorganisms are (1) less DNA content per cell, (2) difficulty in breaking open because of changes in the cell wall that may occur due to carbon or nitrogen starvation or changes in the environmental conditions, and (3) modification of the target gene due to genetic rearrangement. However, Brauns et al. [134] did not attempt to use the hemolysin mRNA as a target for PCR amplification from the nonculturable cells, which could have determined the exact nature of the cells, i.e., alive or dead, and the gene expression of the target.

A study by Mahbubani et al. [85, 86] has shown that mRNA PCR alone is not sufficient to distinguish live from dead *Giardia* cysts, since cysts killed by heat treatment or monochloramination also give positive mRNA PCR amplification. Therefore, in this organism, using the giardin mRNA as a target for PCR amplification, it is necessary to include an mRNA induction step in the procedure to determine the viability of the cysts.

Since the viable but nonculturable stage there may be changes in the gene structure and expression in many microorganisms, a modified approach of the conventional PCR approach may be required for the detection of such state of the microbial pathogens in the environment. A possible approach for the detection of the viable but nonculturable state of many of a microbial pathogen is to use the arbitrarily primed PCR methodology by targeting the genomic DNA or total RNA to generate genomic or RNA fingerprints. Changes in the gene expressions or rearrangement of genomic DNA in viable or nonviable but nonculturable state of a microbial pathogen will be manifested in the fingerprint patterns.

An important issue in environmental microbial molecular genetics is how various genes are regulated and expressed under various environmental conditions. One of the known facts is that some of the environmental microbial pathogens such as *L. pneumophila* and *V. vulnificus* alter their gene expression and remain in a dormant stage as nonculturable organisms in the environment. It has also been predicted that several biodegradative microorganisms may not express their degrading genes in the environment. As a result, one may not be sure whether the released GEMs or indigenous microorganisms are degrading the pollutants at a contaminated site. Using specific mRNA as target for PCR amplification and developing a quantitative assay for such method, it is possible to detect the level of mRNA production with high sensitivity in the environmental samples. A promising method for extraction of specific mRNA from soil seeded with naphthalene-degrading and mercury-resistant bacterial cells has been described [52]. This method can be completed within a few hours and approximately 17 μg of total RNA per gram (wet weight) of soil containing 8.0×10^8 bacterial cells can be purified with a DNA–RNA hybridization detection sensitivity of 160 ng of specific target mRNA. Although this method has potential for studying *in situ* gene expression, the humic acid compounds may precipitate with samples containing high-cation exchange capacity, e.g., some sediments, which will greatly reduce the total RNA recovery efficiency and sensitivity of detection. Application of PCR for detecting specific mRNA extracted from various environmental samples by this method has yet to be evaluated. One very important aspect of PCR gene probe detection of a microbial pathogen in environmental samples is positive amplification signals from nonviable cells providing false-positive results. The ability of the PCR gene probe methodology to detect boiled or UV-treated nonviable bacterial cells in water and other environmental samples have been reported [139]. Similarly, biocide-treated nonviable cells of *S. typhimurium* cells were detected by *Salmonella*-specific primers [79,140]. Identifying this could be a potential problem in the application and regular monitoring of microbial pathogens or understanding the microbial community structures and interactions among themselves. Thus, targeting messages may be the ideal situation. The use of a rapid and efficient cell lysis method followed by capturing and purification of total RNAs from bacterial cells without rapid degradation of mRNA can be achieved (FastRNA kit, Bio101-Savant). Using this RNA extraction approach followed by

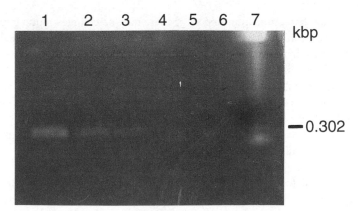

Fig. 10.9. Agarose gel electrophoresis analysis of the reverse transcriptase-PCR (RT-PCR) amplification of a 0.302-kbp segment of the cytotoxin hemolysis mRNA from the viable cells of *Vibrio cholerae*. **Lane 1**, 10^6 cells; **lane 2**, 10^4 cells; **lane 3**, 10^2 cells; **lane 4**, 10^1 cells; **lane 5**, 10^0 cell; **lane 6**, PCR amplification without reverse transcriptase treatment showing no DNA was present in the sample; and lane 7, 123-bp DNA ladder as size standard. All samples were treated with DNAse I prior to RT-PCR.

RT-PCR gene probe methodology, identification of only viable cells of pathogenic *E. coli, Vibrio cholerae*, and *S. typhimurium* has been developed [140] (Fig. 10.9).

10.4.11. Detection of Airborne Microorganisms

A number of air-borne microbial pathogens are fastidious, and it takes a considerable amount of time for their detection by a conventional microbiological approach. The use of PCR gene probes has been applied on the nitrocellulose filters (solid phase PCR) following capture of a model microorganism, *E. coli* DH1 (pWTAla5') [141]. This approach has potential for detecting air-borne microbial pathogens and microorganisms that cannot be cultured by conventional microbiological methods. However, the application and sensitivity of this approach are yet to be tested in "real" environmental samples.

10.4.12. Use of Arbitrarily Primed Polymerase Chain Reaction in Environmental Microbiology

Arbitrarily primed polymerase chain reaction (AP-PCR) (also called random amplified polymorphic DNA [RAPD]) is a relatively new, rapid, and simple technique that generates fingerprints of complex genomes by using single, arbitrarily selected primers to direct amplification. AP-PCR is a variation of the standard, conventional PCR method [16], wherein short single-stranded DNA oligomers are allowed to anneal to denatured template DNA under low stringency conditions [142–145]. At some frequency, two primers will anneal to the template relatively close to one another (100–2,000 bases) and onto complementary DNA strands. In the presence of free nucleotides and DNA polymerase the oligomers (primers) are extended to form a new copy of the target DNA. By then increasing the stringency of the reaction only those products formed in the first few low-stringency reactions are amplified. By repeating this process many times, specific fragments of DNA can be amplified, thereby becoming relatively abundant in the resulting mixture of DNA. Different randomly amplified fragments are generated in different bacteria. When the amplified DNA fragments are separated based on size by using electrophoresis, they produce a readily distinguishable pattern of DNA bands. These bands represent a DNA fingerprint that can be used to identify different microbial groups and even various strains within a specific microbial species. The principles behind AP-PCR are described in Figure 10.10.

Since the introduction of AP-PCR, the DNA from a number of environmentally related microbial pathogens such as *L. pneumophila* [146,147], *Clostridium difficile* [148], *Lactococcus lactis* [149], pathogenic *E. coli* [150], *Actinobacillus* [151], *Rhizobium* [152], *Frankia* [153], *Xanthomonas* [154], and *Pseudomonas* [155], have been subjected to the procedure. The application of AP-PCR has great potential for identification and differentiation of microorganisms within and among themselves and understanding the microbial community structures and interactions in an ecosystem.

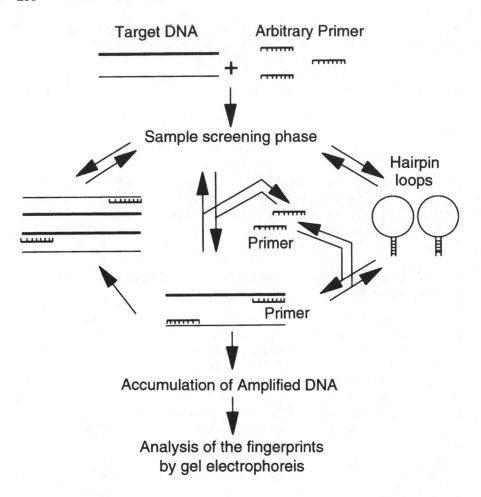

Target DNA   Arbitrary Primer

Sample screening phase

Hairpin
loops

Primer

Primer

Accumulation of Amplified DNA

Analysis of the fingerprints
by gel electrophoreis

(a)

Fig. 10.10a. A model of interactions between molecular species formed during DNA amplification with a single, arbitrary oligonucleotide primer. Following the template "screening"
phase, a set of DNA fragments is synthesized. These first-round amplification products are
initially single-stranded and have palindromic termini that allow formation of hairpin loops.
In subsequent rounds of amplification, the products can be in the form of template–template
and primer template duplexes, as well as in single-strand and hairpin loops. The different
species produced tend to establish an equilibrium while enzyme-anchoring and primer-exten-
sion transform the relatively rare primer–template duplexes into accumulating amplification
products.

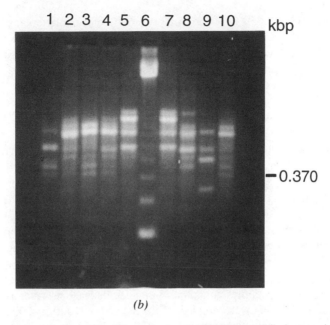

(b)

Fig. 10.10b. Agarose gel electrophoresis analysis of AP-PCR amplified genomic finger-prints from Kanagawa positive (K+) (virulent) and Kanagawa negative (K−) (avirulent) strains of *Vibrio parahemolyticus* using an arbitrarily chosen oligonucleotide primer, RPSE420. **Lanes 1–4** and **10**, AP-PCR-generated genomic fingerprints from *V. parahemolyticus* K+ strains; **lanes 5,7–9**, AP-PCR-generated genomic fingerprints from *V. parahemolyticus* K − strains. Please note that *V. parahemolyticus* K+ and K− strains showed distinct DNA fingerprints patterns especially at an approximately 0.370-kbp fragment area. **Lane 6**, 123-bp DNA ladder as size standard.

10.4.13. Purification of Viral Nucleic Acids from Environmental Samples and Application of PCR Gene Probes for Detection

Outbreaks of enteric viral diseases due to the drinking of contaminated water has increased significantly during the past decade [156,157]. Conventional monitoring for the presence of enteric viruses in the environment and drinking water requires animal cell culture and is technically difficult and time consuming. Since the enteric viruses are relatively resistant to the wastewater treatment and disinfection processes, a rapid and efficient method for periodic, perhaps routine monitoring of the drinking water and finished water sources for possible viral contamination is necessary. Detection of viruses, especially enteroviruses from the environmental water samples, requires sample clean-up for successful PCR DNA amplification.

For identification of viruses in the environmental water samples, the first step is to concentrate the viral particles, followed by processing of the samples to eliminate environmental contaminants that may potentially inhibit the PCR amplification reaction. A model system has been tested for such a purpose by concentrating 1 L 3%

beef extract (BE)–0.05 M glycine to 10–15 ml by biofloculation at pH 3.5 after addition of 0.5 M $FeCl_3$ to a final concentration of 2.5 mM. The concentrated sample was then seeded with hepatitis A or poliovirus 1. The concentrates were then processed by polyethylene glycol precipitation (13% PEG and 0.3 M NaCl) and centrifuged to collect the pellet. The pellet was resuspended in 0.7–1.3 ml of 50 mM Tris.Cl (pH 8.0) and 0.2% Tween 20 buffer and extracted with an equal volume of chloroform. The aqueous phase was purified by using Sephadex G-15, G-25, G-50, G-100, or G-200 spin column chromatography (Pharmacia LKB Biotechnology, Piscataway, NJ) and further concentrated by a Centricon 100 (Amicon, MA) or a Microsep 300 (Filtron, MA) microconcentrator to an approximate volume of 30–60 μl. The retenate was washed once with 1 × PCR reaction buffer and used directly for reverse transcriptase PCR (RT-PCR) amplification detection of the target viruses [157,158]. By following this approach, 97% of the enteroviruses were recovered routinely from the environmental waters [157,158]. This RT-PCR approach, following sample processing, has been shown to detect <1 plaque forming unit (PFU) of most of the known enteroviruses [157–159].

More recently, a related procedure for purification of entroviruses from groundwater samples and samples containing humic acids was described [83]. In this approach, the samples were treated with Sephadex 100 or 200 spin columns (Pharmacia) in combination with Chelex 100 (BioRad Laboratories) following conventional filter elution–adsorption of the viral particles to remove interfering factors for RT-PCR detection. This approach effectively removed the contaminants, including the humic materials, from the water sample, for detection of enteroviruses by RT-PCR with the sensitivity of 0.1 PFU [160].

The presence of human immunodeficiency virus type 1 (HIV-1) in wastewater, sludge, final effluent soil, and pond water released by infected individuals and hospital wastes and their possible role in the spread of disease is a major concern. Total nucleic acids, directly from environmental samples or from viral concentrates from various sources, were extracted by conventional alkaline lysis or by aluminum flocculation, respectively [160]. By using the RT/PCR DNA amplification method, the presence of HIV-1 was detected in several wastewater samples with the viral particle amounts equivalent to 0.04 and 0.4 pg of P24 antigen [160]. However, the study did not show any evidence of the infectivity of these viruses in the wastewaters or their possible role in spreading disease to the human population. More recently, Tsai et al. [161] have described a simple ultrafiltration method to concentrate enteroviruses and hepatitis A viruses (HAV) from sewage and ocean water samples. Using this purification approach, a triplex RT/PCR approach has been developed for simultaneous detection of poliovirus, HAV, and rotavirus in sewage and ocean water samples [161]. Using 1 × Tris·EDTA-NaCl (10 mM Tris·Cl, 1 mM EDTA, 100 mM NaCl) (pH 8.3) equilibrated Sephadex G-50 and Chelex 100 resins columns, the sewage sludge samples were purified for recovery of the viral DNAs that were devoid of PCR-inhibitory substances [162]. By following this purification approach, enteroviruses were detected in 10 different sewage sludge-amended soil samples.

A relatively different approach, called *antigen-capture* PCR (AC-PCR) has been described for the detection of HAV in environmental samples [163]. In this approach, the HAV was captured from seeded liquid wastes by homologous antibody, heat denatured, and RT-PCR performed in a single reaction tube. The AC-PCR amplified products were detected by gene probe hybridization. The detection was specific, with a sensitivity of 4 pfu.

10.4.14. *In Situ* PCR Amplification

In situ PCR is another recent breakthrough in the application of basic PCR methodology (Figure 10.11). Although the *in situ* hybridization approach using

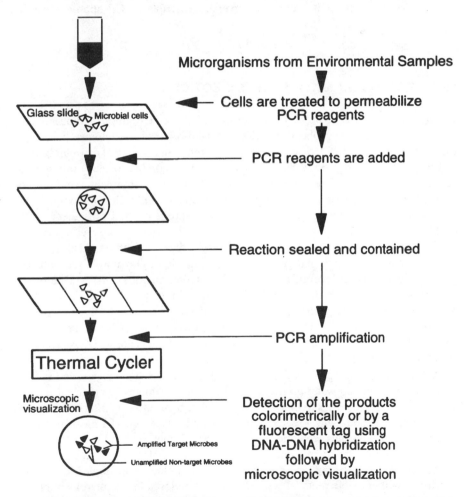

Fig. 10.11. Schematic representation of *in situ* PCR amplification of DNA or RNA within the intact target microorganisms that may have great potential to monitor microbial gene expression in the environment.

oligonucleotide probes has been used to detect microbial pathogens and environmentally significant microorganisms, the application of *in situ* PCR amplification has not been applied to environmental samples to study the presence and activities of specific microorganisms. This approach may have the advantage of detecting single cells in a given environmental sample. Also, the rigorous environmental sample preparation may not be necessary for this approach. The *in situ* PCR amplification approach certainly has applicability in microbiological monitoring of drinking water and viruses in various environmental samples. Also, simultaneous identification of multiple microorganisms in an ecosystem or in an environmental sample by multiplex *in situ* PCR amplification is a possibility. To study the microbial activities, detect a VBNC state of a microbial pathogen, and differentiate live versus dead microbial cells, the *in situ* reverse transcriptase PCR approach can be used.

10.5. DISCUSSION AND FUTURE DIRECTIONS

The applications of gene probe hybridizations alone or PCR amplification in combination with the gene probe DNA–DNA hybridization methodologies have been very useful in answering and solving many environmentally related difficult biological problems, which have remained unsolved for years due to the limitations of conventional methods. One of the important problems in environmental microbiology is monitoring released GEMs in the environment with high sensitivity. It has been shown that the application of the PCR gene probe method can detect 1–100 GEMs per gram of soil or sediment, which is a level of sensitivity several orders of magnitude higher than only the DNA–DNA hybridization method. Also, in many studies it has been shown that gene probe hybridization following PCR amplification provides a higher level of detection of the target microorganisms. One of the drawbacks of this approach is that the PCR method requires purified nucleic acid, which can be achieved from the environmental samples through several rigorous methodological steps. The DNA–DNA hybridization methodology using a labeled gene probe does not necessarily require purification of the genomic DNAs from the target microorganisms.

Application of PCR gene probe technology in monitoring pathogens and indicator microorganisms in water and other environmental samples has reached a stage that one can confidently say that this method can be used as an alternative to the conventional methods with greater specificity and sensitivity. The most important criterion in applying this technology in this area of environmental microbiology is the removal of inhibitors and contaminants from the samples. Although a number of inhibitors, including humic acids from various environmental samples, have been identified with possible removal procedures, it is believed that numerous unidentified inhibitors may exist that are yet to be identified. For detection of microorganisms in soil and sediments, it has been found that the humic and fulvic acid compounds inhibit the polymerase activities and reduce the sensitivity of detection. Although several procedures such as diluting the samples, ion-exchange chromatography, gel filtration chromatography, PVPP treatment, sucrose gradient purification. Chelex 100 resins, etc., have been attempted to remove humic and fulvic acids and other inhibitors from the samples, none of these methods seems to remove them totally. The application of PCR gene probe technology to various environmental samples for detection of pathogens and other microorganisms may be affected severely, if a relatively universal method for removal of inhibitors from the environmental samples is not developed.

Quantitation of microbial populations by conventional methods has several drawbacks. Although there are several reports on the quantitation of PCR-amplified products by using gene probe hybridizations, this aspect requires more investigation to develop a reliable quantitation of a microbial population in a given environment. This will permit the study of microbial succession, completion, and community structure in an ecosystem of interest, including the microbes living in many extreme environments.

Another potential application of PCR gene probe methodology in the area of environmental microbiology is in distinguishing the live cells from the dead ones in a given sample. Although more research needs to be done before applying this approach to actual environmental samples, there have been several reports describing such studies with great promise for the future. However, the concern is the ability of the gene probe hybridization alone or PCR gene probe approach to detect nonviable cells in the environment. Recovery of messages and targeting the mRNA has the potential to solve this problem. Alteration of gene expression in many microbial pathogens and other pollutant-degrading microbes due to various environmental conditions is a growing concern to human health; detection of specific mRNA in the environment by PCR gene probes will give the information of the *in situ* activities of these microorganisms. Much research should be done to target mRNA and optimize the approach in diverse environmental samples.

The recent emergence of AP-PCR has special applications in environmental microbiology and solving many of the microbiological activities that have not been unvailed. The DNA fingerprint analysis will help to trace the specific microbial strains in an ecosystem and to identify their roles in the ecological activities. Differentiating strains that are virulent from avirulent using the RNA/AP-PCR approach has the potential to identify the pathogenic genes in many of the microorganisms that are found in the environment.

Gene probe DNA–DNA hybridization alone or PCR amplification coupled with gene probe hybridization shows promise in cloning genes from environmentally important microorganisms, including those organisms that have not been cultured yet. In the near future, technological improvement and subsequent new development of the PCR method coupled with gene probe DNA–DNA hybridization will solve many unanswered questions in the area of microbial ecology, microbial community structure, environmental health, and environmental analyses of molecular microbiology.

ACKNOWLEDGMENTS

Some of the studies by A.K. Bej and colleagues cited in this chapter are supported in part by (1) National Oceanic and Atmospheric Administration (NOAA), U.S. Department of Commerce under research grant NA16RG155-01, NA56RG0129, NA16RG0155, the Mississippi Alabama Sea Grant Consortium (MASGC), and the University of Alabama at Birmingham; (2) Saltonstall-Kennedy (S-K) Grant Program, National Marine Fisheries Service (NMFS), National Oceanic and Atmospheric Administration (NOAA), U.S. Department of Commerce under the research grant NA37FD0082; and (3) GX BioSystems Inc. Langhorne, PA.

REFERENCES

1. Britten, R.J., Kohne, D.E.: Repeated sequences in DNA. Science 161:529–540 (1968).
2. Ogram, A.V., Sayler, G.S., Gustin, D., Lewis, R.J.: DNA adsorption to soils and sediments. Environ. Sci. Technol. 22:982–984 (1987).
3. Sayler, G.S., Stacey, G.: In Fiksel, J., Covello, V.T. (eds.): Biotechnology Risk Assessment: Issues and Methods for Environmental Introductions. New York: Pergammon Press, 1986.
4. Torsvik, V.L., Goksoyr, J., Daae, F.L.: High diversity of DNA in soil bacteria. Appl. Environ. Microbiol. 56:782–787 (1990).
5. Atlas, R.M., Horowitz, A., Krichevsky, A., Bej, A.K.: Response of microbial populations to environmental disturbance. Microb. Ecol. 22:249–256 (1991).
6. Bej, A.K., Perlin, M.H., Atlas, R.M.: Effect of introducing genetically engineered microorganisms on soil microbial community diversity. FEMS Microb. Ecol. 86:169–176 (1991).
7. Steffan, R.J., Atlas, R.M.: Solution hybridization assay for detecting genetically engineered microorganisms in environmental samples. Bio/Techniques 8:316–318 (1990).
8. Edmond, E., Fliss, I., Pandian, S.: A ribosomal DNA fragment of *Listeria monocytogenes* and its use as a genus-specific probe in an aqueous-phase hybridization assay. Appl. Environ. Microbiol. 59:2690–2697 (1993).
9. Wang, R.F., Cao, W.W., Johnson, M.G.: Development of a 16S rRNA-based oligomer probe specific for *Listeria monocytogenes*. Appl. Environ. Microbiol. 57:3666–3670 (1991).
10. Hodgson, A.L.M., Roberts, W.P.: DNA colony hybridization to identify *Rhizobium* strains. J. Gen. Microbiol. 129:207–212 (1983).
11. Sayler, G.S., Layton, A.C.: Environmental application of nucleic acid hybridization. Annu. Rev. Microbiol. 44:625–637 (1990).
12. Amman, R.I., Krumholz, L., Stahl, D.A.: Fluorescent oligonucleotide probing of whole cells for determinative, phylogenetic and environmental studies in microbiology. J. Bacteriol. 177:762–770 (1990).
13. Amman, R.I., Binder, B.J., Olsen, R.J., Chisholm, S.W., Devereaux, R., Stahl, D.A.: Comparison of 16S rRNA-targeted oligonucleotide probes for flow-cytometry for analyzing mixed microbial population. Appl. Environ. Microbiol. 56:1919–1925 (1990).

14. Mullis, K.B., Faloona, F.A.: Specific synthesis of DNA *in vitro* via a polymerase-catalyzed chain reaction. Methods Enzymol. 155:335–351 (1987).

15. Mullis, K.B.: The unusual origin of the polymerase chain reaction. Sci. Am. 262:56–65 (1990).

16. Saiki, R.K., Gelfand, D.H., Stoffel, S., Scharf, S.J., Higuchi, R., Horn, G.T., Mullis, K.B., Erlich, H.A.: Primer-directed enzymatic amplification of DNA with a thermostable DNA polymerase. Science 239:487–494 (1988).

17. Atlas, R.M., Bej, A.K.: Detecting bacterial pathogens in environmental water samples by using PCR and gene probes. In Innis, M., Gelgand, D.H., Sninsky, J.J., White, T.J. (eds.): PCR Protocols: A Guide to Methods and Applicatons. San Diego: Academic Press, 1990.

18. Atlas, R.M., Bej, A.K.: Polymerase chain reaction. In Gerhardt, P., Murray, R.G.E., Wood, W.A., Krieg, N.R.: Methods for General and Molecular Bacteriology. Washington, D.C.: ASM Press, 1993.

19. Bej, A.K., Mahbubani, M.H., Atlas, R.M.: Amplification of nucleic acids by polymerase chain reaction (PCR) and other methods and their applications. Crit. Rev. Biochem. Mol. Biol. 26:301–334 (1991).

20. Bej, A.K., Mahbubani, M.H.: Applications of the polymerase chain reaction in environmental microbiology. PCR Method Appl. 1:151–159 (1992).

21. Bej, A.K., Mahbubani, M.H.: Detection of microbial pathogens in the gastrointestinal tract by PCR and gene probe methods. In Ehrlich, G.D., Greenburg, S.J. (eds.): PCR-Based Diagnostics in Infectious Disease. Boston: Blackwell Scientific, 1993.

22. Erlich, H.A., Gelfand, D., Sninsky, J.J.: Recent advances in the polymerase chain reaction. Science 252:1643–1651 (1991).

23. Grunstein, M., Hogness, D.S.: Colony hybridization: A method for the isolation of cloned DNAs that contain a specific gene. Proc. Natl. Acad. Sci. U.S.A. 72:3961–3965 (1975).

24. Hanahan, D., Meselson, M.: Plasmid screening at high colony density. Gene 10:63–67 (1980).

25. Sayler, G.S., Shields, M.S., Tedford, E.T., Breen, A., Hooper, S.W.: Application of DNA–DNA colony hybridization to the detection of catabolic genotypes in environmental samples. Appl. Environ. Microbiol. 49:1295–1303 (1985).

26. Echeverria, P., Seriwatana, J., Chityothin, O., Chaicumpa, W., Tirapat, C.: Detection of enterotoxigenic *Escherichia coli* in water by filter hybridization with three enterotoxin gene probes. J. Clin. Microbiol. 16:1086–1090 (1982).

27. Fitts, R., Diamond, M., Hamilton, C., Neri, M.: DNA–DNA hybridization assay for the detection of *Salmonella* spp. in foods. Appl. Environ. Microbiol. 46:1146–1151 (1983).

28. Hill, W.E., Payne, W.L., Aulisio, C.C.G.: Detection and enumeration of virulent *Yersinia enterocolitica* in food by DNA colony hybridization. Appl. Environ. Microbiol. 46:636–641 (1983).

29. Miliotis, M.D., Galen, J.E., Kaper, J.B., Morris, J.G. Jr.: Development and testing of a synthetic oligonucleotide probe for the detection of pathogenic *Yersinia* strains. J. Clin. Microbiol. 27:1667–1670 (1989).

30. Datta, A.R., Wentz, B.A., Hill, W.E.: Detection of hemolytic *Listeria monocytogenes* by using DNA colony hybridization. Appl. Environ. Microbiol. 53:2256–2259 (1987).

31. Falkenstein, H., Bellemann, P., Walter, S., Zeller, W., Geider, K.: Identification of

Erwina amylovora, the fiberlight pathogen, by colony hybridization with DNA from plasmid pEPA29. Appl. Environ. Microbiol. 54:2798–2802 (1988).

32. Nortermans, S., Chakrobarty, T., Leimeister-Wachter, M., Dufrenne, J., Heuvelan, K.J.: Specific gene probe for detection of biotype and serotype *Listeria* strains. Appl. Environ. Microbiol. 55:902–906 (1989).

33. Atlas, R.M., Sayler, G., Burlage, R.S., Bej, A.K.: Molecular approaches for environmental monitoring of microorganisms. Bio/Techniques 12:706–717 (1992).

34. Fredrickson, J.K., Bezdicek, D.F., Brickman, F.J., Li, S.W.: Enumeration of *Tn5* mutant bacteria in soil by using a most-probable number-DNA hybridization and antibiotic resistance. Appl. Environ. Microbiol. 54:446–453.

35. Jain, R.K., Sayler, G.S., Wilson, J.T., Houston, L., Pacia, D.: Maintenance and stability of induced genotypes in groundwater aquifer material. Appl. Environ. Microbiol. 53:996–1002 (1987).

36. Blackburn, J.W., Jain, R.L., Sayler, G.S.: Molecular microbial ecology of a naphthalene-degrading genotype in activated sludge. Environ. Sci. Technol. 21:884–890 (1987).

37. Barkay, T., Fouts, D.L., Olson, B.H.: The preparation of DNA gene probe for the detection of mercury resistance genes in gram-negative communities. Appl. Environ. Microbiol. 49:686–692 (1985).

38. Barkay, T., Olsen, B.H.: Phenotypic and genotypic adaptation of aerobic heterotrophic sediment bacterial communities to mercury stress. Appl. Environ. Microbiol. 52:403–406 (1986).

39. Barkay, T.: Adaptation of aquatic microbial communities to Hg^{2+} stress. Appl. Environ. Microbiol. 53:2725–2732 (1987).

40. Barkay, T., Liebert, C., Gilman, M.: Hybridization of DNA probes with whole community genome for detection of genes that encode microbial responses to pollutants: *mer* genes and Hg^{2+} resistance. Appl. Environ. Microbiol. 55:1574–1577 (1989).

41. Amy, P.S., Hiatt, H.D.: Survival and detection of bacteria in an aquatic environment. Appl. Environ. Microbiol. 55:788–793 (1989).

42. Datta, A.R., Moore, M.A., Wentz, B.A., Lane, J.: Identification and enumeration of *Listeria monocytogenes* by nonradioactive DNA probe colony hybridization. Appl. Environ. Microbiol. 59:144–149 (1993).

43. Pettigrew and Sayler, G.S.: The use of DNA:DNA colony hybridization in the rapid isolation of 4-chlorobiphenyl degradative bacterial phenotypes. J. Microbiol. Methods, 5:205–213 (1986).

44. Bej, A.K., Perlin, M.H., Atlas, R.M.: Model suicide vector for containment of genetically engineered microorganisms. Appl. Environ. Microbiol. 54:2472–2477 (1988).

45. Bej, A.K., Molin, S., Perlin, M.H., Atlas, R.M.: Maintenance and killing efficiency of conditional lethal constructs in *Pseudomonas putida*. J. Ind. Microbiol. 10:79–85 (1992).

46. Bej, A.K., Mahbuabani, M.H., Atlas, R.M.: Detection and molecular serogrouping of *Legionella pneumophila* by polymerase chain reaction amplification and restruction enzyme analysis. *Legionella*: Current Status and Emerging Perspective. American Society for Microbiology, Washington, D.C., 1993.

47. Molin, S., Boe, L., Jensen, L.B., Kristensen, C.S., Givskoy, M., Ramos, J.L., Bej, A.K.:

Suicidal genetic elements and their use in biological containment of bacteria. Annu. Rev. Microbiol. 47:139–166 (1993).

48. Holben, W.E., Jansson, J.K., Chelm, B.K., Tiedje, J.M.: DNA probe method for the detection of specific microorganisms in the soil bacterial community. Appl. Environ. Microbiol. 54:703–711 (1988).

49. Ogram, A., Sayler, G.S., Barkay, T.: The extraction and purification of microbial DNA from sediments. J. Microbiol. Methods 7:57–66 (1988).

50. Steffan, R.J., Goksoyr, J., Bej, A.K., Atlas, R.M.: Recovery of DNA from soils and sediments. Appl. Environ. Microbiol. 54:2908–2914 (1988).

51. Tsai, Y., Olson, B.H.: Rapid method for direct extraction of DNA from soil and sediments. Appl. Environ. Microbiol. 57:1070–1074 (1991).

52. Tsai, Y., park, M.J., Olson, B.H.: Rapid method for direct extraction of mRNA from seeded soils. Appl. Environ. Microbiol. 57:765–768 (1991).

53. Tsai, Y., Olson, B.: Rapid method of separation of bacterial DNA from humic substances in sediments for polymerase chain reaction. Appl. Environ. Microbiol. 58:2292–2295 (1992).

54. Tsai, Y., Olson, B.H.: Detection of low numbers of bacterial cells in soils and sediments by polymerase chain reaction. Appl. Environ. Microbiol. 58:754–757 (1992).

55. Pillai, S.D., Josephson, K.L., Bailey, R.L., Gerba, C.P., Pepper, I.L.: Rapid method for processing soil samples for polymerase chain reaction amplification of specific gene sequences. Appl. Environ. Microbiol. 57:2283–2286 (1991).

56. Smalla, K., Cresswell, N., Mendonca, L.C., Wolters, A., van Elsas, J.D.: Rapid DNA extraction protocol from soil for polymerase chain reaction-mediated amplification. J. Appl. Bacteriol. 74:78–85 (1993).

57. Young, C.C., Burghoff, R.L., Keim, L.G., Minak-Bernero, V., Lute, J.R., Hinton, S.M.: Polyvinylpyrrolidone-Agarose gel electrophoresis purification of polymerase chain reaction-amplifiable DNA from soils. Appl. Environ. Microbiol. 59:1972–1974 (1993).

58. Picard, C., Ponsonnet, C., Paget, E., Nesme, X., Simonet, P.: Detection and enumeration of bacteria in soil by direct DNA extraction and polymerase chain reaction. Appl. Environ. Microbiol. 58:2717–2722 (1992).

59. Bruce, K.D., Hiorns, W.D., Hobman, J.L., Osborn, A.M., Strike, P., Ritchie, D.A.: Amplification of DNA from native populations of soil bacteria by using polymerase chain reaction. Appl. Environ. Microbiol. 58:3413–3416 (1992).

60. Selenska, S., Klingmüller, W.: DNA recovery and direct detection of *Tn5* sequences from soil. Lett. Appl. Microbiol. 13:21–24 (1991).

61. Tebbe, C.C., Vahjen, W.: Interference of humic acids and DNA extracted directly from soil in detection and transformation of recombinant DNA from bacteria and a yeast. Appl. Environ. Microbiol. 59:2657–2665 (1993).

62. Sommerville, C.C., Knight, I.T., Straub, W.L., Colwell, R.: Simple, rapid method for direct isolation of nucleic acids from aquatic environments. Appl. Environ. Microbiol. 55:548–554 (1989).

63. Paul, J.H., Cazares, L., Thurmond, J.: Amplification of the *rbc*L gene from dissolved and particulate DNA from aquatic environments. Appl. Environ. Microbiol. 56:1963–1966 (1990).

64. Weller, R., Ward, D.M.: Selective recovery of 16S rRNA sequences from natural microbial communities in the form of cDNA. Appl. Environ. Microbiol. 55:1818–1822 (1989).

65. Fuhrman, J.A., Comeau, D.E., Hagstrom, A., Cham, A.M.: Extraction from natural planktonic microorganisms of DNA suitable for molecular biological studies. Appl. Environ. Microbiol. 54:1426–1429 (1988).

66. Lee, S., Fuhrman, J.A.: DNA hybridization to compare species compositions of natural bacterioplankton assemblages. Appl. Environ. Microbiol. 56:739–746 (1990).

67. Zehr, J.P. McReynold, L.A.: Use of degenerate oligonucleotides for the amplification of the nifH gene from the marine cyanobacterium *Trichodesmium thiebautii*. Appl. Environ. Microbiol. 55:2522–2526.

68. Oyofo, B.A., Rollins, D.M.: Efficacy of filter types for detecting *Campylobacter jejuni* and *Campylobacter coli* in environmental water samples by polymerase chain reaction. Appl. Environ. Microbiol. 59:4090–4095 (1993).

69. Bej, A.K., Steffan, R.J., DiCeasre, J.L., Haff, L., Atlas, R.M.: Detection of coliform bacteria in water by polymerase chain reaction and gene probes. Appl. Environ. Microbiol. 56:307–314 (1990).

70. Bej, A.K., Mahbubani, M.H., Miller, R., DiCesare, J., Haff, L., Atlas, R.M.: Multiplex PCR amplification and immobilized capture probe for detection of bacterial pathogens and indicators in water. Mol. Cell. Probes 4:353–365 (1990).

71. Bej, A.K., Mahbubani, M.H., DiCesare, J.L., Atlas, R.M.: PCR–gene probe detection of microorganisms using filter-concentrated samples. Appl. Environ. Microbiol. 57:3529–3534 (1991).

72. Bej, A.K., McCarty, S.C., Atlas, R.M.: Detection of coliform bacteria and *Escherichia coli* by multiplex polymerase chain reaction: Comparison with defined substrate and plating methods for water quality monitoring. Appl. Environ. Microbiol. 57:2429–2432 (1991).

73. Bej, A.K., Mahbubani, M.H., Atlas, R.M.: Detection of a viable *Legionella pneumophila* by using polymerase chain reaction and gene probes. Appl. Environ. Microbiol. 57:597–601 (1991).

74. Bej, A.K., DiCesare, J.L., Haff, L., Atlas, R.M.: Detection of *Escherichia coli* and *Shigella* spp. in water by using polymerase chain reaction (PCR) and gene probes for *uid*. Appl. Environ. Microbiol. 57:1013–1017 (1991).

75. Bej, A.K., Mahbubani, M.H.: Genetically engineered microorganisms: Monitoring and containing. In Corn, M. (ed.): Handbook of Hazardous Materials. San Diego: Academic Press, 1993.

76. Bej, A.K., Mahbubani, M.H.: Thermostable DNA polymerases for *in vitro* DNA amplifications. In Griffin, H., Griffin, A. (eds.): *PCR Technology: Current Innovations*. San Diego: Academic Press, 1993.

77. Lang, A.L., Tsai, Y.L., Mayer, C.L., Patton, K.C., Palmer, C.J.: Multiplex PCR for detection of the heat-labile toxin gene and Shiga-like toxin I and II genes in *Escherichia coli* isolated from natural waters. Appl. Environ. Microbiol. 60:3145–3149 (1994).

78. Southworth, J., Bej, A.K.: Species-specific identification of *Escherichia coli* by polymerase chain reaction (PCR) for monitoring drinking water quality. 94th General Meeting of American Society for Microbiology (ASM), Las Vegas, Nevada.

79. Graves, S., Bej, A.K.: Use of polymerase chain reaction (PCR) in distinguishing live

Salmonella typhimurium from biocide-treated dead cells in water. 94th General Meeting of American Society for Microbiology (ASM), Las Vegas, Nevada, 1994.

80. Jones, D.D., Law, R., and Bej, A.K.: Detection of *Salmonella* spp. in contaminated oysters using polymerase chain reaction (PCR) and gene probes. J. Food. Sci. 58(6):1191–1197, 1202 (1993).

81. Way, J.S., Josephson, K.L., Pillai, S.D., Abbaszadegan, M., Gerba, C.P., and Pepper, I.L.: Specific detection of *Salmonella* spp. by multiplex polymerase chain reaction. Appl. Environ. Microbiol. 59:1473–1479 (1993).

82. Abbaszadegan, M., Gerba, C.P., Rose, J.B.: Detection of *Giardia* cysts with a cDNA probe and applications to water samples. Appl. Environ. Microbiol. 57:927–931 (1991).

83. Abbaszadegan, M., Huber, M.S., Gerba, C.P., Pepper, I.L.: Detection of enteroviruses in groundwater with the polymerase chain reaction. Appl. Environ. Microbiol. 59:1318–1324 (1993).

84. Mahbubani, M.H., Bej, A.K., Perlin, M.H., Schaeffer, F.W., Jakubowski, W., Atlas, R.M.: The differentiation of *Giardia duodenalis* from other *Giardia* spp. based on the polymerase chain reaction and gene probes. J. Clin. Microbiol. 30:74–78 (1992).

85. Mahbubani, M.H., Bej, A.K., Perlin, M.H., Schaeffer, F.W., Jakubowski, W., Atlas, R.M.: Detection of *Giardia* using the polymerase chain reaction and distinguishing live from dead cysts. Appl. Environ. Microbiol. 57:3455–3461 (1991).

86. Mahbubani, M., Bej, A.K., DiCesare, J., Miller, R., Haff, L., Atlas, R.M.: Detection of bacterial mRNA using polymerase chain reaction. Bio/Techniques 10;48–49 (1991).

87. Mahbuabani, M.H., Jones, D.D., and Bej, A.K.: Species-specific detection of *Shigella* spp. by polymerase chain reaction (submitted 1996).

88. Mahbubani, M.H., Shaeffer, F.W. III, and Bej, A.K.: Detection of *Giardia lambia* cysts in environmental water using polymerase chain reaction and immuno-magnetic capture beads. (submitted 1996).

89. Starnbach, M.N., Falkow, S., Tompkins, L.S.: Species specific detection of *Legionella pneumophila* in water by DNA amplification and hybridization. J. Clin. Microbiol. 27:1257–1261 (1990).

90. Jain, R.K., Sayler, G.S.: Problems and potential for *in situ* treatment of environmental pollutants by engineered microorganisms. Microbiol. Sci. 4:59–63 (1987).

91. Sayler, G.S., Harris, C., Pettigrew, C., Pacia, D., Breen, A., Sirotkin, K.M.: Evaluating the maintenance and effects of genetically engineered microorganisms. Dev. Ind. Microbiol. 27:135–149 (1987).

92. Yates, M.V., Yates, S.R., Warrick, A.W., Gerba, C.P.: Use of geostatistics to predict virus decay rates for determination of septic tank setback distances. Appl. Environ. Microbiol. 52(3):479–483.

93. Kuritza, A., Shaughnessy, P., Salyers, A.A.: Enumeration of polysaccharide degrading *Bacteroides* species in human feces by using species-specific DNA probes. Appl. Environ. Microbiol. 51:385–390 (1986).

94. Pasculle, A.W., Veto, G.E., Krystofiak, S., McKelvey, K., Vrsalovic, K.: Laboratory and clinical evaluation of a commercial DNA probe for the detection of *Legionella* spp. J. Clin. Microbiol. 27:2350–2358 (1989).

95. Drake, T.A., Hindler, J.A., Berlin, G.W., Bruckner, D.A.: Rapid identification of *mycobacterium avium* complex in culture using DNA probes. J. Clin. Microbiol. 25:1442–1448 (1987).

96. Jiang, X., Estes, M.K., Metcalf, T.G., Melnick, J.L.: Detection of hepatitis A virus in seeded estuarine samples by hybridization with cDNA probes. Appl. Environ. Microbiol. 52:711–717 (1986).

97. Pace, N.R., Stahl, D.A., Lane, D.J., Olsen, G.J.: The analysis of natural microbial populations by ribosomal RNA sequences. Adv. Microb. Ecol. 9:1–55 (1986).

98. Giovannoni, S.J., Delong, E.F., Olsen, G.J., Pace, N.R.: Phylogenetic group-specific oligonucleotide probes for identification of single microbial cells. J. Bacteriol. 170:720–726 (1988).

99. Ward, D.M., Weller, R., Bateson, M.M.: 16S rRNA sequences reveal numerous uncultured microorganisms in a natural community. Nature 345:63–65 (1990).

100. Reysenbach, A.L., Wickham, G.S., Pace, N.R.: Phylogenetic analysis of the hyperthermophilic pink filament community in Octopus Spring, Yellowstone National Park. Appl. Environ. Microbiol. 60:2113–2119 (1994).

101. Kane, M.D., Poulsen, L.K., Stahl, D.A.: Monitoring the enrichment and isolation of sulfate-reducing bacteria by using oligonucleotide hybridization probes designed from environmentally derived 16S rRNA sequences. Appl. Environ. Microbiol. 159:682–686 (1993).

102. Raskin, L., Stromley, J.M., Rittmann, B.E., Stahl, D.A.: Group-specific 16S rRNA hybridization probes to describe natural communities of methanogens. Appl. Environ. Microbiol. 60:1232–1240 (1994).

103. Kopczynski, E.D., Bateson, M.M., Ward, D.M.: Recognition of chimeric small-subunit ribosomal DNAs composed of genes from uncultivated microorganisms. Appl. Environ. Microbiol. 60:746–748 (1994).

104. Burggraf, S., Mayer, T., Amann, R., Schadhauser, S., Woese, C.R., Stetter, K.O.: Identifying members of the domain *Archaea* with rRNA-targeted oligonucleotide probes. Appl. Environ. Microbiol. 60:3112–3119 (1994).

105. Trebesius, K., Amann, R., Ludwig, W., Muhlegger, K., Schliefer, K.H.: Identification of whole fixed bacterial cells with nonradioactive 23S rRNA-targeted polnucleotide probes. Appl. Environ. Microbiol. 60:3228–3235 (1994).

106. Ludwig, W., Dorn, S., Springer, N., Kirchhof, G., Schleifer, K.H.: PCR-based preparation of 23S rRNA-targeted group-specific polynucleotide probes. Appl. Environ. Microbiol. 60:3236–3244 (1994).

107. DiChristina, T.J., DeLong, E.F.: Design and application of rRNA-targeted oligonucleotide probes for the dissimilatory iron- and manganese-reducing bacterium *Shewanella putrefaciens*. Appl. Environ. Microbiol. 59:4152–4160 (1993).

108. Walia, S., Khan, A., Rosenthal, N.: Construction and applications of DNA probes for detection of polychlorinated biphenyl-degrading genotypes in toxic organic soil environments. Appl. Environ. Microbiol. 56:254–259 (1990).

109. Steffan, R.J., Breen, A., Atlas, R.M., Sayler, G.S.: Application of gene probe methods for monitoring specific microbial population in freshwater ecosystems. Can. J. Microbiol. 35:681–685 (1989).

110. Mancini, P., Fertels, S., Nave, D., Gealt, M.A.: Mobilization of plasmid pHSV106 from *Escherichia coli* HB101 in a laboratory-scale waste treatment facility. Appl. Environ. Microbiol. 53:665–671 (1987).

111. Bentjen, S.A., Fredrickson, J.K., Van Vorris, P., Li, S.W.: Intact soil-core microcosms

for evaluating the fate and ecological impact of the release of genetically engineered microorganisms. Appl. Environ. Microbiol. 55:198–202 (1989).

112. Steffan, R.J., Atlas, R.M.: DNA amplification to enhance the detection of genetically engineered microorganisms in environmental samples. Appl. Environ. Microbiol. 54:2185–2191 (1988).

113. Chaudhry, G.R., Toranzos, G.A., Bhatti, A.R.: Novel method for monitoring genetically engineered microorganisms in the environment. Appl. Environ. Microbiol. 55:1301–1304 (1989).

114. Wimpee, C.F., Nadeau, T.L., Nealson, K.H.: Development of species-specific hybridization probes for marine luminous bacteria by using *in vitro* DNA amplification. Appl. Environ. Microbiol. 57:1319–1324 (1991).

115. Hwang, I., Farrand, S.K.: A novel gene tag for identifying microorganisms released into the environment. Appl. Environ. Microbiol. 60:913–920 (1994).

116. Greer, C.W., Beaumier, D., Bergeron, H., Lau, P.C.K.: Polymerase chain reaction isolation of a cholorocatechol dioxygenase gene from a dichlorobenzoic acid degrading *Alcaligenes dentrificans*. 91st General Meeting of the American Society for Microbiology. Abstract Q-99:292 (1991).

117. Neilson, J.W., Josephson, K.L., Pilliai, S.D., Pepper, I.L.: Polymerase chain reaction and gene probe detection of the 2,4-dichlorophenoxyacetic acid degradation plasmid, pJP4. Appl. Environ. Microbiol. 58:1271–1275 (1992).

118. Ka, J.O., Holben, W.E., Tiedje, J.M.: Use of gene probes to aid in recovery and identification of functionally dominant 2,4-dichlorophenoxyacetic acid-degrading populations in soil. Appl. Environ. Microbiol. 60:1116–1120 (1994).

119. Ka, J.O., Holben, W.E., Tiedje, J.M.: Genetic and phenotypic diversity of 2,4-dichlorophenoxacetic acid (2,4-D)-degrading bacteria isolated from 2,4-D treated field soils. Appl. Environ. Microbiol. 60:1106–1115 (1994).

120. Herrick, J.B., Madsen, E.L., Batt, C.A., Ghiorse, W.C.: Polymerase chain reaction amplification of naphthalene-catabolic and 16S rRNA gene sequences from indigenous sediment bacteria. Appl. Environ. Microbiol. 59:687–694 (1993).

121. Amann, R.I., Stromley, J., Devereux, R., Keryl, R., Stahl, D.A.: Molecular and microscopic identification of sulfate-reducing bacteria in multispecies biofilms. Appl. Environ. Microbiol. 58:614–623 (1991).

122. Voordouw, G., Shen, Y., Harrington, C.S., Telang, A.J., Jack, T.R., Westlake, D.W.S.: Quantitative reverse sample genome probing of microbial communities and its application to oil field production waters. Appl. Environ. Microbiol. 59:4101–4114 (1993).

123. Moran, M.A., Torsvik, V.L., Torsvik, T., Hodson, R.E.: Direct extraction and purification of rRNA for ecological studies. Appl. Environ. Microbiol. 59:915–918 (1993).

124. Jeffrey, W.H., Nazaret, S., Von-Haven, R.: Improved method for recovery of mRNA from aquatic samples and its application to detection of *mer* expression. Appl. Environ. Microbiol. 60:1814–1821 (1994).

125. Olson, B.H.: Tracking and using genes in the environment. Environ. Sci. Technol. 25:604–611 (1991).

126. Ogunseiten, O.A., Delgado, I.L., Tsai, Y.L., Olson, B.H.: Effect of 2-hydroxybenzoate on the maintenance of naphthalene-degrading pseudomonads in seeded and unseeded soil. Appl. Environ. Microbiol. 57:2873–2879 (1991).

127. Hofle, M.G. In Hattori, T., Ishida, Y., Maruyama, Y., Morita, R.Y., Ucida, A. (eds.): Recent Advances in Microbiol Ecology. Tokyo: Japan Scientific Press, 1989.

128. Cleuziat, P., Baudouy-Robert, J.: Specific detection of *Escherichia coli* and *Shigella* species using fragments of genes coding for beta-glucuronidase. FEMS Microbiol. Lett. 72:315–322.

129. Mahbubani, M.H., Bej, A.K., Miller, R., Haff, L., DiCesare, J., Atlas, R.M.: Detection of *Legionella* with polymerase chain reaction and gene probe methods. Mol. Cell. Probes 4:175–187 (1990).

130. Bej, A.K., Mahbubani, M.H., Boyce, M.J., Atlas, R.M.: Detection of *Salmonella* in shellfish by using polymerase chain reaction. App. Environ. Microbiol. 60:368–373 (1994).

131. Bej, A.K., Jones, D.D.: Differentiation of total *Salmonella* spp. from pathogenic *Salmonella typhimurium* in shellfish by polymerase chain reaction (in preparation, 1995).

132. Chamberlain, J.S., Gibbs, R.A., Ranier, J.E., Nguyen, P.N., Radolf, A.: Deletion screening of Dunchenne muscular dystrophy locus via multiplex DNA amplification. Nucleic Acids Res. 16:1141–1156 (1988).

133. Beasley, L., Jones, D.D., Bej, A.K.: A rapid method for detection and differentiation of KP+ and KP– *Vibrio parahemolyticus* in artificially contaminated shellfish by *in vitro* DNA amplification and gene probe hybridization methods. 94th General Meeting of American Society for Microbiology (ASM), Las Vegas, Nevada, 1994.

134. Brauns, L.A., Hudson, M.C., Oliver, J.D.: Use of the polymerase chain reaction in detection of culturable and nonculturable *Vibrio vulnificans* cells. Appl. Environ. Microbiol. 57:2651–2655 (1991).

135. Colwell, R.R., Brayton, P.R., Grimes, D.I., Roszak, D.B., Huq, S.A., Palmer, L.M.: Viable but nonculturable *Vibrio cholerae* and released pathogens in the environment: implication for release genetically engineered microorganisms. Bio/Technology 3:817–820 (1985).

136. Hussong, D., Colwell, R.R., O'Brien, M.O., Weiss, E., Pearson, A.D., Einer, R.M., Burge, W.D.: Viable *Legionella pneumophila* not detectable by culture on agar media. Bio/Technology 5:947–950 (1987).

137. Oliver, J.D.: In Kjelleberg, S. (ed.): Formation of viable but nonculturable cells. Starvation in Bacteria. New York: Plenum Press, 1993.

138. Roszak, D.B., Colwell, R.R.: Survival strategies of bacteria in the natural environment. Microbiol. Rev. 51:365–379 (1987).

139. Josephson, K.L., Gerba, C.P., Pepper, I.L.: Polymerase chain reaction detection of nonviable bacterial pathogens. Appl. Environ. Microbiol. 59:3513–3315 (1993).

140. Bej, A.K., Mahbubani, M.H., Atlas, R.M.: Bacterial detection using PCR and colorimetric gene probe methods. (submitted 1995).

141. Alvarez, A.J., Buttner, M.P., Toranzos, G.A., Dvorsky, E.A., Toro, A., Heikes, T.B., Mertikas-Pifer, L.E., Stetzenbach, L.D.: Use of solid-phase PCR for enhanced detection of airborne microorganisms. Appl. Environ. Microbiol. 60:374–376 (1994).

142. Caetano-Anolles, G., Bassam, B.J., Gresshoff, P.M.: Primer-template interaction during DNA amplification fingerprinting with single arbitrary oligonucleotides. Mol. Gen. Genet. 235:157 (1992).

143. Caetano-Anolles, G.: Amplifying DNA with arbitrary oligonucleotide primers. PCR Methods Appl. 3:85 (1993).

144. Welsh, J., McClelland, M.: Fingerprinting genomes using PCR with arbitrary primers. Nucleic Acids Res. 18:7213 (1990).

145. Welsh, J., McClelland, M.: Genomic fingerprinting using arbitrarily primed PCR and a matrix of pairwise combinations of primers. Nucleic Acids Res. 19:5275 (1991).

146. Gomez-Lus, P., Fields, B.S., Benson, R.F., Martin, W.T., O'Conner, S.P., Black, C.M.: Comparison of arbitrarily primed polymerase chain reactions, ribotyping, and mono-clonal antibody analysis for subtyping *Legionella pneumophila* serogroup 1. J. Clin. Microbiol. 31:1940–1942 (1993).

147. van Belkum, A., Struelens, M., Quint, W.: Typing of *Legionella pneumophila* strains by polymerase chain reaction-mediated DNA fingerprinting. J. Clin. Microbiol. 31:2198–2200 (1993).

148. McMillin, D.E., Muldrow, L.L.: Typing of toxic strains of *Clostridium difficile* using DNA fingerprints generated with arbitrary polymerase chain reaction primers. FEMS Microbiol. Lett. 92:5–10 (1992).

149. Cancilla, M.R., Powill, I.B., Hillier, A.J., Davidson, B.E.: Rapid genomic fingerprinting of *Lactococcus lactis* strains by arbitrarily primed polymerase chain reaction with ^{32}P and fluorescent labels. Appl. Environ. Microbiol. 58:1772–1775 (1992).

150. Lett, P., Jones, D.D., Bej, A.K.: Multiplex PCR DNA amplification and gene probe methods for species-specific simultaneous detection of enteropathogenic/toxigenic, en-terohemorrhagic, and enteroinvasive *Escherichia coli* in artificially contaminated ground beef. 94th General Meeting of American Society for Microbiology (ASM), Las Vegas, Nevada, 1994.

151. Hennessy, K.J., Iandolo, J.J., Fenwick, B.W.: Serotype identification of *Actinobacillus pleuropneumoniae* by arbitrarily primed polymerase chain reaction. J. Clin. Microbiol. 31:1155–1159 (1993).

152. Bruijn, de F.J.: Use of repetitive (repetitive extragenic palindromic and enterobacterial repetitive intergenetic concensus) sequences and the polymerase chain reaction to fin-gerprint the genome of *Rhizobium meliloti* isolates and other soil bacteria. Appl. Environ. Microbiol. 58:2180–2187 (1992).

153. Sellstedt, A., Wullings, B., Nystrom, J., Gustafsson, P.: Identification of Casuarina-Frankia strains by the use of polymerase chain reaction (PCR) with arbitrary primers. FEMS Microbiol. Lett. 93:1–6 (1992).

154. Manulis, K.S., Valinsky, L., Lichter, A., Gabriel, D.W.: Sensitive and specific detection of *Xanthomonas campestris* pv. pelargonii with DNA primers and probes identified by random amplified polymorphic DNA analysis. Appl. Environ. Microbiol. 60:4094–4099 (1994).

155. Louws, F.J., Fulbright, D.W., Stephens, C.T., Bruijn, F.J.: Specific genomic fingerprints of pathogenic *Xanthomonas* and *Pseudomonas* pathovars and strains generated with repetitive sequences and PCR. Appl. Environ. Microbiol. 60:2286–2295 (1994).

156. Craun, G.F.: Surface water supplies and health. J. Am. Water Works Assoc. 80:40–52 (1988).

157. DeLeon, R., Shieh, C., Baric, R.S., Sobsey, M.D.: Detection of entroviruses and hepati-tis A virus in environmental samples by gene probes and polymerase chain reaction. In

Proceedings of the 1990 Water Quality Technology Conference, American Water Works Association, Denver, CO, 1990.

158. Schwab, K.J., DeLeon, R., Baric, R.S., Sobsey, M.D.: Detection of rotaviruses, enteroviruses, and hepatitis A virus by reverse transcriptase-polymerase chain reactions. In Proceedings of the 1991 Water Quality Technology Conference. American Water Works Association, Denver, CO (in press).

159. Chapman, N.M., Tracy, S., Gauntt, C.J., Fortmueller, U.: Molecular detection and identification of enteroviruses using enzymatic amplification and nucleic acid hybridization. J. Clin. Microbiol. 28:843–850 (1990).

160. Ansari, S.A., Farrah, S.R., Chaudhry, G.R.: Presence of human immunodeficiency virus nucleic acids in wastewater and their detection of polymerase chain reaction, Appl. Environ. Microbiol. 58:3984–3990 (1992).

161. Tsai, Y.L., Tran, B., Sangermano, L.R., Palmer, C.J.: Detection of poliovirus, hepatitis A virus, and rotavirus from sewage and ocean water by triplex reverse transcriptase PCR. Appl. Environ. Microbiol. 60:2400–2407 (1994).

162. Straub et al. 1994.

163. Deng, M.Y., Day, S.P., Cliver, D.O.: Detection of hepatitis A virus in environmental samples by antigen-capture PCR. Appl. Environ. Microbiol. 60:1927–1933 (1994).

INDEX